T0296587

IMPERIAL BOTANICAL CONFERENCE
LONDON, JULY 7–16, 1924
President: SIR DAVID PRAIN, C.M.G., C.I.E., F.R.S.

Report of Proceedings

Photo.Russell.London.

Yours sincerely
D. Prain

IMPERIAL BOTANICAL CONFERENCE

LONDON   JULY 7–16, 1924

President: SIR DAVID PRAIN, C.M.G., C.I.E., F.R.S.

Held at the Imperial College of Science and Technology, South Kensington, by kind permission of the Governing Body

Report of Proceedings

EDITED BY

F. T. BROOKS

Hon. Secretary

CAMBRIDGE
AT THE UNIVERSITY PRESS
1925

CAMBRIDGE
UNIVERSITY PRESS

University Printing House, Cambridge CB2 8BS, United Kingdom

Cambridge University Press is part of the University of Cambridge.

It furthers the University's mission by disseminating knowledge in the pursuit of education, learning and research at the highest international levels of excellence.

www.cambridge.org
Information on this title: www.cambridge.org/9781107464193

First published 1925
First paperback edition 2014

A catalogue record for this publication is available from the British Library

ISBN 978-1-107-46419-3 Paperback

Cambridge University Press has no responsibility for the persistence or accuracy of URLs for external or third-party internet websites referred to in this publication, and does not guarantee that any content on such websites is, or will remain, accurate or appropriate.

....

Every effort has been made in preparing this book to provide accurate and up-to-date information which is in accord with accepted standards and practice at the time of publication. Although case histories are drawn from actual cases, every effort has been made to disguise the identities of the individuals involved. Nevertheless, the authors, editors and publishers can make no warranties that the information contained herein is totally free from error, not least because clinical standards are constantly changing through research and regulation. The authors, editors and publishers therefore disclaim all liability for direct or consequential damages resulting from the use of material contained in this book. Readers are strongly advised to pay careful attention to information provided by the manufacturer of any drugs or equipment that they plan to use.

CONTENTS

FOREWORD

By SIR DAVID PRAIN, C.M.G., C.I.E., F.R.S.

PRESIDENT

A DECADE ago an event in European history prevented botanists throughout the globe from carrying into effect a resolution they had formulated five years earlier. Finding themselves, five years later, no longer empowered to comply with the letter of that international resolution, representatives of botany in this country decided that they would, if possible, fulfil its spirit. The steps taken and the results attained are recorded in the Report now published of the proceedings of the Imperial Botanical Conference held in London in 1924.

While the proceedings recorded here were in progress, the hope became general that this Conference might prove to be but the first of a series of similar reunions of botanists at work in the various Dominions and dependencies of the Empire. The feeling was entertained that the permanent record, in an accessible form, of the proceedings of this first Conference, might serve as a means to the end desired. At the close of the Conference, those botanists who had undertaken the tasks of organising its convocation and arranging its deliberations were accordingly requested by their colleagues to consider, before demitting office, the feasibility of editing and publishing an account of its proceedings. This request has received the careful consideration of the Executive Committee and, as a result, it has been found possible to prepare a Report which, it is hoped, may meet the wishes of those who took part in the Imperial Botanical Conference of 1924.

LONDON
1 *November*, 1924

PRESIDENTIAL ADDRESS AND OPENING OF CONFERENCE

By Sir David Prain, C.M.G., C.I.E., F.R.S.

Ladies and Gentlemen,

I have a pleasant duty to perform this morning. On behalf and in the name of the Executive Committee that has organised this Conference I have to welcome all the botanists who are here to-day. I have also, on behalf of the botanists of this country, to thank those botanists from our overseas possessions who, by their attendance, have made this Conference Imperial in fact as well as in name. I feel too that I may, on behalf of all here, welcome those foreign botanists whose interest in our proceedings has induced them to honour us by being with us as our guests.

I have been asked oftener than once to explain how and why this Conference has come to be summoned. Many here know; some do not. It may therefore be of use if the circumstances be reviewed: some of the things that have happened may have a bearing on some of the conclusions to be reached when we assemble again in this room at noon on 12th July.

In 1900 a quinquennial International Botanical Congress met in Paris. An official invitation from Austria was then received and accepted; the ensuing Congress was therefore held in Vienna in 1905. At the Vienna Congress it was a general wish that the next Congress should take place in London in 1910. The British botanists present at Vienna had come unprovided with official authority to tender the necessary invitation. They were unable to obtain the requisite authority before the Congress of 1905 was over, and, as a consequence, the Congress of 1910 met in Brussels. British representatives at the Brussels Congress took with them official authority to invite the International Congress to meet in London in 1915. This invitation was accepted and after the Brussels Congress was over, an Organising Committee was duly appointed in this country to take over charge from the Belgian Organising Committee. Ten years ago to-day everything save minor details had been arranged by the British Organising Committee for the London Congress due in 1915.

In the beginning of August, 1914, events over which the British Committee had no control, affected international relationships so adversely that the International Botanical Congress due to meet in London in 1915 could not be held. The constitution of these quinquennial Congresses made provision for such a contingency. It had been prescribed in advance that if, for whatever reason, a duly constituted Organising Committee should be prevented from holding its particular Congress, the duty of summoning the next Congress must devolve upon the Association Internationale des Botanistes. When it was announced that the International Botanical Congress due in London in 1915 must be abandoned, the British Organising Committee were reminded by the Association Internationale des Botanistes what the terms of the constitution of these International Congresses are.

The British Organising Committee recognised the right of the Association Internationale to summon the next Congress. But, until a new Organising Committee had been duly appointed it was the duty of the British Committee to remain in being. The Association Internationale had not felt the effects of war when it intimated to the British Organising Committee in 1916 its intention to arrange for an International Botanical Congress in 1920. But in 1919 the Association Internationale was so seriously affected by the blessings of peace that its own continued existence had become precarious; it was no longer in a position to arrange for an International Botanical Congress. A new situation had been created. The Association which alone had authority to initiate an International Botanical Congress was unable to do this: the Organising Committee which might have been able to summon an International Botanical Congress had no authority to do this. Having considered this new situation, the British Organising Committee decided to convene, if possible, a Botanical Conference which should be Imperial instead of International.

The possibility of summoning an Imperial Botanical Conference depended on two considerations. The first question to be considered was that of the funds in the custody of the Organising Committee. These funds had been contributed by subscribers in order to meet expenses to be incurred in connection with the International Botanical Congress proposed to be held in 1915. Having been subscribed for a definite purpose they could not be used by the Organising Committee for a different purpose. Individual subscribers to the International Botanical Congress fund had therefore to be asked whether they desired their subscriptions to be returned or if they could agree to their subscriptions being used to meet expenses to be incurred in connection with an Imperial Botanical Conference. The response to this enquiry was extremely gratifying and it is largely owing to this fact that we are here to-day.

When it was clear that the Organising Committee could, without impropriety, use the funds subscribed on account of an International Congress in meeting the expenses of an Imperial Conference, it became possible to investigate the further consideration whether the attendance of botanists from overseas could be hoped for. Individual botanists resident in our various dominions and dependencies were therefore asked to say whether, in the event of such a Conference being summoned, they would be able to be present. The response to this second enquiry proved as gratifying as the response to the former one. The replies however were so often coupled with a suggestion and, indeed, a hope that the Conference be deferred until the British Empire Exhibition should take place, that it was decided that the Exhibition year should be also the year of this Imperial Conference.

During the final phase of their work the members of the Organising Committee benefited by welcome and unlooked for help. That Committee was appointed immediately after the Brussels Congress of 1910. Since then a new and vigorous generation of botanical workers has grown up. The idea that had been explored by the Organising Committee in 1920, suggested itself independently to active members of the younger generation, and at the meeting of the British Association held at Hull a Committee was appointed to consider the possibility of holding an Imperial Botanical Conference in 1924. The new Committee came into touch with the older Committee, and the Executive Committee, which has drawn up the programme for this Conference and in whose name it is my privilege to bid you welcome now, came into existence through the amalgamation of the two bodies.

I am glad to be able to add, by way of completing this history, that the British Organising Committee, which came into existence in 1910, has at last been freed from its responsibilities. Recently the welcome information has reached this country that a Committee has been constituted to organise an International Botanical Congress to be held in the near future in the United States of America.

I should like now to invite your attention to a collection of interesting exhibits which has been displayed for inspection in the rooms of the Botany Department of this College. These exhibits will repay examination and I venture to suggest that we now adjourn so as to have an opportunity of seeing them and, at the same time, of renewing old and forming new acquaintanceships. From this afternoon onwards you will find the programme before you a full one. You are not only to be kept busy all day and every day during the rest of this week; you will have, during much of the time, to consider that insoluble problem—the possibility of being in two places at once.

I now declare this Conference open.

———————

During the Conference the Acting Agent-General for Victoria conveyed his best wishes for the success of the Conference and delivered a message from the Government of Victoria that the latter was willing to co-operate in any technical directions the Conference might advise.

During the discussion on "The Best Means of Promoting a Complete Botanical Survey of the Different Parts of the Empire" a unanimous vote of appreciation for his services to Botany was accorded to Mr J. H. Maiden, F.R.S. on the occasion of his retirement from the Directorship of the Sydney Botanic Garden, accompanied by the best wishes of all members of the Conference for continued health and prosperity during his retirement.

PLANT PHYSIOLOGY

THE PHYSIOLOGY OF CROP YIELD: A SURVEY OF MODERN METHODS OF ATTACK

(CHAIRMAN: DR F. F. BLACKMAN, F.R.S.)

Dr F. F. Blackman, F.R.S. GENERAL INTRODUCTION
(not given orally)

Crop yield has been an age-long pursuit of mankind. Striking were the achievements of agriculture before the dawn of history—tilth—irrigation—manuring—rotation of crops. With the beginning of the scientific age came fertilisers, glass houses, etc.

The modern problem is to produce the greatest yield of crop per unit area. But still it is an applied problem, not one of pure science. Its dominating limitation is that this must be done at the least possible expense.

How different would be the outlook on plant physiology were the goal, instead of the largest yield per area of natural soil, the largest possible yield from each single plant individually, and were it pursued regardless of expense! Then plant physiology might be in the favoured position of human physiology and our field of research be endowed and pushed forward in the way that falls to the lot of medical researches which aim at improving the efficiency of the individual human frame without thought of cost.

The financial setting of our problem prevents the exploitation of wholly artificial conditions of plant development and concentrates attention only on the use of semi-natural soil and semi-natural conditions of growth.

This financial limitation of research to such variable conditions introduces immense complexity and makes very difficult the task of establishing fundamental truths of wide application; but on the other hand it gives great importance to small margins of difference.

In this session our object is to bring together the different chapters of modern scientific investigation which combine to illuminate the general problem. We may distinguish five such chapters.

CHAP. I. *The experimental study of the factors and conditions affecting Growth*

Crop yield is a special case of plant growth, and underlying each special crop problem is the general physiological one of the augmentation of growth, and its control by such factors as nutriment, aeration, light and dark, temperature, CO_2-content of the air, and expense, together with other factors that may yet be brought into account such as traces of special chemical elements, the electrical setting of the plant, etc. It would have been impossible to investigate the single factors in the field, and fundamental studies have been made in the laboratory where control of factors is possible. In the field these uncontrolled factors are generally grouped as weather and left to natural variation. With intensive cultivation in glass houses, temperature and humidity are brought under human control and we get nearer to laboratory conditions. Light can now be added to the controllable factors; using modern powerful electric illuminants it is possible to supply artificial light adequate for plant growth, either continuously or discontinuously, and the study of the interaction of these "weather factors" can be pushed further in the laboratory.

The first communication, by Dr F. G. Gregory, illustrates the study of plant growth in the laboratory, with controlled artificial light; the second, by Professor V. H. Blackman, elucidates the mixed results of electro-culture in the field by laboratory studies in controlled conditions.

CHAP. II. *The Ontogeny of the Crop and the Duration of the Development Sequence*

In this chapter we stress that complexity of the whole situation which is due to the organisation of the plants we work upon. We realise that crop yield covers a wide range of special cases. The desired crop may be the fleshy parts of the whole plant or its wood, fibre or bark. On the other hand it may be some special morphological part, as the flower buds of the caper, the petals of the rose, the stamens of the crocus, the fruits of the apple, the seeds of the pea, the seed-hairs of cotton, not to mention various tubers and roots.

The desired part comes at a late stage of a long development sequence. All other parts except this one morphological unit may be valueless, and cultivation has to work upon the morphological plasticity that most plants exhibit, and endeavour to produce the maximum of the crop parts and the minimum of the antecedent or alternative parts. For the mastery of this aspect of crop yield we need detailed studies of the normal sequence of development of each crop plant combined with an exploration of its plasticity under the influence of natural and artificial variations of environment, including surgical operations such as pruning.

A further set of problems arises out of the effect on yield of varying duration of the development sequence in the individual crop plants. In most plants this sequence is not a closed one, but development of crop may be prolonged by suitable external conditions and yield thus increased. Under completely controlled laboratory conditions the development sequence of an annual plant could be carried through at the ideal rate found to give the maximum duration of the crop-yielding phase of the plant's ontogeny. In the field it is generally the weather cycle which initiates and closes the conditions for the development sequence. With unknown weather before him the agriculturist has to decide year by year when he will start the development sequence—to fix the sowing date for each crop. Ages of past experience have accumulated much empirical wisdom on this matter, and it is clear that for the purposes of crop yield it is possible to improve on the natural determination of sowing date which coincides with the end of the previous growth cycle.

This chapter will be illustrated by Dr W. L. Balls' account of his method for analysing the development of the plant by collating records of the significant stages antecedent to the final yield. An analysis of the effect of different sowing dates on the cotton crop in Egypt will be presented by the same method.

CHAP. III. *The quantitative relations of Factors and Yield*

No plant can develop except in the presence of innumerable "factors" the magnitudes of which are capable of affecting its rate of development and also its ultimate yield.

As a crude analysis these factors may be distinguished into those which concern the supply of materials of growth and those that concern the conditions of growth. Some factors, conspicuously light, act in both categories.

A line along which investigation has been pushed is to endeavour, with a series of cultures, to keep all factors but one constant and then to compare the effect, on growth and yield within the series, of varying the one selected factor over a wide range. Data for single mineral constituents of the soil of pot cultures have been accumulated and similarly experiments with varying area of natural soil for each plant to draw upon for field cultures. We are faced with the problem of relating the grades of increasing yield with the grades of increasing factor, in such series of experiments, on some empirical or rational basis.

Mr G. E. Briggs will discuss the "general law" put forward by Mitscherlich, Baule and others, as governing universally the relation between plant yield of any type and the magnitude or intensity of every outside factor, be it a mineral constituent of the soil, light or water supply.

CHAP. IV. *The complexity of the plant's Spatial Environment—Soil*

Were finance not a governing factor no doubt each crop would be grown in an ideal standardised soil, artificially prepared. As it is, a given crop cannot even be grown only in the one locality which has the nearest natural soil to the ideal. Economic reasons demand that each crop shall be grown on many natural soils.

The complexity of natural soil is enormous. We know it is a microcosm in itself with its own internal fauna and flora as well as wide variations in chemical nature and physical properties. The addition of one and the same given amount of fertiliser to two different types of soil may produce quite different reactions, directly or indirectly, when the equilibrium of the microcosm is thus disturbed. To establish whether or not small margins of favourable effect do actually occur for a given crop by shifting the existing equilibrium of the soil towards some new equilibrium requires extreme critical care in experimentation. All the experiments must be adjacent and also simultaneous, as no weather sequence during development is ever exactly repeated. Everything turns on the critical use of control plots of soil as standards. There has grown up a whole series of investigations aimed at securing the most trustworthy method of experimentation on this subject.

Mr E. J. Maskell will give an account of the methods of critical plot-culture from the point of view of experimental work at Rothamsted.

CHAP. V. *The complexity of the plant's Temporal Environment—Season*

Natural atmospheric conditions and weather lack even the element of stability that the soil possesses. With sufficient trouble and expense it would be practicable to prepare a large area of really uniform soil, but uniform seasons are here beyond our powers and our finance. Definitive studies of crop yield must range over a series of years and cannot escape the action of this immensely variable set of factors. The components of weather have a considerable independence so that the task of disentangling the effects of duration of sunshine, mean temperature, humidity, rainfall and air movement is one of great difficulty. Given a sufficient mass of data and adequate records of the components of weather we can add to the experimental method the methods of statistical analysis. At Rothamsted crops have been grown and weather recorded for the same plots for many years in succession.

Dr R. A. Fisher will show how significant relations between weather factors and yield can be computed by modern statistical methods from data of this type.

Dr F. G. Gregory. EXPERIMENTS ON PLANT GROWTH WITH CONTROLLED LIGHT AND TEMPERATURE

(*Abstract*)

Since plants growing under natural conditions are exposed to the action of many varied external factors, it becomes necessary for the purpose of analysing the effect of any one factor to ascertain in what manner the actions of two or more single factors combine in determining growth; to ascertain, for example, whether the effects of single factors are directly additive or more complexly related.

This can only be done in experiments under controlled conditions. A series of eight such experiments has been completed with Cucumber plants grown with continuous artificial light of two different constant intensities, with constant humidity and soil conditions, and covering a temperature range from 62° F. to 95° F.

The measures of growth studied have been the growth in area of single leaves and of total leaf surface, and the dry weight increase. From these the nett assimilation rate has been calculated. These assimilation data are unique in that the assimilating system has been allowed to develop under the same set of conditions in which the assimilation rate has been determined.

For all the experiments at the lower light intensity (approx. that of average winter sunlight) the curves of leaf surface increase were found to be of the same type, although a very marked optimum was found at 77° F. (25° C.). This temperature was found to be optimum also for assimilation rate, respiration rate, and for such morphological characters as leaf size, and date of appearance of the first foliage leaf.

At all temperatures over the range studied with lower light intensity the relative growth in leaf area fell off in a regular manner from the beginning of growth onwards. This falling off in growth became very rapid above the optimum temperature.

To account for the detrimental factor at work three possibilities presented themselves:

(1) That continuous light was in itself harmful.

(2) That respiration was kept at a high level by the high temperatures, whereas the low light intensity limited the assimilation rate, resulting in progressive starvation.

(3) That the low light intensity exerted a detrimental effect on the plant.

Direct experiments with continuous light as compared with discontinuous showed results in favour of continuous light, thus disposing of the first suggestion.

The experiments at supra-optimal temperatures showed a steady nett assimilation rate in spite of the fact that the rate of increase in leaf area was rapidly falling. This is incompatible with the suggestion of progressive starvation and suggests that the growth rate of the leaf surface is not directly dependent on the assimilation rate. This is corroborated by the temperature coefficients of assimilation and relative leaf growth rate. The value of both coefficients was found to be 2·3 per 10° C. rise in temperature from the lowest temperature studied up to the optimum, but above this point the coefficient of assimilation rate falls off more rapidly than for relative leaf growth rate, indicating that assimilation rate does not directly control leaf growth, but that other factors also are concerned. There is some evidence, from the study of the growth of single leaves, to show that cell division is inhibited at the higher temperatures.

If the third suggestion were correct it ought to be possible to remove the detrimental factor by increasing the light intensity. This has been effected by increasing the light intensity three times. A similar state of things had been previously noted with plants growing under greenhouse conditions in the winter, and here the detrimental effect disappeared in the spring as the intensity and duration of light increased.

The interaction of the two factors light and temperature on growth may be represented by an optimum curve, and there is evidence to show that with increasing light intensity the optimum point for temperature rises towards some unknown limit.

The complexity of the whole problem is reflected in the changing morphological characters of the plants grown under the varying external conditions. Such characters include varying leaf size and shape, leaf thickness, and relative growth of leaf and stem.

Professor V. H. Blackman, F.R.S. THE ELECTRICAL CONDITIONS OF PLANT GROWTH

(Abstract)

Plants growing in the open and unscreened by taller buildings or other plants are subjected normally to an "air-earth" current which is of the order of 2×10^{-16} amp. per cm.^{2}, the current passing to the whole surface of the globe being only about 1000 amperes. Whether this minute current is a favourable factor in plant growth—as has been claimed for nearly 200 years—is still uncertain. The favourable effect, if any, cannot be very large, otherwise it would be impossible to grow plants satisfactorily in greenhouses. The question of the favourable effect of atmospheric electric currents artificially produced starts with the work of Lemström, who first carried out experiments about 50 years ago by stretching above field crops wires charged to a few thousand volts. He claimed that marked increases of yield could be obtained by this means, but his results did not carry conviction to most agriculturists. Since 1915, field experiments have been carried out with overhead installations consisting of thin wires (5–15 feet apart) placed at a height of 7 feet and charged to 50,000 volts. The current was at the rate of 0·5–1·0 milliamp. per acre for six hours a day during the growing season, the wires being in all cases charged positively. Of eighteen experiments carried out between 1915 and 1920 there were fourteen positive results and four negative ones. Of twelve experiments with spring-sown cereals ten were positive and two negative, the mean increase being of the order of 22 per cent.

Pot-culture experiments, in which the current passing through the plants is measured, have shown that increased yields both of dry weight and of grain yield can be obtained both with maize and barley using currents of the order of 10^{-9} amp. They have also shown that currents of higher intensities (10^{-7} amp.) are injurious, and that favourable results can be obtained with wires charged negatively as well as with those charged positively, and also with alternating current. Furthermore, such experiments have shown that application of the discharge for the second month only may be as effective as, or more effective than, one applied for the whole growing season; they also indicate that the effect on grain yield may be greater than that on vegetative growth.

Laboratory experiments have been undertaken in which barley seedlings have been subjected to the discharge from an electrified point at a potential of a few thousand volts, the current passing through the plant being 0.5×10^{-10} amp. Three experiments were carried out, two in which the point was charged positively and the discharge was given for three and one hour respectively, and one in which the wire was charged negatively and the discharge given for three hours. With the point charged positively a favourable effect on the rate of growth is produced during the first hour and continues during the period of the discharge. On the cessation of the discharge a remarkable *after-effect* is to be observed, for during the fifth hour an increased rate of growth of 15 per cent. is shown. The after-effect is thus greater than the direct effect, and it is greater with a short period than with a long period of the discharge. A discharge with the needle point negatively charged gave an increased rate of growth during the first hour, but the effect during the second hour is slight and during the third hour negligible. With the cessation of the discharge the after-effect appears as before, but it is much less intense than with the positive discharge.

Converging evidence from field experiments, from pot-culture experiments and from the laboratory thus leaves no doubt as to the favourable effect of minute electric discharges on the growth of plants. For the elucidation of the action of the discharge much further work is necessary; the occurrence of the remarkable after-effect which is greater than the direct effect and also greater with short than with a long period of discharge indicates that the interpretation will not be an easy one. The field and pot-culture experiments show clearly, however, that the effect produced is out of all proportion to the energy supplied. In the field experiments the energy of the discharge (1 milliamp. at 50,000 volts for six hours a day), not all of which is available to the plant, is less than 0.5 per cent. of that fixed from sunlight by a growing barley crop, yet the increased yield is of the order of 20 per cent. Again, in one pot-culture experiment the energy of the current passing to the plants was only 30 calories, while the additional dry material fixed had a combustion value of about 15,000 calories. The effect may thus be classed as of the nature of a stimulation. It is obvious that much further work is required before the proper conditions (such as strength of current, period of discharge, etc.) required for the optimal effect can be determined; these conditions will no doubt vary with different plants and with different climatic and edaphic factors.

Dr W. Lawrence Balls, F.R.S. DEVELOPMENTAL ANALYSIS OF CROP YIELD

(*Abstract*)

The expression of Yield as a single unanalysed final value is unsatisfactory, seeing that this final value has been reached by a process of building, stage by stage and organ by organ. Opinions are readily available in agricultural practice as to the particular stages which have contributed most to the making or marring of the final value; but for scientific study it is necessary to substitute statements of observed fact for mere expressions of opinion. Provided that the day-to-day facts about the antecedent stages are known, there is a reasonable prospect of tracing their bearing on the ultimate yield; incidentally, information on the physiology of the plant concerned is obtained under field conditions.

The author developed a method for cotton, which consists in selecting such cardinal features and points of development of the plant as may be convenient for observation. The detail to which this observation is carried is limited only by the number of observers—skilled or unskilled—and computers available. The essential feature of the observations made is that they shall be continuous from day to day throughout the growing season, wherever observable material exists, and that the results shall be presented in this day-to-day form. The latter condition is rather important, since it leads to the presentation of differentiated curves, in contradistinction to the integrated presentation employed by, *e.g.*, Mitscherlich. In the latter case differences and peculiarities, which are conspicuous in differentiated curves, may be largely obscured.

The method had its genesis in the detailed examination of individual cotton plants for genetical study, in the course of which it had been realised that physiological characters and time relations were as important as morphological ones. Systems for recording them were therefore developed. In the light of such records we found that even the skilled farmer was extraordinarily unaware of what happened in the normal crop; and the method was therefore turned to making sample observations on masses of similar plants, *i.e.* the normal field crop. From such observations we computed the behaviour of the Average Plant, and so were able for the future to regard the crop as a mere multiple of the Average Plant[1].

---

[1] The graphs drawn from these data have been termed "Plant Development Curves," an inconvenient expression; and a shorter title would be useful, but has not been invented, the word "Phytography" having been taken up already.

The advantages of the method are great, in that we substitute for the single final value a continuous biographical statement (of definable accuracy at every step) which (in the case of cotton) takes cognisance of the rate of stem growth, the number of flowers, number of fruits, weight of fruit contents, and so forth. Yield thus analysed into its components can be further analysed into problems of Yield Causation. It may be objected that cotton is a particularly suitable plant for such analysis, as shown by subsequent workers in Egypt (Keeling, Hurst,. Hughes, Bailey and Trought), in the Sudan (Massey), in the West Indies (Harland), etc., since the yield is built up over a period of several weeks; and it would appear from Engledow's preliminary studies that cardinal points for observation are by no means so obvious in the cereals, although Prescott has used the method very successfully for maize. Additional difficulty in observation does not invalidate the value of the records when once obtained. The observations also have practical possibilities, in that they enable statements of Crop Condition to be made objectively and precisely, in place of their present dependence upon personal opinion; while an obvious extension of this is to an improvement of Crop Forecasts.

The comparison of varieties and the fitting of each locality with its most suitable strain, on the importance of which Mr Engledow has rightly insisted elsewhere in this Conference, are much facilitated when the reaction of each variety to each environment is thus represented biographically. In particular, the seasonal peculiarities, which ordinarily necessitate the repetition of variety tests for several years, are largely eliminated, since the effect of each component of the environment can be—as knowledge accumulates—identified and separated from other components in the process of analysis. Thus, a test in a wet year may give quite good indications of what would happen in a dry year. Again, it is not merely possible, but usual, for different varieties to arrive at the same final yield by different biographical paths; when these paths are known the variety taking that which best suits the usual local conditions can be chosen, even though the total final yield of that variety may not be quite the best during the particular seasons of experiment.

Lantern slides were shown to demonstrate the application of the method to the study, in cotton, of the spacing of the crop, of the time of sowing and of the effect of manure. The data for the last named were drawn from the work of Harland; and particular attention was directed to the form of these manurial experiment curves, which illustrate in a striking manner the curious fact that the earliest application of science to agriculture by Liebig is the one point about which least is known physiologically when we apply this test of Yield Analysis.

Mr G. E. Briggs. PLANT YIELD AND INTENSITY OF EXTERNAL FACTORS; MITSCHERLICH'S "WIRKUNGSGESETZ"

(Abstract)

Mitscherlich's law. General form

$$y = A \left(1 - e^{-c_1 x_1}\right)\left(1 - e^{-c_2 x_2}\right)\ldots\ldots\left(1 - e^{-c_n x_n}\right),$$

y is yield, A maximum yield, c_1 etc. the *Wirkungsfaktoren* of the factors x_1 etc., e the base of the natural logarithms.

Considering one variable x_1 the expression becomes

$$y = A_1\left(1 - e^{-c_1 x_1}\right),$$

where $$A_1 = A\left(1 - e^{-c_2 x_2}\right)\left(1 - e^{-c_3 x_3}\right)\ldots\ldots\left(1 - e^{-c_n x_n}\right),$$

and is the yield when x_1 is increased until further increase has no effect on y. The equation expressed in other forms

$$y = A_1\left(1 - e^{-0.7\frac{x_1}{h_1}}\right) = A_1\left(1 - \tfrac{1}{2}^{\frac{x_1}{h_1}}\right),$$

where h_1 is the value of x_1 when y is $A_1/2$ [Baule's form]. When $x_1 = 2h_1$, $A_1 - y = A_1/4$ and $A_1/8$ when $x_1 = 3h_1$ etc.,

or $$\frac{dy}{dx_1} = c_1\left(A_1 - y\right),$$

that is, that the small increase in yield for a small increase in the value of x_1 is proportional to the deficit from the maximum $(A_1 - y)$

or $$\log_e\left(A_1 - y\right) = \log_e A_1 - c_1 x_1.$$

In practice the increase in x_1, and not x_1 itself, is measured. Calling this x,

$$\log_e\left(A_1 - y\right) = \log_e A_1 - c_1\left(b + x\right),$$

where b is the unknown value of x_1 when $x = 0$, and then $y = a_1$,

$$\log_e\left(A_1 - a_1\right) = \log_e A_1 - c_1 b;$$

hence $$\log_e\left(A_1 - y\right) = \log_e\left(A_1 - a_1\right) - c_1 x.$$

This is the original form of the equation put forward by Mitscherlich in 1909 (he uses logarithms to the base 10 which are 0.434 times those to the base e).

Mitscherlich claims that the relation between yield and the intensity of the external factor (water supply, mineral nutrient, light, etc.) is expressed by this equation, *i.e.* c_1 is independent of x_1 and A_1. Moreover, he claims that his equation holds whatever the nature of the plant, portion of the plant (corn or straw) and the time of harvest, c_1 being the same in all cases for the same factor; *i.e.* no matter what the plant, soil or weather, the yield, expressed as a fraction of the yield when x_1 is relatively very great, is the same in all cases for a given value of x_1.

A consideration of the flexibility of the equation in application, and of the available data shows that his claims are scarcely justified. The relative yield for example does vary from plant to plant and from year to year.

In making use of the equation to determine the amount of nutrient already in a sample of soil different values are obtained with different plants. Owing to uncertainty as to the depth of soil available in field cultures it is impossible to utilise data from pot cultures for forecasting the effect of manure in field cultures with the same soil. The ability to determine the amount of nutrient present and the effect of further supplies is not dependent upon the validity of Mitscherlich's equation.

He also applies his equation to spacing.

There are exceptional cases in which Mitscherlich recognises that the equation does not hold—interaction of the nutrient added with those already present.

There is a possibility of errors in the technique.

Physiological considerations: effect of variation of a factor on the relation of corn to straw; effect of intensity of factors at different stages of the life cycle (Mitscherlich uses amount of nutrient added at the commencement as a measurement of the intensity of the factor); effect of manure on the soil bacteria.

Other possible expressions for the yield-factor relation.

Mr E. J. Maskell. THE TECHNIQUE OF PLOT EXPERIMENTS

(*Abstract*)[1]

An accurate estimate of the effect of treatment upon yield depends upon a satisfactory solution of the problem of differences in yield which arise not from differences in treatment but from differences in the positions of the experimental plots. The problem is to devise some "control" in the arrangement of the plots which shall reduce the magnitude of the error arising from this cause, and to secure a valid estimate of the error arising from causes which remain beyond control.

[1] Full paper appears on pp. 373 ff.

Uniformity trials have shown that heterogeneity (the association in yield of adjacent plots) is a universal feature of soil variation. This element of regularity must therefore be taken into account in devising the arrangement of treatments. Beaven has met the problem for the simple case of two varieties (or treatments) by the arrangement shown below:

$$ab \ ba \ ab \ ba \ ab \ ba, \text{ etc.}$$

The relative positions of the treatments a and b are reversed in each successive pair of plots, and the estimate of experimental error is based upon the series of differences between adjacent plots ("Students'" method). In general this estimate of error is considerably below the estimate based upon a random arrangement—an indication of the increased accuracy of comparison secured by the method.

Two principles are involved in the method; first the comparison of adjacent plots; second the regular alternation of the direction of comparison thus—$a \, b \, b \, a$, etc.—which makes the whole arrangement balanced. The unbalanced arrangement—$a \, b \, a \, b$, etc.—is unsatisfactory and may give misleading results.

In experiments involving the comparison of several treatments, comparisons cannot regularly be made between adjacent plots, but the principle of maximum contiguity may be secured by subdividing the experimental area into as many units (trials) as there are to be replications of the treatments in each of which units each treatment occurs once. Most chessboard arrangements are of this type, and a method of analysis due to Dr Fisher is available for the estimate of the experimental error of such arrangements.

Not all chessboards are, however, balanced and in consequence in some cases a correction for position has been made. It is better to obviate the need for correction by making the arrangement a balanced one.

Where the number of replications can be made equal to the number of treatments balance may be secured by making the plan form a square, the number of rows and of columns being equal to the number of treatments. Each treatment then is made to recur once in each row and once in each column. Subject to those restrictions the arrangement is a random one.

Considerable increase in accuracy may be obtained with this method of arrangement.

Considerations similar to those outlined for field experiments should govern the technique of sampling for growth studies.

Dr R. A. Fisher. THE ANALYSIS OF WEATHER-CROP DATA

(*Abstract*)

In considering, as a whole, the variation observable in farm yields, and the causes of that variation, it is clear that the weather is by far the most influential single factor which we can name. Very little is at present known as to what features of the weather sequence are responsible for the observed large variations in yield; the study of the causal *nexus* connecting meteorological instrumental observations on one side, with crop yields on the other is one of great difficulty.

The data which at first sight seem most available are the official returns made in different countries showing the estimated average yield of each crop over an area, such as a county. Numerous attempts have been made to discover, from such data, *critical periods* in the life of the crop during which the meteorological influences are especially important. Such attempts have often used only short crop records, and by the methods commonly employed "critical periods" would have been found, even if dummy data, entirely unrelated to the weather, had been substituted for the crop record. The number of meteorological variates in the weather experienced by a single crop is enormous, especially if short periods are treated separately, and no crop record in the world is long enough to give reliable information on these lines.

A second difficulty of official data respecting yields obtained under industrial, as opposed to experimental, farming is that the accuracy of the yield estimates is unknown. Great care is undoubtedly taken by the departments concerned, but two important sources of bias seem not to have been overcome. In the first place it would appear that county averages show much less variation from year to year, than one would expect from the records of experimental yields; it seems probable that the reporters unconsciously tend to make their estimates approximate more or less to the known average of their county, so overestimating the yields in bad years, and underestimating them in good years. Secondly, the land under the several crops is not a constant quantity. In new countries, where the cultivated area is increasing, the best land is usually taken up first, and an apparent falling off in yield occurs as worse land is brought into cultivation. In England, one effect of a wet autumn is to diminish the area under wheat next year, for the farmers get behind with their autumn work, and leave a proportion of land to be sown in spring with oats or barley. I believe this last effect is the cause of the discrepancy between the results obtained by Hooker using Ministry of Agriculture crop returns, and those found from the Rothamsted crop records. For most of the year the results are in agreement, but for rain in October, which Hooker found to be markedly associated with bad wheat yields, I find an effect, never more than slightly harmful, and on many plots positively beneficial.

The long period experimental data upon which I have been working at Rothamsted appear to be capable of yielding reliable results, but unfortunately they are practically unique, and naturally cover only a few crops on a single small area. For the future we must look to data obtained under experimental conditions, obtained by parallel experiments at a large number of different centres. Only so can we avoid having to wait thirty or forty years until data parallel with those at Rothamsted have been accumulated for other crops and at other centres. The conditions for the success of such research may to a large extent be laid down in advance.

The accuracy of the individual experiments must be assured by adequate replication, and, what is equally important, each experiment must be so planned that its accuracy is capable of valid estimation. The principles of the arrangement of plots are quite definite and somewhat stringent; we have only recently come to understand them clearly, or to realise the high degree of accuracy obtainable on quite small areas of land.

Next the different factors, Weather, Variety, Soil and Manure, should not be studied in isolation. The Rothamsted data show conclusively that the weather response of the different plots is much affected by the manurial treatment. This interaction of the weather and manure is of great agricultural importance, for it shows that a manurial treatment which is unprofitable in some seasons may be profitable in others; it points the way to methods of enabling the farmer to adapt his manurial treatment to season and climate.

Probably of even greater importance are the inter-relations of Weather and Variety. It can scarcely be questioned that varieties differ greatly in their weather responses. This has proved a great stumbling block in variety trials, for the results of one year's experiments are often, one might say usually, contradicted by the results of repeating the experiment next year. Only by very prolonged trials could we be sure that our experiment had been carried out under average weather conditions, and then our choice of variety will only be a safe one for the climatic district of the experiment. On the other hand, we should set ourselves deliberately to evaluate the differences in the weather response of varieties, and plan our experiment so that this can be done. The main weather features are now known all over the country, and if in testing varieties we were to evaluate the differences in weather response, we could assign to each variety the region in which it was the best. It would in fact be possible systematically to adapt variety to climate.

Mr F. L. Engledow. ANALYSIS OF CEREAL YIELD
(*Abstract*)

An analysis of cereal yield similar to that made for cotton by Dr Balls is difficult owing to the tillering habit. The tillers (axillary shoots from basal nodes) develop their own root systems so that the cereal plant is an aggregation of more or less interdependent units. Spacing increases the number of tillers, but as all the late tillers do not ripen simultaneously this is in practice a disadvantage. Seed rate is therefore important, but as in the field many plants die, the actual seed rate must exceed that which would be calculated by observation on individual plants.

In a yield investigation on the basis of the plant as a unit certain characters indicative of the degree of development of the individual were found to be useful, *e.g.* the maximum length of the first green leaf and the time at which it was attained. A large population of plants growing under "uniform" conditions was observed systematically and the degree of development of each individual at various stages of its life was noted; from the data it was possible to select "modal" plants for each developmental phase. At harvest certain plants with parallel life histories were picked out, but it was found that there were great differences in the way in which their dry weight was distributed among their tillers and between grain and straw in the individual tiller. This case is eloquent of the difficulties encountered in dealing with the cereal plant as a unit.

The "modal plant" method is the most practicable one for small scale yield trials. In the past insufficient stress has been laid on testing varieties in different localities under different climatic conditions. Future yield trials conducted on "modal plant" lines simultaneously in several localities should facilitate the selection of improved varieties suited to particular districts.

Dr H. M. Leake. Rainfall and Crop Yield in the Tropics

(*Abstract*)

In tropical countries, and especially in those subject to a monsoon type of rainfall, owing to the elimination of temperature as a limiting factor to plant growth, the relation of rainfall to plant growth becomes very marked. For the determination of the inter-relation between rainfall and crop yield it has been customary to divide the season into periods and to work with the aggregate of rainfall during these periods. A complex problem illustrating the use of partial correlations has to be worked out. One of the best known examples of investigation along these lines is that of Hooker.

By a simple method, to be described, the rainfall is here recorded as a continuous graph. Applications of the method seem to indicate that a particular crop will have a characteristic graph. If this prove to be true it may become possible to forecast from the rainfall graph of a tract the crops most likely to succeed in that tract. Further, it may be possible to classify varieties on the basis of the graph.

This varietal graph would take the form of a band, greater or less in depth, and the suitability of the climate would be determined by the extent to which the seasonal rainfall graphs fall within this band.

Since crop returns are, in the majority of cases, approximate estimates only, and since experimental results lack varietal continuity over a sufficiently long series of years, the necessary data are not available. The most rapid method of collection would be a series of co-operative experiments conducted with standard varieties and under a series of divergent climatic conditions.

PLANT PHYSIOLOGY

THE BIOLOGICAL PROBLEMS OF THE COLD STORAGE OF APPLES

(CHAIRMAN: PROFESSOR V. H. BLACKMAN, F.R.S.)

Dr F. F. Blackman, F.R.S. GENERAL INTRODUCTION
(not given orally)

The aim of this session is to present the biological and physiological problems that have been encountered in the scientific study of the keeping properties of apples and the best methods of storing them, which is being carried out by a staff of botanical workers supported by the Food Investigation Board of the Department of Scientific and Industrial Research of which Mr W. B. Hardy, Sec. R. S., is Director. The contributions to be made to this session do not represent all the workers or all the matters under investigation, but a selection has been arranged to illustrate the fundamental physiology of the apple.

The problems are of many types; some are narrow and biochemical, others broad and biological. In practice, no one problem can be investigated in isolation, or presented quite independently in discourse. It may therefore be advantageous to provide, as general introduction, a brief dissection of the interwoven types of problems that are involved in the subject matter of the four sections that make up this session.

Heterogeneity of apple cells in time. The tissues of a stored apple exhibit a steady drift in their condition with time, labelled by the consumer as unripe, ripe and rotten. Considering this physiologically we find we have to deal with successive stages in what may be called the *ontogenetic metabolic drift*, which characterises the progress of cells from adolescence through maturity to death. Work on stored apples concentrates our attention on the later or senescent stages of this drift, with which is associated the colour change from green to yellow to brown. This drift in apples is well marked by changes in chemical composition and in metabolic activity, as in respiration. We have to realise that a state of equilibrium at no time exists, and that actually the metabolism of each cell is never quite the same to-day as it was yesterday.

Heterogeneity of apple cells in space. Another set of problems arises from the fact that the unit of all our physiological investigations—the whole apple—cannot be regarded as homogeneously constituted. Even if all the cells of the flesh of one apple were inherently identical, physiologically, they are not all under identical metabolic conditions so that at any moment of investigation they may well be at different stages of their individual ontogenetic metabolic drifts. On analysis it is found that the distribution of organic acids and other substances in the regions of the flesh is not uniform but tends to show radial gradients, distinguishing the centre of the flesh from the periphery. When moderately injurious substances or conditions of storage are applied to apples, it is rare for all regions of the flesh to be affected equally; thus with excess of carbon dioxide the whole centre may become brown and killed while the periphery remains in perfect condition—"brown heart" of apples. Other injurious effects may be quite superficial—as "superficial scald."

The difficulty which this heterogeneity creates for the investigator is this, that if the cell-population of an apple is not practically homogeneous, then the observed progress of respiratory activity of a whole apple, during senescence, does not give a true picture of the course of respiration of a single constituent cell, but tends towards being only a statistical curve of the distribution of states of greater or less activity amongst the cell-population.

The combination of these two heterogeneities is a considerable obstacle to progress, for it makes the exact repetition of experiments in sequence almost impossible, and the utilisation of simultaneous control experiments very difficult.

In the first communication to this session it will be shown that senescent leaves present a metabolic drift with the same characteristics as that of the apple, and the problem of the essential causation of the drift will be discussed.

Rate of progress of senescence. The control of this rate is the part of our general enquiry that bears directly on economic practice. The rate of senescence is made evident by the quicker or slower succession of the series of symptoms associated with ripeness and rottenness. It is not in itself a progressive change in an amount of matter, and so is not to be treated on physico-chemical lines. It may possibly be brought about by some change in concentration of cell constituents, but at present the rate of the drift has to be treated empirically.

The arrival of the last phase alone, that of brownness, can be timed by inspection. At the surface of apples this is usually associated with the germination of the ever-present spores of rot-fungi, but in the sterile centre of the apple this can be reached without fungal invasion. In commercial practice, when 10 per cent. of the apples have reached the last

2–2

phase, the contents of a store have to be marketed at once; but a statistical study of the ultimate fate of the whole store-population, with a determination of the invading fungi, has been carried through by Mrs Kidd.

Internal evidence of the rate of the senescent drift in all stages is obtained by chemical analysis of apples in different storage conditions. The drift of the various carbohydrates, of organic acids and acidity, of the nitrogenous substances, and of pectin changes in the cell-walls are being followed systematically.

Such analyses involve the destruction of the apple; at present we have but one method for studying the living apple continuously throughout its drift, and that is by measuring its respiration.

Our knowledge of the biochemical drift will be presented by Dr Haynes for the cell-contents and by Miss Carré for the cell-wall in the second section of this session, while the evidence from respiration studies will be given by Dr Kidd in the third section.

At least three methods of retarding the rate of senescence seem possible: two of them, the use of cold and of "gas storage," which turns on the use of gas-mixtures poorer in oxygen and richer in carbon dioxide than ordinary air, have been investigated. A third possibility is that of lowering the water-content of the apple by partial drying. This has not been yet investigated because of the economic obstacle, that no salesman will look at shrivelled apples however good they may be to the taste.

Predetermination of keeping power by conditions during growth. Considered biologically our enquiry is one into the longevity of a population, not of human beings but of apples, and may be treated on parallel lines. In contrast to the environment of a human population that of a cold store is extremely uniform. But growing on the tree the apples, in adolescence, have had more varied and adventurous careers, and we have examined the extent to which the circumstances of early growth affect the duration of life after picking. The effect is considerable and complicates very much our lines of physiological enquiry. The fruits of the different varieties of apples, of course, differ largely in longevity and keeping power, but it has now been made clear that, for any one variety, the longevity varies greatly according to the age of the tree, the soil of the orchard, and the season in which the crop was grown.

In the last section of this session Dr West will present our knowledge of this biological side of the whole investigation.

Dr F. F. Blackman, F.R.S. The Drift of Metabolism
through Senescence to Death

(*Abstract*)

The intensive study of the behaviour of apples in storage widens the
outlook of the plant physiologist because it extends his knowledge of
metabolism beyond the phase of maturity of the plant, into that of
senescence. It is now clear that metabolic activity does not merely tail
away towards a zero point at death, but that there are specific changes
in the living system marking the transition from maturity to senescence.

It seems that we may profitably formulate the general conception of
a continuous *ontogenetic metabolic drift* throughout the life-history of the
cells and tissues of the various component parts of the higher plants.
Some drift of this nature might be found in the short life-history of the
individual cells of a unicellular plant, but probably much of it, in the
higher plant, is the outcome of mutual relationships between the parts
and the whole.

The intensity of respiration supplies the most general key to the
magnitude of metabolism, and a comparison of available data on the
respiration of plants in germination, adolescence, maturity, senescence
and death suggests a formulation of the cause of the drift as one with
two peaks, one initial and the other at the onset of senescence. Though
respiration studies give only a summary measure of metabolism they
have the advantage that they can be carried on without destroying the
organ investigated. Unfortunately, there is no convenient standard by
which to express magnitude of respiration throughout ontogeny, as the
amount of inactive skeletal matter in the tissues is not constant.

Attention is drawn to two sets of experiments from which an idea of
the course of the respiratory drift can be obtained.

The initial and mature phases are illuminated by measurements of
the respiration of Sunflower plants[1], and the later stages by the writer's
unpublished experiments on the respiration of leaves of Cherry, Laurel
and Tropaeolum. It is interesting to find that apples show the same
increase of respiration at about the stage when they pass from green
to yellow as do isolated leaves; and this is held to mark the change from
maturity to senescence.

The problem of the causation of these phases of respiratory activity
is of deep physiological interest. Quantitative variations of a metabolic

[1] Kidd, Briggs and West, *Proc. Roy. Soc.* B, 92, 1921.

process might be caused by alterations of three different types, (1) variations of the amount of effective catalyst in the cells, (2) variations in the concentration of the effective substrate, and (3) alterations of the physical organisation of the protoplasm.

Theoretically the whole course of the respiratory drift might be due to the first cause, but there are not yet direct experiments on this matter. For only one cell-catalyst—amylase—have we data on alterations of amount with the different stages of development. Sjoberg's data show that the alterations are considerable, and their drift can be brought into harmony with the form of the respiratory drift.

In some stages respiration can be clearly related to the second cause, represented by sugar concentration, but it would not be possible to interpret the whole drift in terms of substrate.

With regard to the third suggested cause, increase of protoplasmic permeability seems to be the dominant factor in the death of the cell and it may well be significant in the senescent stage.

Researches are in progress aimed at disentangling and evaluating these factors.

Dr D. Haynes. THE METABOLIC DRIFT IN APPLES. BIOCHEMICAL CHANGES

(Abstract)

The most obvious chemical change in stored apples is loss of acidity. In an acid apple such as Bramley Seedling acids may run down from $1\frac{1}{2}$ to 0·2 per cent. or less if they are kept till late in the season. Cold storage keeps up the acid level and this is one of its more important functions, since rate of loss of acid appears to determine the rate of progress of disintegrative chemical changes in the senescent apple. Too great a degree of cold causes injury, and the damaged condition is shown by a rapid increase in rate of loss of acid.

It has been found that acidity plotted against time in store gives a logarithmic curve. Rate of loss of acid is thus proportional to concentration of acid throughout the period of storage. It is probable that acid is continuously produced by the oxidation of sugar at the same time that it is consumed; if this be so, for the logarithmic relation to hold, the rate of production of acid must also be proportional to the acid concentration. Injury from overcooling, which is shown by a breakaway from the logarithmic curve in the downward sense, is not accompanied by an increase in the rate of respiration; therefore what is apparently a more rapid rate of consumption of acid must on the above hypothesis be interpreted as a less rapid rate of production, and injury by cold must involve an interference with the normal respiration processes.

Deviation from mean acidity tends to be greater in cold stored apples than in those kept at higher temperatures. This indicates that in cold store the more acid apples are those which keep up their acidity best. More investigation of the behaviour of apples in this respect is necessary, but it may be suggested that it is a consequence of the structure of the cells forming the tissue of the apple since a definite relation has been found in Bramley Seedling apples between acidity and nitrogen content (determined by Miss Helen Archbold), which suggests that the more acid apples are those whose content of protoplasm is low. A somewhat similar relation has also been found to hold between acidity and the cellulose-pectin residue (the cell-wall stuff) of the apple. A comparison of nitrogen content with rate of respiration (determined by Dr F. Kidd and Dr C. West) shows a marked correlation between these two factors, which justifies the assumption, made by Miss Archbold on chemical grounds, that the nitrogen content is a measure of the protoplasm.

There is some evidence that the substance of the protoplasm itself is slowly attacked during the whole storage period and that this process is accelerated when "internal breakdown" sets in. The investigation of these changes is difficult owing to the very small proportion of nitrogen in the apple and the very large differences between the nitrogen content of individual apples. The cell-wall undergoes a much more marked disintegration, the proportion of substance insoluble in alcohol steadily decreasing as acidity falls. This process takes place more rapidly as temperature rises. It is probable that the disintegration affords respiratory material to the cell and it may supply salts; on both these counts it may not be entirely to the advantage of the apple to keep its cell-wall intact.

The proportion of total sugar changes very little during storage, as loss in respiration is largely balanced by the increase of concentration produced by transpiration, but the proportion in which different sugars are present alters considerably. Miss H. M. Judd has found in particular that sucrose becomes rapidly inverted. There is some evidence that a low concentration of sugar is a factor favouring "internal breakdown" in cold storage.

Miss H. M. Carré.   THE METABOLIC DRIFT IN APPLES. THE PECTIC CHANGES

(Abstract)

The pectin compounds of fruits exhibit variation as ripening proceeds, and the following account of a series of experiments carried out with apples, shows that pectin substances are closely connected with the ripening of the fruit, and the metabolic drift from maturity to senescence.

Three forms of pectic compounds have been distinguished in apple tissue by chemical means, pectin, pectose, and pectic acid, or its salts. *Pectin* is present in the juice and can be readily washed out from the tissues since it is soluble in water. *Pectose* forms part of the cell-wall and is insoluble in water, but it can be hydrolysed into the soluble Pectin by treatment with dilute acids. The acid simultaneously decomposes any salts of Pectic Acid into the free acid which is insoluble in water and in acids, but is readily dissolved out by weak alkalis.

The distribution in the tissues of these three pectic constituents was examined by microchemical methods, and conclusive evidence was obtained that the middle lamella is composed of pectic compounds, and that the cell-wall is not homogeneous, but is an intimate mixture of the pectic substances in combination with cellulose from the very earliest stages of fruit formation. This widespread distribution of pectic compounds in the apple suggests that they play an important part in metabolic processes, and this view is confirmed by chemical observations of the relative proportions of pectose, pectin, and pectic acid, and of the changes which they undergo during development and senescence of apples.

It was found that the soluble *Pectin* is negligible in amount in unripe fruits, but develops in the juice during ripening at the expense of the Pectose till it reaches a maximum amount. At this stage the apples exhibit the condition known as "mealiness," and the fruit is over-ripe from a market point of view.

As over-ripeness develops the pectin content begins to fall, presumably by hydrolysis into its decomposition products.

The increase of pectin during ripening is balanced by a corresponding decrease in the *Pectose* content of the cell-wall. It follows, therefore, that Pectose is the precursor of the soluble pectin which it apparently gives rise to by hydrolysis.

Finally as the cell-walls become depleted of pectose and the over-ripe soft condition becomes more advanced, the pectic acid and pectates of the middle lamella also begin to decrease in amount by a process of decomposition. This gradual process continues until in the last stages of senescence there is a total disruption of the tissues, and ultimately a marked decrease in the total pectic constituents.

Microscopic observations made on the tissue periodically bear out the results arrived at from the chemical side.

It was observed that as the season advances the walls get thinner and contain less pectose material, and that the soft over-ripe condition occurs simultaneously with a gradual separation of the cells, owing to the breakdown of the middle lamella pectic substance. In extreme cases of senescence, the whole framework of the cells is observed to have disappeared, and the cells separate from one another on the slightest pressure.

From the above results we may conclude that the ripening of the apples and the metabolic drift, which the tissues undergo leading to death, are closely associated with a series of gradual changes in the pectic compounds which occur in the cell-walls and cell-framework.

We cannot yet say whether the pectic changes in the cell-walls are secondary to the protoplasmic changes expressing the metabolic drift or whether they are primary bringing about fatal disorganisation of the tissues.

Dr Franklin Kidd. THE RATE OF SENESCENCE AND THE CONTROL OF IT BY CONDITIONS

(*Abstract*)

(1) *Evidence of Senescence Rate by Study of Fungal Invasion in Storage (Researches of Mrs M. N. Kidd)*

The course of senescence in apples has been studied by investigating the changes in susceptibility to fungal disease with advancing age. This work, the results of which are now incorporated in three papers[1], has been carried out by Mrs Kidd. The susceptibility to disease has been studied statistically and is expressed as "disease rate," that is to say the

[1] M. N. Kidd and A. Beaumont, "Apple Rot Fungi in Storage," *Trans. Brit. Mycol. Soc.* Vol. x. 1924, pp. 98–118.

M. N. Kidd and A. Beaumont, "An Experimental Study of the Fungal Invasion of Apples in Storage, with particular Reference to Invasion through the Lenticels," *Ann. Applied Biology* (in the press).

M. N. Kidd, "Variations in the Rate of Senescence and Fungal Invasion in Apples." Thesis, deposited in the University Library, Cambridge.

number of invasions taking place per unit time per unit surface. The surface of apples is thickly covered with spores of various apple rot fungi at the time of gathering. Those that are important from the point of view of subsequent invasion are those which lie in the lenticel cavities. The change in disease rate with age can be expressed as a curve. Either a single curve can be obtained without discrimination as to species of fungus invading, or a number of curves can be constructed one for each individual species of invading organism. In either case the curves are of the same type (Fig. 1). Susceptibility steadily increases with age.

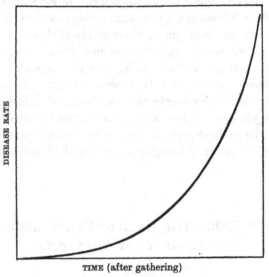

TIME (after gathering)

FIG. 1. Fungal Invasion.

The next stage in the enquiry was to attempt to explain the increasing susceptibility with age in more detail in terms of the condition of the apple, the fungus and the environment. As the result of an experimental study of fungal invasion of apples in storage, with particular reference to invasion through the lenticels, the following conclusions have been tentatively reached: (i) that the fundamental controlling factor is the exosmosis equilibrium of the general mass of the apple tissue in supplying moisture and nutrient outside the living cell, (ii) that the critical stage of the fungus conditioning invasion is not necessarily the germination of the spore, but the attainment of a certain critical vigour of development.

The results obtained in this study have indicated the probable explanation of the phenomenon of lenticel spotting in apples in storage, about which there has been disagreement as to whether the disease is functional or fungal in origin. Experiments indicate that the disease is functional in origin and that a local physiological breakdown of the apple cells in the neighbourhood of the lenticel supplies sufficient moisture and nutrient for fungi present in the lenticels to germinate, but not enough to enable it to reach a critical vigour of development necessary for active invasion.

Probably a competition for the moisture and nutrient liberated by the disorganised cells round the lenticels takes place between the underlying apple tissue and the fungus.

(2) *Drift of Respiration with Senescence (Researches of Dr Franklin Kidd and Dr Cyril West)*

During growth on the tree the carbon dioxide production by apples per unit fresh weight decreases. A mature apple has a respiratory activity of about one-tenth that of the young apple fruit. This change in rate of

Generalised Curve of Respiration (CO$_2$ production) of an Apple

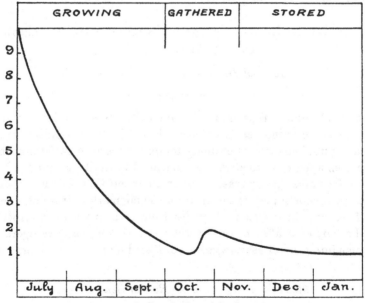

FIG. 2. Respiration of Apples.

carbon dioxide production is correlated fairly closely with a corresponding decrease in the concentration of acid in the apple. The concentration of sugars in the apple on the other hand increases during growth to maturation on the tree.

About the time that apples are normally gathered the respiratory activity reaches a minimum. It then increases again, approximately doubling itself, and then finally falls off to values about equal to the original minimum which are attained at the end of the life of the apple.

The rise from the minimum is a general phenomenon which has not been previously recorded. It is not obviously correlated with any change in acid or sugar concentration in the fruit and is not conditioned by gathering. Fig. 2 gives the generalised curve for the respiration of an apple throughout its life on the tree and in storage.

When apples are gathered somewhat under-ripe the respiration continues to fall in storage. The minimum is reached after gathering and the subsequent rise and fall towards the end point then occur. When these phenomena are studied in relation to temperature it is found that the time taken to reach the minimum is scarcely affected by temperature, likewise the final fall off to the end point. In contrast to this the time taken to rise from the minimum to the maximum is markedly influenced by temperature; for example, 800 hours at 8° C., 80 hours at 22·5°.

Dr Cyril West. LONGEVITY AND VARIATION OF SENESCENCE RATE AMONG APPLES IN STORAGE

(*Researches of Drs Franklin Kidd and Cyril West*)

(*Abstract*)

The world's apple crops are produced locally and over relatively short seasons. Owing to inherent differences the cultivated varieties of apples have a natural longevity at ordinary temperatures after gathering which varies from a few days to about six months. But during recent years the demand for apples has increased to such an extent that in order to maintain a continuous supply of this fruit for the big markets it is both stored out of season and transported by sea from regions where the ripening period occurs at a different time of the year. In both cases refrigeration has come into general use as a means of retarding the ripening processes (*i.e.* rate of senescence).

Methods of Retarding the Rate of Senescence

(*a*) *Temperature.* While cold storage as compared with ordinary storage has been found as a rule to extend the storage life of the apple (see Table I) this is by no means invariably the case. In certain cases ordinary storage has proved superior to cold storage owing to the premature onset at the low temperature of a functional disease known as "internal breakdown."

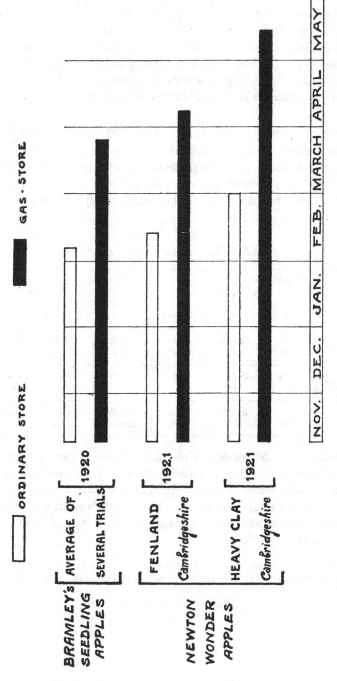

FIG. 3.

The horizontal bands indicate the length of the commercial storage life in each case.

(*b*) *Composition of the Storage Atmosphere.* During the past four years it has been found that by storing apples in atmospheres containing 5–8 per cent. oxygen as compared with 21 per cent. in normal air, and 12–15 per cent. carbon dioxide as compared with a minute trace in normal air, the storage life of the fruit can be materially lengthened (see Fig. 3).

(*c*) *Humidity.* Very little is known at present as to the effect upon the longevity of apples of the humidity of the storage atmosphere.

Pre-determination of Storage Life by Conditions during Growth

A series of storage trials carried out for a number of years under carefully controlled conditions with a single variety of apples, both at ordinary temperatures and at low temperature, have shown that the storage life of the fruit depends not only upon the storage conditions, but is largely

Table I

End of Commercial Storage Life[1]

Season	In cold storage, with Bramley's Seedlings off			In ordinary storage, with Bramley's Seedlings off		
	Fenland	Chalk soil	Silt soil	Fenland	Chalk soil	Silt soil
1920–21	Early Jan.	Early March	Late May	Middle Dec.	Middle Dec.	Early March
1921–22	Late March	Middle April	Late May	Early Jan.	Late Jan.	Middle Jan.
1922–23	Late Jan.	Middle Feb.	Late Feb.	Middle March	Early March	Early April

pre-determined by the conditions obtaining during growth on the tree, the rate of senescence being materially affected by such orchard factors as locality or soil, season, climate, etc. The variation in longevity and keeping power of a single variety, namely, Bramley's Seedling, according to the soil of the orchard, and the season in which the crop was grown, is clearly shown in Table I.

[1] The end of the "commercial storage life" of the fruit is an arbitrary point, and it has been judged as ended when the loss due to fungal rots or functional diseases amounts to 10 per cent.

GENETICS

THE ECONOMIC POSSIBILITIES OF
PLANT BREEDING

(CHAIRMAN: MR W. BATESON, F.R.S.)

Mr F. L. Engledow

In the experience of most of us, this occasion must be almost unique. Imperial gatherings are rare: perhaps even more rare are formal assessments of the economic possibilities of plant breeding. But five-and-twenty years of breeding has been done in the more orderly way which Mendel's principles opened up, so that a discussion of possibilities is not now premature. The reticence of plant breeders has not kept the achievements and potentialities of their art in obscurity. Popular journals have for years, with insistence, with optimism, but not with much truth or understanding, stirred public expectancy. A few months before Mendel's work was disinterred, the *Strand Magazine* had an article on "Wonders in Wheat Growing." On evidence mainly photographic, it assured its readers that from breeding had sprung achievements just as wonderful as making two blades of grass grow where only one grew before. The journalists are no more quick to improve than our crops, for only a few years ago, in *Chambers' Journal*, dated, happily enough April 1st, were the following words: "The vast possibilities of plant breeding can hardly be estimated. It would not be difficult for one man to breed a new wheat which would produce one more grain to each head....What would be the result? In five years in the U.S.A. alone, the inexhaustible forces of nature would produce annually without effort and without cost 15 million extra bushels of wheat...etc. But these vast possibilities are not alone for one year or for our own time and race but are beneficent legacies for every man, woman or child who shall ever inherit the earth." Behind the amusement we derive from such glimpses of ourselves through others' eyes, there lurks an uneasy feeling. Has much been done to bring the obvious possibilities to maturity and have we clearly focussed the possibilities that remain? To satisfy that feeling we must go over the past, dealing solely with facts: we must look ahead, admitting reasonable speculation.

For comprehensive discussion there are two bases. First, for the crop plants in turn, might be set out a list of economically important plant characters—these are the units conceivably available for building. Logically, all plant characters are equally important, for no one can be defined without reference to others. But plant breeding is an art, not a science, so we need not hesitate to describe as plant characters those attributes which, effectively, do count in agricultural practice. Take wheat. There are degrees of susceptibility to diseases, gradations in growth period, tall and short straws and the like. Various combinations of these things are theoretical possibilities of improvement. Heritability and all its implications enter into these possibilities and thus constitute the second basis of our discussion. Interwoven everywhere runs the question of practicability. It is useless to speculate on what is impracticable, and equally unwise to neglect anything practicable which may lead to improvement. In so big a field, it would be rather doctrinaire to try to implement this comprehensive logical policy in making our assessment. It will serve as a guide but it must not lead us far away from agricultural practice.

My purpose is to deal only with the crop plants of England, in particular, with wheat. Selection is to be separately discussed so that our essential concern is hybridisation. Of the heritability of plant characters and, in a wide sense, the procedure which is entailed, it will be best to speak first. Manipulative difficulties are few. For oats we still need better hybridisation technique, and with the naturally out-pollinating crops there are considerable difficulties especially in the provision of large stocks of pure seed. I think it would be of general interest if those who are constantly faced with these difficulties would give us their views. They are a definite limitation to breeding possibilities. What has been called "crossability" interests us here. Some inter-species crosses used to appear to hold great promise. Rivet wheat (*T. turgidum*) × Bread wheats (*T. vulgare*) is full of theoretical possibilities but in practice has yielded, mainly, morphological freaks. The chromosome difference of the parents, 14 and 21, involves great sterility, and there appears little promise of economic success. I imagine that the effective infertility of *Brassica napus* × *B. rapa* must similarly have occasioned disappointment and that, generally speaking, we cannot look very hopefully to inter-species crosses in English crop plants. Linkage threatens us at some points. There is no *a priori* reason why it should, but it was feared long before it received a name. Strength and yield in wheat were once declared incompatible. That, emphatically, was wrong: but winter hardiness and high yield in oats like free tillering and the six-row habit in barley: or high yield and the absence of awns, still appear non-associable. But geneticists are rapidly advancing the knowledge of linkage and as time goes on it appears to be less and less of a fundamental obstacle.

It may seem a digression to take into account the machinery for testing and marketing new hybrid forms. But in this lie some of the prime factors limiting the economic fulfilment of breeding possibilities. In testing, especially in yield testing, great statistical accuracy is now attainable. But the tests do not disclose some points of field behaviour. A wheat that characteristically sprouts in the stook or lodges on a very rich soil, may excel in a formal test. Or a new form, under average in a test in Kent, might be valuable in Shropshire. We cannot go further into this great question now. More opportunity for informal observation over a wide range of conditions is required for the products of breeding. It would enhance the possibilities of reaping the full benefit of breeding results though, naturally, it is not to be regarded as anything but a supplement to formal statistical tests.

Here we may draw the line under this list of difficulties of procedure. The total is not formidable and, as far as our considerations have carried us, it seems permissible to look hopefully at the possibilities of breeding.

Now we must turn to the essence of our discussion—the fundamental possibilities, the possibilities of building new aggregates of desirable plant characters. What has already been done is an unassailable index to possibilities, so let us start with a salutary lesson from history. The cereal yield in England can be traced from about ten bushels per acre in Manorial times to about thirty-two bushels in 1883. In that year the Board of Agriculture began to collect definite statistics. These, covering the forty years to the present time, show for wheat, oats, and barley no noteworthy increase and that is a disquieting fact, for yield, of all things, is the sign of crop improvement. In Ireland, Holland, Denmark and Germany, during the same forty years, increases of from 20 per cent. or more in cereal yields are recorded. New forms, some of hybrid origin, have, beyond any question, brought part of this increase. Now in England there has been no lack of new forms of cereal plants nor has modesty constrained the propaganda which heralded them. Old seed catalogues abound in forgotten names and, omitting the decency of obituary notices, introducers are still marketing yet newer forms. The dismal constancy of the national yield makes us ask if the highest possible yielding capacity has been attained. At the least it makes us speculate on the possibility of evolving an altered system of breeding for yield. Other plant characters and other parts of the Empire may be expected to furnish a happier story, for centuries of skilful and incessant endeavour have gone to the improvement of cereal yielding capacity in England. But it is the outstanding concern in this small country and if there appear possibilities of yield improvement, we may be happy about betterment along other lines. It is for these reasons that, in what follows, considerable attention is given to cereal yield.

The economic possibilities of plant breeding are measured by the prospects of breeding new forms of crop plants which will bring the grower a better financial return than the old. Enhanced returns may accrue from higher yield, better quality, and from a number of other amenities. Under these headings it is proposed to conduct the enquiry into plant characters or the available units in breeding.

Let us begin with quality and, although it is familiar, examine the story of Yeoman wheat. All other wheats suited to England are weak and, for flour milling, have to be blended with strong imported wheats, the English forming 20–50 per cent. of the blend. Inland millers are being crippled by the rail freights on the foreign wheat they bring in from the ports. Their plight is desperate and port mills, grinding no English wheat at all, are absorbing the whole milling industry. Correspondingly, the mill demand for English wheat is falling and the farmer suffers. Yeoman thrives over most of our wheat area and its strength is such as to make a blend containing from 60 to even 100 per cent. Yeoman, suitable for bread flour. It can be bought at the door by many inland millers and it is already—I quote the Linlithgow Committee—producing a revival in inland milling. I am anxious not to overstate the case. Yeoman is not the heaviest cropper all over our wheat area: but there is no reason why other equally strong wheats should not be specially bred for every part of that area. Again, the strength of Yeoman is not the same in all seasons and localities: but it is always greater than that of any other English wheat. There is one thing more. We used to look on Fife as the world's strongest wheat. That is possibly not the case. From yet stronger parents may be bred heavy cropping forms of higher strength than Yeoman—a consideration opening up big possibilities. All this is well known, but its basis is solid fact and it is well to fasten on to the successes, for in breeding we meet failures in abundance.

A financial consideration arises here. Millers are loth to pay more for Yeoman than for other English wheats and that is quite natural. But it has made pessimists ask if breeding for quality or strength is really going to help the farmer. Candidly and briefly, these are the salient facts. Some farmers do not recognise a Yeoman sample when they see it: others market any good red sample as Yeoman. Some millers cannot recognise Yeoman and thus may market flour from other wheats as Yeoman flour. The bakers still find it hard to believe—on "principle"—that English flour is fit for bread. These are only transient difficulties and there is no doubt that improved quality is finding its market and its price.

We may reasonably conclude that in the improvement of quality in cereals are definite possibilities. Oats are as yet untouched but the lines of action are visible. Among the barleys the scope is probably not so great, but the latest successful hybrid—Spratt Archer—in which high yield and excellent quality are combined, ought to be the forerunner of a considerable series. Keeping-quality in onions, sugar-content and non-bolting in mangolds, flavour and cooking capacity in a number of vegetables, cumulatively represent untouched breeding possibilities.

We can devote only a little time to what have been called other amenities. Directly or indirectly, most of them affect yield or quality and all influence financial return. Disease resistance comes first. Little Joss wheat is a living testimony to the practicability and value of uniting yielding capacity and disease resistance. Now it is a significant fact that among the wheats—and there are probably no less than 3000 forms—are a considerable number specifically resistant to one or more fungoid diseases. Quite recently a wheat from remote China has furnished a hybrid which crops well and is resistant to *Puccinia triticina* in the Argentine. Exhaustive trial can scarcely fail to furnish several other wheat areas with resistant forms and once such a form is found there is practical certainty of a successful breeding sequel. South Africa—to name but one potential wheat territory—needs rust resistant forms and could find them. Bunt, smut, mildew, and *Helminthosporium* are other baneful cereal diseases and, save mildew, we may say of them all, that the beginnings of breeding for resistance are demonstrable.

In the habit of early ripening are considerable potentialities. Harvest may linger even till Christmas Day in parts of north Yorkshire, an unhappy outcome of the lack of early ripening forms. Yet among the cereals, growth period has a very wide range so that the building unit is available and the heredity of the case fairly clear. What has been done, notably with beans and hearting broccoli, can be done with cereal crops.

Most needful and most intractable of this group of attributes is winter hardiness. Scarcely an English crop can be named whose growing would not be remuneratively extended by the production of more hardy forms.

In some cases hardy or half-hardy forms already exist, but invariably in association with relatively low yield. The genetics of hardiness still remains unknown. Let us consider oats. A full hardy, white, oat would be a great asset. It has been much sought, even prematurely proclaimed, but never found. Efforts to breed it have brought fresh knowledge and even encouragement. Still more knowledge must be added and with that, it is believed, will come success. Sugar-content of the plant tissues and morphological habit may now be said to offer a practicable index to hardiness. What we still lack is an alternative to the dreary waiting for a hard testing winter such as comes to us but once in five or six years. This we may find, before long, in the low-temperature stations. Even if distant, the great possibilities are by no means unapproachable.

Liability to lodging may conceivably set the limit to cropping capacity for it increases with the yield. Its real causes are only now beginning to be traced, but some of the most recent wheats are proof that it can be "handled" by the breeder. Swedish Iron wheat, though of low quality, crops very abundantly on many soils and stands magnificently. Barleys, the six-row forms in particular, offer special difficulties. "Neck-breaking" has been almost "bred out" of the two-row barleys without detriment to yield and malting value. In the six-row barleys corresponding improvements have begun and, proof of another possibility, more free tillering has been implanted in them. Some of the best Continental six-row barleys stand witness to this. No one familiar with an area such as our Fens where half the fields are too laid to be cut with the binder, can have any doubts about the big possibilities that remain in this direction.

There is a good market now for poultry wheat and the poultrymen are paying top prices. When more is known of the feeding value of the different wheats it may become a sound economic proposition to breed a poultry wheat. For biscuits too, the production of special wheats may soon prove remunerative. Fodder oats for green-feed and ensilage, are beginning to arouse interest and some of the forms abounding in foliage but producing little grain, which appear almost inevitably in breeding for grain production, may prove valuable. Then again, England still lacks a good spring wheat. All these requirements rank as reasonable possibilities, for they appear to call for breeding achievements of an order already attained.

Yield, because it is the prime concern of crop husbandry and the greatest difficulty in breeding, has been left till last. What may be called the ancillary yield characters, non-lodging, rust-resistance and the like, have been considered. There remains the elusive complex best described, perhaps, as the intrinsic yield characters. We have to consider the possibilities of identifying and transposing them in such fashion as to breed forms of greater intrinsic yielding capacity. By crossing two high-yielding forms and meticulously selecting and testing the segregates, there have been produced, in the last twenty years, cereal forms which, in certain localities, outyield all pre-existing ones by 3–5 per cent. This, I think, is a fair statement of the case. Such advances, if extended to the whole cereal area, would augment the growers' net returns by some £4,000,000 annually. It is estimated that £40,000 spent on breeding and testing in Ireland has brought an increase equivalent to £250,000 per year. We may thus approximately gauge in money the ultimate possibilities of the existing methods of breeding for higher cereal yield. We cannot foretell all the difficulties which will be involved.

When, for any locality, a 3–5 per cent. increase has been obtained by breeding a new form, further advance of the same order is exceedingly difficult to obtain. Of that we may be sure. Now diligent search all over the world for new parent forms, though of proved value in regard to disease resistance, will not directly aid us in breeding for higher yield in England. Forms of superlative intrinsic yielding capacity are not to be found. Of that, also, we may be sure. It may be that, generally speaking, the maximum of intrinsic yielding capacity has been reached. If that be so, disease resistance and the other ancillary yielding characters hold all the remaining possibilities. Many are genetically tractable and, as a body, they still offer great scope to the breeder.

But, theoretically, there is yet another line of possibility. It is far removed from practice but discussion of possibilities is our purpose: and behind us is the driving force of an industry which existing cereal forms are not at this moment preserving from decay. Craving indulgence for arbitrary phraseology, we may thus propound the possibility. We ought to try to *analyse* intrinsic yielding capacity, to identify the *plant characters* which mainly govern yield: then by suitable hybridisations, attempt to breed optimum combinations of these characters. The italics signalise an arbitrariness of wording adopted so that we may deal in comfort with the elusive conception, "yield."

If in England we grew, side by side, Squarehead's Master and any Durum wheat, we should in all circumstances find the same result. The Squarehead's Master would give the higher yield and, correspondingly, tiller more freely, have larger ears, and better filled grains. Finding so constant an association of high tillering, etc., with high yield we may, retaining our arbitrary language, say that such *plant characters* are the *cause* of high yield. But if we compare say, Squarehead's Master and White Victor in a district where they commonly differ by about 5 per cent. in yield, we cannot thus name the causal plant character differences. But such differences there must be. Upon the possibility of identifying them with a precision which will serve in genetics, depends this ambitious scheme of *synthesising* higher yielding forms.

Let us try to estimate whether we are in a position to approach the analysis of yield experimentally. Primarily, it involves environmental factors and plant characters. Factors which produce mechanical injury to crops, thereby lowering the yield, need receive no further notice. Soil and climate factors interest us most here and some examples of "adaptation" to these factors will quickly lead us to the heart of the problem. Yeoman wheat yields very big crops on the silt of the Holbeach area. Near by, in parts of the black fen, with the same climate, it is not thus outstandingly good. Again, over much of our wheat area the best forms—probably Squarehead's Master, Swedish Iron, and Yeoman—all crop heavily. But in any one locality their order of merit varies from year to year. As the soil is constant it can simply be said that facts of this kind reflect certain *adaptations* of the wheat forms concerned to certain *climatic factors*. So it is in the case of soil adaptations. In these different wheat forms must be a number of character combinations variably affected by environment. Orderly synthetic breeding for higher yield may become possible if we can make progress in the identification of these character combinations. We can no longer hope, as breeders once did, for the new form which everywhere and in all years will excel. Our hope is of breeding for every locality the form best adapted to the environment it offers.

Now we must deal with plant characters: not with the hypothetical ones in terms of which we have described the phenomenon of adaptation, but with characters we can observe and whose statistical relation to yield we can evaluate. The attributes we deal with in genetics and related studies and which we call plant characters are not fundamental things. They are resultants or manifestations of the interplay of vital plant processes and environment. These vital processes are the real plant characters. Analysis of yield in terms of them would imply the solution of most of the major visible problems of plant physiology. While we wait for physiological advances we may speak of tillering and the like as plant characters, and we must employ them as our experimental variables if we are to make the investigations proposed.

Attempts have often been made to resolve cereal yield statistically into such components as survival rate of the plants, tillering, ear size, and grain size. By simple analytical extensions such resolution may gain considerable value. For instance, it has been shown that in some forms —particularly of spring oats—there are characteristically produced a well-developed main axis with a considerable number of small and nearly valueless side tillers. Other forms produce a few side tillers each bearing a considerable amount of grain. Further study on these lines will cast much light on grain production in the different cereal forms. But the tendency has been to limit observation to the mature plant and to seek in its attributes alone, the *characters which govern yield*. Yield or average weight of grain per acre is an end product. It necessarily reflects in some complex way every vital process of the plant and every day in its life. We must infer, therefore, that though we cannot study the vital processes, it is incumbent on us to watch the developmental stages. And, correspondingly, climatic factors cannot be gauged from yearly or even weekly means. Climatic factors are effective not always in measure of actual intensity but largely according to their incidence in physiological time upon the growing plant. The cold dry spell of early May last year produced on some spring cereal forms a very adverse effect in a few days. The characteristic susceptibilities of cereal forms to these obscure climatic influences—baneful and beneficial alike—are responsible for the facts of "adaptation." It is suggested that by studying a number of forms in several localities and over a span of years, knowledge might be gained upon the nature of these susceptibilities or at any rate upon their distribution among different cereal forms. Daily meteorological data would be required and among the plant characters to be observed should be early developmental stages. Of these latter early tillering is believed to be of the first importance.

This very short outline has necessarily been imperfect. It is offered in the belief that form-adaptation is an approachable problem and in the conviction that the very fact of adaptation is proof of untouched possibilities. Certainly they are distant ones but the ultimate policy of breeding for higher yield must take the form of an orderly synthesis for separate localities. By a corollary we are reminded of a procedure already suggested. As far as possible, promising new forms should, early on in breeding, be tested over a wide range of localities. This may preserve for suitable districts forms which at the breeding station are not very outstanding. Little Joss went into cultivation as a light land wheat and still does well on many such lands. But it has been a great success in many parts of the very fertile black fen. From facts of this kind we may be emboldened to believe that, even before breeding for higher yield, lie great economic possibilities, the road to them being the study of adaptation.

For some of the possibilities which have been mooted the basis of fact is small. But in what have been grouped as quality, and ancillary yield characters, lie possibilities firmly implanted in the facts of cropping practice. While we cannot be unmindful of the reservations and the difficulties we may, and we ought, to dwell on the achievements. Plumage Archer and Spratt Archer among the barleys: Yeoman, Swedish Iron, and Little Joss among the wheats: Victory and, where it stands the frost, Marvellous among the oats—these, in the cereals alone, have justified themselves and justify, on several grounds, belief in future possibilities.

Dr Redcliffe N. Salaman. THE INHERITANCE OF CROPPING IN THE POTATO

The investigations of which the following is a brief and but partial account, were begun in 1911 and deal with some 25,000 seedling potatoes contained in over 400 families. In order to estimate cropping, weighing is not sufficient, for it fails to give any information in regard to the relationship between the mass of tuber and the mass or area of the foliage. No satisfactory method of measuring this relationship accurately by some objective method could be devised, so that recourse was had to a purely subjective one. In addition to weighing the crops, an estimation by which they were valued by the eye and placed into one of five classes was adopted. The classes are (see Fig. 7, p. 48):

Crop 1, in which the proportion of the tuber mass to the above-ground haulm is very high. Such crops would be described as *very good*.

Crop 2, in which the proportion is less, and in which the crop would be described as *fair* or *medium*.

Crop 3, in which the proportion is still less and the crop, though an indifferent one, is in no sense negligible, but would be described as *poor*.

Crop 4, where the proportion sinks still further and the crop, almost negligible, may be described as *extremely poor*.

Crop 5, where there is no tuber formation, or at most a few thickened bulbous swellings on the stolons. Such plants are designated *zero* croppers.

An essential preliminary requisite to such a grouping as is here described is that the soil on which the trials are grown shall be such that it is suitable for the growth of the potato; a soil markedly deficient in Potash will give quite erroneous results: on the other hand, a heavy clay soil and a light gravel soil, both in "good heart," will allow of identical estimations of similar material being made.

By plotting the percentages of the numbers of individuals in any seedling family, which fall into these five classes respectively, a curve is obtained. Repeated examination in the same and different years of families derived from the same matings or selfings, both by myself and, independently, by my friend and late assistant Mr J. W. Lesley, gave rise to curves so similar as to be in many cases superimposable (Figs. 1 and 2). In addition, curves illustrating the actual weights and the actual heights of the plants were made and it was further found that repeated cultures of the same family gave the same weight and height curves when grown in Barley or in Cambridge respectively, but that whilst the cropping curve, estimated as already described, by any given selfing or cross, was the same whether grown in Barley or in Cambridge, the curve of actual weights and the curve of actual heights was invariably different, and always in the sense that the plants in Barley were larger and gave heavier crops than those in Cambridge. There seems no reason to doubt that the method, though a subjective one, when once fully grasped and employed by the same persons, is highly efficient, and is reliable on all soils which are in a good average farming condition.

It is thus possible to determine for any individual a curve representing the crops of its offspring derived from its own self-fertilization. Such a curve is constant in shape for the same individual wherever it is grown or in whatever year, so long as the environmental conditions lie within the range normal to potato culture.

It is similarly true, that the curve representing the crop of a family derived from any given cross is a constant for that particular cross. Reciprocal crosses (Fig. 3), moreover, produce identical results, so that it is permissible to conclude that whatever the factors may be which control cropping, they are not sex-linked.

An examination of a large number of curves shows that they may be divided into six classes (Fig. 4):

Curve I *a* composed of about 75 % Crop 1 bearing individuals
15 Crop 2 " "
10 Crop 3 " "
no Crop 4 " "
no Crop 5 " "

Curve I composed of about 50 % Crop 1 " "
25 Crop 2 " "
15 Crop 3 " "
10 Crop 4 " "
no Crop 5 " "

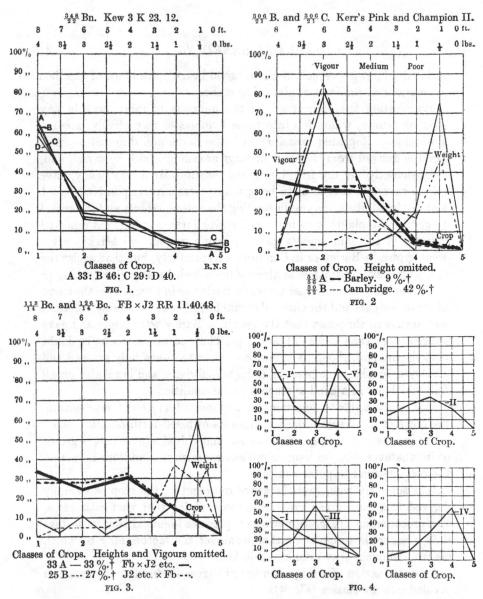

$\frac{348}{22}$Bn. Kew 3 K 23. 12.

Classes of Crop. R.N.S
A 33: B 46: C 29: D 40.

FIG. 1.

$\frac{306}{21}$ B. and $\frac{306}{21}$ C. Kerr's Pink and Champion II.

Classes of Crop. Height omitted.
$\frac{51}{56}$ A — Barley. 9 %.†
$\frac{30}{52}$ B --- Cambridge. 42 %.†

FIG. 2.

$\frac{113}{14}$ Bc. and $\frac{126}{14}$ Bc. FB × J2 RR 11.40.48.

Classes of Crops. Heights and Vigours omitted.
33 A — 33 %.† Fb × J2 etc. —.
25 B --- 27 %.† J2 etc. × Fb ---.

FIG. 3.

FIG. 4.

FIG. 1. Curves representing the cropping of four separate families, A, B, C, and D, all derived from the selfing of the same mother plant (Kew 3 K 23.12). The four families were grown in 1922 in Barley. The figures 33, 46, etc., represent the number of individuals in each family which survived throughout the season. The similarity of the curves needs no comment.

FIG. 2. Curves of two families, both being crosses of Kerr's Pink × Champion II, "A," containing 52 individuals, was grown at Barley by R. N. Salaman, and "B," containing 56 individuals, was grown by J. W. Lesley at Cambridge. The crop curves are almost identical, as are also the vigour curves. The weights are greater, as were also the heights, in the Barley family.

FIG. 3. Reciprocal crosses of Flourball and a seedling, J2 RR 11.40.48. "A" represents the curve of the family in which Flourball is the mother, "B" in which it is the pollen parent. There were 33 and 25 individuals respectively in each family. Both families were grown in Barley, and the cropping curves are almost identical; the weights are very similar.

FIG. 4. The six types of curve which may be distinguished, are represented. They must be onsidered as the average form of each type.

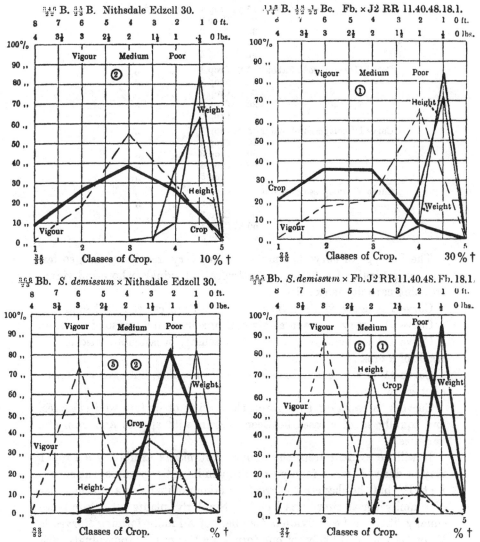

FIG. 5. The curve in the left upper corner results from the selfing of the seedling Nithsdale, Edzell 30, and is of type 3. The curve immediately beneath it is the result of mating this seedling with *S. demissum*, and represents a family of great vigour but of type 5 as regards crop. The curve in the right-hand top corner similarly represents the selfed family of the seedling Fb. J2 RR 11.40.48.18.1, and is of type 2. The curve immediately beneath is the result of mating this seedling with *S. demissum* and gives rise, as in the former case, to a very vigorous family with a very poor crop of types 4 to 5. The numbers in the left lower corners of the diagrams represent the numerical strength of the family. The nominator is the number of seedlings which survived the season; the denominator, the total number planted out. The numbers in the right lower corners of the diagrams represent the percentage mortality of the seedlings in each family.

Curve II composed of about	15 %	Crop 1	bearing individuals	
	25	Crop 2	,,	,,
	35	Crop 3	,,	,,
	25	Crop 4	,,	,,
	no	Crop 5	,,	,,
Curve III composed of about	5 %	Crop I	,,	,,
	20	Crop 2	,,	,,
	55	Crop 3	,,	,,
	20	Crop 4	,,	,,
	no	Crop 5	,,	,,
Curve IV composed of about	5 %	Crop 1	,,	,,
	10	Crop 2	,,	,,
	30	Crop 3	,,	,,
	55	Crop 4	,,	,,
	no	Crop 5	,,	,,
Curve V composed of about	no	Crop 1	,,	,,
	no	Crop 2	,,	,,
	no	Crop 3	,,	,,
	60 %	Crop 4	,,	,,
	40	Crop 5	,,	,,

The I *a* type of curve has not been met very frequently and, so far, only as a result of a first cross between individuals whose individual cropping capacity is excellent.

Curve I is fairly common both as the result of crossing and selfing.

Curve II is very common as a result of selfing, and less so of crossing.

Curve III is frequent as a result of selfing, occasionally also as the result of a cross.

Curve IV is very common as the result of selfing, but rare as a result of crossing.

Curve V is constant as regards the selfing of certain wild varieties, and equally so of crosses between certain wild species and domestic varieties.

It will be evident from the above, that plants exhibiting a Crop 1 may be derived from a family exhibiting in its entirety any of the curves I *a* to IV, but the chances are largely in favour of its being a unit member of a family whose curve was I or I *a*. Similarly, plants with Crop 2 are equally likely to be individual members of a family forming Curves I, II, or III and, less likely, of one forming a Curve IV. Individuals with Crop 3 may possibly be derived from a Curve I, or with about equal probability from a Curve II or IV family, but with greater probability from Curve III family. Individuals with a Crop 4 are theoretically as likely to be derived from Curve IV as Curve V, but, in fact, their derivation from a Curve V family may be almost excluded so long as purely domestic varieties are being dealt with.

Conversely, individuals with a Crop 1 on selfing should, on the average, give a higher rate of curves of type I than a similar number of individuals whose personal crop was, say, 3 or 4.

The following table, giving the results of the selfing of one hundred and forty-six individuals, whose own individual crops were known, gives ample support to this contention:

	10 families forming Curve		I=17 %		
58 individuals of Crop 1,	29	„	„	„	II=50
when selfed, gave rise to	9	„	„	„	III=16
	10	„	„	„	IV=17
	1	„	„	„	I= 3 %
35 individuals of Crop 2,	14	„	„	„	II=40
when selfed, gave rise to	12	„	„	„	III=34
	8	„	„	„	IV=23
	2	„	„	„	I= 8 %
25 individuals of Crop 3,	3	„	„	„	II=12
when selfed, gave rise to	8	„	„	„	III=32
	11	„	„	„	IV=44
	1	„	„	„	V= 4
	0	„	„	„	I= —
16 individuals of Crop 4,	4	„	„	„	II=25 %
when selfed, gave rise to	0	„	„	„	III= —
	11	„	„	„	IV=69
	1	„	„	„	V= 6
12 individuals of Crop 5,	0	„	„	„	I= —
all members of wild va-	0	„	„	„	II= —
rieties or their deriva-	0	„	„	„	III= —
tives, when selfed, gave	0	„	„	„	IV= —
rise to	12	„	„	„	V=100 %

Similarly, by growing on selected individuals of selfed families whose curve has already been determined, and determining the curve of the selfed families of these individuals, it is found that of the five curves, Curve V and Curve IV are almost certainly produced by a homozygous combination of factors influencing cropping. Curve III is probably not in all cases the result of a homozygous condition. Curve II represents probably always a heterozygous state, and it is very probable that Curve I may do the same.

It is suggested tentatively that as far as the domestic varieties are concerned, we have to deal with at least two cumulative cropping factors —A and B—and that $AaBB$ possibly rather than $AABB$ represents the highest genetic composition of any individual, the latter form being perhaps non-vital. The genetic composition representing the lowest crop would be $aabb$, one type of "zero cropper."

The effects of crossing. In general when the curves derived from the selfing of each parent are known, then the higher the type of curve each may possess, the higher the type of curve exhibited by the F_1 family.

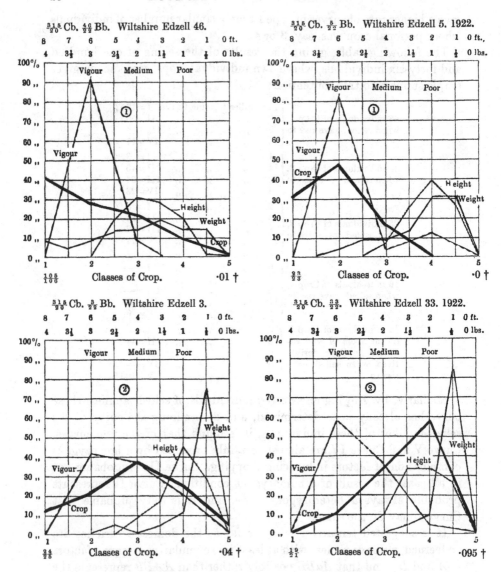

FIG. 6. Four family curves are shown, all being derived from F_2 families obtained by selfing various F_1 individuals of the cross Edgecote Purple × Edzell Blue. The numbers in the left-hand bottom corner of the graphs have the same significance as those in Fig. 5. The F_1 curve, not shown here, was of type 1. The two upper curves of Fig. 6 emanate from "selfed" individuals whose crop was 1. They have produced families of the type 1 curves. The two lower curves are of type II and IV respectively and the selfed parents of each possessed a crop 2. The series illustrates the segregation of cropping curves.

This is very well exhibited by the material in hand, and it has thus been possible to select a parent which, when used in crosses, with domestic varieties will invariably produce an F_1 with a high curve. In a general sense, the curve of an F_1 family bears a strong resemblance to the theoretical curve obtained by summating the selfed curve of each of the parents. Crossing between domestic varieties and the wild species *S. demissum*[1], whose cropping curve is a Curve V, produces an F_1 of extraordinary vigour but whose Family crop is typical of that of the Curve V (Fig. 5). In other words, *S. demissum* is possessed of crop inhibitors, and these inhibiting factors are dominant to the A and B factors which cause cropping in the domestic varieties.

A cross has been made between a zero cropping plant derived as an F_2 from a wild species cross which, when mated to a domestic, produced only partial inhibition of the crops. It is, therefore, probable that each of the cropping factors A and B has its own specific inhibitor represented in the wild varieties. These latter considerations have a very suggestive bearing on the possible origin of the domestic potato.

The segregation of curve types. In crosses between the domestic varieties there is abundant evidence of subsequent segregation to specific curve types in the various F_2 families (Fig. 6). The same appears to be true with regard to the crosses between the wild type *S. demissum* and the domestic potato, but here the evidence is at present rather scanty.

Relation of vigour to crop. A few facts in regard to this subject may be stated, leaving their discussion to some future time. Whilst the relation between actual crop weights and the vigour of the plants producing them is a close one, there is no definite relation between the cropping curve of a family and the vigour of the plants which give rise to it. A high curve may be associated with low vigour and a low one with high vigour, or *vice versa*. On the other hand, zero croppers are very frequently accompanied by the highest vigour. A considerable number of brother-sister matings have been made in highly inbred families, with a result that the vigour may be considerably restored by such mating, whilst the crop is not necessarily affected. Further examination of the material is required before any more definite statement can be made. It would appear, however, that not only is the genetic basis of cropping independent of vigour, but that there is probably no relationship between the genetic factors for cropping and such others as may conduce towards the production of vigour. On the other hand, the total quantity of tuber material

[1] It is not certain whether *S. demissum* is the correct name for this species. It bears a close resemblance to the description of the variety *S. utile*.

formed by any given plant, is largely determined by the vigour of that plant, and that that vigour may well be dependent on the heterosis

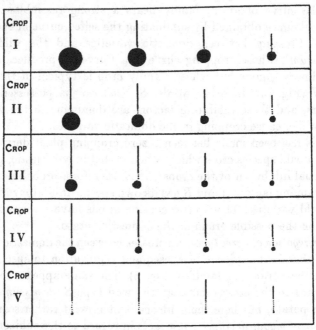

FIG. 7. The diagram represents the relation between the crop, the circular area, and the haulm as represented by the vertical line above it. It is the proportion between these two quantities which forms the criterion for the estimation of the crop of any plant into one or other of the five classes. It will be observed that the tuber mass constituting the crop of any plant might, considered apart from the haulm, entitle it to a position in almost any class, but it is only when considered in relation to the mass of foliage above it, that a correct estimate is obtained.

exhibited by the individual as determined by such factors as those for colour and shape, would appear to be highly probable from the analysis of the material so far undertaken.

Mr M. A. Bailey

When the Secretary asked me to join in this discussion he asked me to approach the question with special reference to the subject of cotton. I accepted his suggestion with considerable diffidence; firstly because my experience in Cotton Breeding is limited to one country, Egypt, and that country no longer comes within the scope of this Conference;

secondly because the cotton-plant offers a rather unfavourable medium for what may be called "constructive" plant-breeding.

On the other hand, Egypt is a country which, given a suitable and efficient administration, could be made to profit to the greatest possible extent from any improvements effected by the scientific agriculturist. A short account of the conditions which exist there may therefore be of some interest if only for purposes of comparison with other countries.

The cultivated area of Egypt is very small indeed—only about 12,000 square miles, or less than half the area of Ireland—but the population is very dense (over 1000 people per square mile), affording an ample supply of cheap labour.

Even with the present diminished yield of cotton, the return per acre is still well above that obtained in most other cotton-growing countries. The average yield may be taken at not much below 400 lbs. of ginned lint per acre for all except the poorest lands, and the average price obtained at Alexandria during the months January–April, 1924, was nearly £8 10s. per 100 lbs., representing about £32 per acre.

These few figures serve to indicate what might be described as the "accessibility" of the country for administrative control, and also the importance of the crop to the country.

One word as to the present condition of the crop in Egypt viewed from the standpoint of the plant breeder. It is sufficient to sit in a train travelling from Alexandria to Cairo and to look out of the window, to be impressed by the great irregularity of the general crop. Rogue plants of exceptional height stand out prominently and are often very numerous indeed.

A closer inspection reveals the presence of many more rogue plants of less obvious types.

These plants are almost invariably characterised by lint of very inferior quality, and their presence in the crop reduces very considerably the value of the lint in general, and paves the way for further deterioration in succeeding crops.

The exact mode of origin of these rogues is uncertain, but it has been clearly established that they *can* be eliminated by bulk selection methods.

Bulk selection was first practised on a *wide* scale in Egypt on the lands of the State Domains Administration, where the maintenance and improvement of type of Sakellarides Cotton has been the object. The results have been very satisfactory indeed and large quantities of this selected seed are distributed to cultivators every year.

B   4

Bulk selection, both positive and negative, has also been practised for many years on the lands of the Ministry of Agriculture's own Experimental Farms.

Here again, using the simplest and most direct methods known to plant breeders, improvements have been effected both in quality and in yield.

Various types of cotton have been dealt with in this manner and it may be of interest to consider the economic possibilities of the results obtained.

Owing to the present unsettled conditions of the world markets, and the temporary shortage of short-staple cotton, the distribution in space of the different varieties of cotton over the extent of Egypt has become very involved. For simplicity, therefore, we will go back a couple of years, say to 1922, when the situation was more clearly defined. In that year the distribution of varieties was, very roughly, as follows:

$$\begin{array}{ll} \text{Sakel} & \tfrac{3}{4} \text{ of whole area} \\ \left.\begin{array}{l}\text{Ashmuni}\\ \text{Zagora}\end{array}\right\} & \tfrac{1}{4} \quad \text{,,} \quad \text{,,} \end{array}$$

On such a basis, what would such relatively small improvements as those already indicated mean to the country as a whole?

First as regards yield, the results of trials over a number of years would seem to indicate that average increases in the neighbourhood of $2\tfrac{1}{2}$ per cent., in lbs. of lint per acre, had been obtained in the case of short-staple cottons, and smaller increases in the case of other cottons.

As regards quality, the best economic test available at the moment would seem to be the premiums offered by actual buyers for certain selected strains compared with the bulk crop of Egypt.

In the 1922–23 season certain samples of strains selected in this way were proffered to the buyers at Alexandria and the premiums offered on that occasion amounted to something as follows:

$$\begin{array}{lll} \text{Selected Sakel} & \dots \quad \dots & \cdot 75 \text{ pence per lb.} \\ \text{,,} \quad \text{Upper Egyptian} & \cdot 39 & \text{,,} \quad \text{,,} \end{array}$$

This is not very much to go on by itself, but it tallies with the obvious superiority of the selected lint and with opinions expressed by expert graders on other occasions.

If now we imagine all the Sakel and Ashmuni grown in Egypt in 1922 entirely supplanted by such improved strains, what difference will this make in the national revenue?

The two factors, increased yield and improved quality, taken together would appear to increase the national wealth in one year by nearly 1¼ million pounds.

It may be objected that if the *whole* country were put under these new strains, then the buyers would no longer be prepared to pay the premium suggested, and that the general level of prices would fall to where they were before. This might have been true some years back, but Egypt is no longer the sole purveyor of Sakel cotton. To take one instance, and that one near at hand; Sakel from the Tokar region of the Sudan has already attained a high reputation, and higher prices have actually been paid for last season's growth than for Egyptian Sakel.

Admitting the roughness of these calculations and the fact that it would take from seven to eight years to bring about such a complete transformation, yet I think that the figure is sufficiently near the mark to emphasise my point, which is that organised plant breeding combined with organised distribution of improved seed cannot fail to repay its own cost many times over in a country like that which we are now discussing.

If only a hundredth part of the general extra profit indicated above is recovered by taxation the sum realised would be more than sufficient to pay for the maintenance of the plant breeding staff, farm and laboratories.

I have mentioned the question of the organisation of *seed distribution*.

Without a properly organised and controlled system of seed distribution, much of the economic value of plant breeding will be lost at once through delay. It is not sufficient to produce improved strains—they must be distributed in the shortest possible time. Also they must be distributed without their purity becoming impaired.

In Egypt as also in many parts of the British Empire where cotton is grown entirely for consumption abroad, the question of *Seed-control* is one of the highest importance as a standardised product is obviously desirable.

By "seed control" I mean two things: firstly, the taking of special precautions to prevent the accidental mixing of seed of different varieties and also the intentional adulteration of good seed with bad, and *secondly*, the formulation of a definite policy with regard to the variety to be grown in any one particular district.

Seed control goes hand in hand with seed distribution and both should in such circumstances be operated by the Government. They are, to my mind, of an importance equal to that of the plant breeding itself. Seed distribution should pay for itself—the cost of seed-control would easily be covered by the early results of plant breeding work.

I have dwelt on the question of bulk selection in its relation to cotton in Egypt—what of the improvements which can be effected by other methods?

These methods are:

(1) Introduction of new types from other countries.

(2) "Pure-lining" of existing varieties.

(3) Direct selection of new and desirable types in the field, *i.e.* the selection of useful rogue plants.

(4) Hybridisation.

As regards the first of these—the introduction of new types from abroad—this method is not likely to be of much value to Egypt which, with the exception of the West Indian Islands, stands pre-eminent amongst cotton-growing countries as regards both yield and quality.

In other less fortunate countries, however—and this includes most of the countries within the British Empire where cotton can be grown—this method assumes the highest importance.

Not only are some introduced varieties eminently suited for growth in their new country, but in some cases the acclimatised type is even superior to the original.

An outstanding example of this is the "Mesowhite" of Irak originally introduced from India as "Webber 49," a long-staple American Upland cotton.

Passing now to the *pure-line* method. Pure-lining is valuable both as affording reliable material for constructive hybridisation, and also as insuring the preservation of the best features of existing varieties. Not merely this, it has the same effect as is obtained in bulk selection, but with the advantage that the poorer types of plants are eliminated once and for all.

From the economic standpoint, the final results to be obtained from the pure-line method in Cotton are likely to be somewhat better than those obtained from bulk selection, but they take longer to achieve.

What now of the method of direct selection of desirable rogue plants in the field?

This is a possibility that is always with us. Success in this direction depends on two things, luck and continued application.

All the great improvements in the past in Egypt have been dependent on this method. The Assili variety arose in 1906–7 by the selection of a single unusually early plant from the bulk of the Mit Afifi crop.

In 1911, M. Sakellarides rose to fame as the originator of the now famous Sakel Cotton—again a chance selection of a single plant, chosen by a discerning eye.

This method is likely to lead to better and more striking results than the methods referred to above, but it has one drawback from the economic viewpoint—uncertainty. It may be fifty years or more before it becomes self-supporting.

This type of work, however, can conveniently be combined with the normal routine of bulk selection and its expenses paid in this manner, until the lucky stroke has been achieved.

Finally we come to the question of improvement by hybridisation.

In this respect cotton, without doubt, presents great difficulties.

A wide field of original parents are available and in most instances the initial cross can be made without difficulty.

In the second generation segregation undoubtedly takes place, but, in our experience at any rate, the various characters of economic significance merge into each other without perceptible gradation.

Especially is this so in regard to lint length, one of the most important of all characters in the case of cotton. To produce the results observed it is probable that as many as five separate cumulative factors are concerned.

The problem of improvement of lint length is hard enough in itself. But lint length is not all—high yield is almost as important, and obviously depends not merely on a single series of cumulative factors but on a whole array of totally unrelated characters.

Clearly the fixation of new types by this method is not merely difficult, but is bound to take a very long time. In many instances it may be said that the economic botanist "has not the time" to carry it through at all.

Here again we have a method of improvement which, if somewhat more certain in its results than that depending on the selection of useful off-type plants, is much handicapped by the question of time.

Where the staff is small and the plant breeder almost single-handed, it were best to leave hybridisation severely alone—*it will not pay*.

If the staff is sufficient it should certainly be undertaken, the initial cost being covered by the results of bulk selection work or acclimatisation of foreign types.

Before leaving the subject of the economic possibilities of plant breeding, I should like to express an opinion on the relative economic importance of plant breeding and *plant physiology* in the case of the cotton crop.

In a small experimental station in a cotton-growing country, I should recommend that attention should first be given to the improvement of existing varieties by straightforward bulk selection. This work may be expected to give results of economic value which will enable work in other directions to be attempted. Next, I should recommend that attention be given to the physiology of the plant.

In Egypt it would seem that the cotton plant rarely bears more than two-fifths of the full number of bolls, which it should be capable of carrying. This is principally due to the action of three separate factors—the premature shedding of unopened buds, insect attack, and the shedding of unripe bolls. I have mentioned them in the order which appears to be their order of importance.

Of these, the first and the last are dependent almost entirely on physiological reactions. In part, no doubt, they are affected by the resistance shown by individual plants to the particular conditions of the environment which bring them about, and they are thus capable of amelioration by selective breeding.

In the main, however, their successful control may be assumed to depend on the discovery of the exact nature of the unfavourable environmental factors which produce them and on the elaboration of methods which will modify these environmental factors or counteract their effects.

If these factors could be eliminated greater progress would have been registered in increasing the yield of cotton, than could be expected from any of the methods of plant breeding referred to above, even if applied over a long interval of years.

The analysis of "yield" is at least as important in the case of cotton as with other crops. This analysis has to be made before serious work on the improvement of yield by constructive breeding can be undertaken. This problem is to a large extent physiological.

In cotton our most useful weapon in this attack is the "Flowering Curve" method introduced by Dr Lawrence Balls, but much still remains to be learnt as to the meaning of results when expressed in this form. It is becoming quite clear that Flowering Curves represent the resultant of a variety of different reactions and we are only now beginning to realise what these reactions are.

Mr J. M. F. Drummond

In the crop of cereals, in which the most striking improvement has already been effected by scientific breeding, there still remain important problems which should prove soluble by standard methods, *e.g.* the isolation of quickly maturing oat varieties suitable for upland districts.

As regards potatoes, the outlook seems less hopeful, chiefly owing to the unexpected complexity and elusiveness of the virus disease problem. Cropping records of a large number of seedlings grown during the years 1921 to 1924 at the Scottish Plant Breeding Station clearly show a progressive deterioration of the stocks accompanied by increasing infestation by Leaf-roll and Mosaic. Quite a fair proportion of the seedlings in question appeared very promising in their first season, but so far not a single one of them has fulfilled its early promise in later years. Further information regarding virus diseases is required by the breeder, particularly with regard to the following questions:

1. Are any or all of these diseases transmitted through the (true) seed?

2. Are they spread from plant to plant in any other way than by insect agency?

3. Are any of the mosaic-like diseases of other plants transferable to the potato?

Meanwhile it is difficult to escape from the unwelcome conclusion that it is futile to attempt the raising of new potato varieties except in localities which are as far as possible free from virus diseases. A fuller knowledge of the inheritance of Wart Disease immunity would be of great value to the practical breeder; the most promising line of advance here seems to be the isolation of homozygous immunes.

As far as swedes and turnips are concerned, the genetically mixed condition of most if not of all commercial stocks is an obstacle to progress, aggravated in the case of turnips by rather pronounced self-sterility. A further difficulty is the lack of any definite criterion of feeding-value. The Scottish Plant Breeding Station is attempting the isolation of homozygous swede strains, and the feeding-value problem is being attacked by a co-ordinated scheme of research, in which all the Scottish Agricultural Colleges and Research Stations are taking part. The development of strains of swedes and turnips truly resistant to Finger-and-Toe is greatly to be desired; but in view of the generalised parasitism of the organism concerned, the chances of success are probably not very great. A satisfactory form of plot-technique for determining the cropping power of root-crops on a small scale (50 to 100 roots) has still to be worked out.

The line of work which possibly offers the brightest prospect of economic results is the exploitation of indigenous forage grasses, especially with a view to the improvement of pastures. It is now clear that "Rye-grass," "Cocksfoot" and "Timothy" are "Sammelarten," each comprising a great variety of types differing in habit, leaf-production, seed-production, duration, etc. The raw material for the artificial evolution of pasture and other types of agricultural grasses is thus abundantly available. Unfortunately there are serious obstacles—both inherent and incidental—in the way of rapid progress. The prevalence of self-sterility (notably in the rye-grasses) obviously complicates the breeder's task at the outset by rendering the isolation of pure lines difficult if not impracticable; it is particularly unfortunate that the ease with which a large clone of a grass can be built up for the purpose of rapid multiplication should be so largely discounted by the low degree of seed-production that ensues if the clone is thoroughly protected from the access of foreign pollen. The successful isolation of mass cultures of desirable strains for multiplication is another problem presenting considerable difficulty. Resort has to be made to localities not only far distant from hay-fields and permanent pastures but also free from any appreciable flora of indigenous plants of the grass concerned. Generally speaking such places are to be found on elevated moorland where the conditions are somewhat unfavourable to seed-production. A difficulty of a very different kind in connection with the development of pasture types arises out of the divergent demands, on the one hand of the farmer, who requires a leaf-producing type, and on the other of the seed-grower, who naturally prefers the heavy-seeding type. There are several alternative possible solutions of this economic problem, but it is not easy at present to decide which of them is to be preferred. In any case the work of isolating pedigree strains of the divergent types of each important grass, which is being actively carried on both at the Welsh and at the Scottish Plant Breeding Stations, must necessarily constitute the first step in any scheme of grass-improvement. As a matter of fact, evidence recently obtained at the Scottish Plant Breeding Station indicates that, as far as perennial rye-grass is concerned, the leafy and long-lived types are not necessarily poor seeders.

At the risk of repetition, emphasis may once more be laid on the fundamental importance of the self-sterility phenomenon. It is significant that the outstanding successes of modern crop breeders have been achieved with those cereals (wheat, oats and barley) in which natural selfing is the prevalent method of fertilisation to so pronounced an extent that pure lines can be isolated, even from a large number of varieties grown side by side without any artificial isolation, with the greatest ease. Although self-fertility and self-sterility are doubtless essentially constitutional characteristics, there is evidence to show that they are subject, within limits, to modification by environment. Whether adaptability in this respect is sufficiently great and of such a kind as to admit of artificial control in the interests of the breeder is a question urgently demanding investigation. There is here a fascinating field of enquiry for the plant physiologist, who is in general better equipped for an investigation of this particular subject than the average crop breeder.

In conclusion, reference may be made to the desirability of more effective co-ordination in plant breeding research. In a separate paper ("The Formation of Herbaria of Crop-plants") the writer has drawn attention to the crying need for a properly organised scheme of classification and nomenclature of crop plants. The standardisation of yield-trial plot-technique for the various crops and the unification of technical terminology are other aspects of co-ordination urgently demanding attention.

Mr T. S. Venkatraman. SUGAR-CANE BREEDING IN INDIA

Within the last half a dozen years plant breeding carried on with sugar-cane in India has achieved some very striking results. From the results already obtained, it would not be too much to say that breeding work on this crop appears to be one of the most fruitful in results for India.

Certain special features of sugar-cane breeding

The difficulty and uncertainty of getting sugar-canes to flower and seed and the comparatively greater care and time that are needed to grow seedling canes to maturity, coupled with the ease with which sugar-canes can be grown from cuttings, have made the latter the universal method for planting canes for cultivation. The growing of canes from seed is confined to special breeding stations where it is done with the definite object of evolving new seedling canes.

Breeding in sugar-cane differs from that for most other crops in certain important respects. Whereas in most other crops the parents breed more or less true when selfed, in the case of the cane, even a batch of selfed seedlings shows wide variations in botanical, agricultural and chemical characters, none of the descendants resembling the parent; so wide is this variation that the seed-produced crop is often of little value from the crop point of view. This wide variation is an asset of some value on account of the consequent rapidity with which new seedling canes can be obtained. As the subsequent multiplication of the new cane is carried out by the vegetative method, the seedling continues to maintain, more or less, the original characters for fairly long periods; there is not the need for the elaborate fixing of characters or for precautions against the new cane degenerating through undesirable cross-pollination, as in the case of crops grown entirely from seed.

Cane breeding is not, however, without certain serious handicaps. Firstly, the absence of any easily ascertainable law governing the inheritance of characters in the sugar-cane renders attempts at building up a desired combination, or repeating it, less certain and more difficult. Secondly, the new seedling canes often deteriorate under cultivation after some years; this, it is believed in certain quarters, is due to the vegetative method of propagation.

It is thus seen that, whereas the early results in the cane may be comparatively rapid, continuous work is needed to keep up the level of the canes under cultivation by periodic introductions of fresh seedlings.

The poor quality of Indian canes

Though India must have been growing canes long before many of the other countries of the world, yet, today, her canes are some of the most primitive in existence. The bulk of the Indian area—situated in temperate North India—grows a type of cane which many of the factories in the tropics may not care to handle. The acre yields in sugar for India are consequently much less than those for the more favourably situated tropical countries; the yields are less than one-third that of Cuba, one-sixth that of Java and one-seventh that of Hawaii.

Indian growth conditions for sugar-cane

The poor quality of the canes grown is largely due to the unfavourable climatic and cultural conditions obtaining over the bulk of the Indian area. Over this area the growing period for cane is often limited to less than nine months on account of the very cold winters—frost not being uncommon in the Punjab and parts of the United Provinces. In India the cane plant is often required to develop its sugar in about half the period that is available to it in most tropical countries. Again, this limited growing period and the rigors of the North Indian winter render it difficult, if not impossible, to improve the cultivation or manuring beyond a certain limit.

Cane breeding for India

Whereas in the other countries a certain amount of success has been obtained through raising seedling canes without a deliberate attempt at cross-pollination, such work has been found unsuitable to solve the Indian problem. The tropical canes generally give rise to seedlings more or less of the tropical type and are, therefore, unsuitable for growing in temperate India. The Indian canes, on the other hand, have been found to yield seedlings generally poorer than the parents. The most satisfactory results have been secured by a well-planned scheme of cross-pollinations between the hardy Indian and the rich tropical canes. Besides a marked increase in vigour of growth and thickness of cane, the cross-bred seedlings have often shown a deeper and more vigorous root system—a factor of considerable importance for success under conditions obtaining in Northern India.

Results obtained

The results obtained through cane breeding in India have been very marked and well beyond the expectations which were entertained at the founding of the station. The Coimbatore-bred canes have been found to yield as much as two to three times that of the indigenous kinds and, in certain parts of Bihar, the crops from these seedlings have been pronounced to be not inferior to those of Cuba or Java. When it is remembered that India possesses within her confines half the world's sugar-cane area it will be readily conceded that sugar-cane breeding for India has mighty potentialities in the future. In spite of the large cane area in India she had to import in the year 1923 refined sugar valued at over seventeen crores of rupees.

GENETICS

THE VALUE OF SELECTION WORK IN THE IMPROVEMENT
OF CROP PLANTS

(CHAIRMAN: MR W. BATESON, F.R.S.)

Prof. J. Percival. THE METHODS AND VALUE
OF SELECTION

Plant breeders, and all who are engaged in the cultivation of farm and garden crops, should periodically transfer their attention to the wild flora of their country.

Nowhere among the natural vegetation are found the large and luscious fruits of the garden, the big edible roots of the carrot and beet, or the long well-filled pods of the pea and bean. The all-important cereal grains are also conspicuously absent.

The contrast between the crops of the field and garden and the natural flora is arresting, and when it is realised that the cultivated crops are the descendants of plants once wild, the imagination of the dullest of us is enlivened and stimulated.

The time and exact mode of origin of our cultivated plants is lost for ever. Our cereals, beans and peas, the best of our root and vegetable crops such as turnips, carrots, beet and cabbages, and the majority of our fruits, apples, pears, plums and grapes, have been cultivated for not less than three or four thousand years: not a single food plant of first class importance has been added to our fields or gardens within historic time. Nevertheless, with a few exceptions the wild prototypes from which the cultivated crops have been derived can be more or less clearly recognised.

On comparing both, it is seen that the peculiar characters which rendered the wild plant attractive or useful to man have been greatly increased; in the language of the plant breeder the wild plant has been enormously improved. Such improvement is entirely dependent on the natural variability of plants and their responses to the environment of our cultivated fields and gardens.

In the progeny raised from a single plant no two individuals are exactly alike; in regard to their height some are taller than others; the colour of the flowers often varies, and in chemical composition some are richer in starch, sugar, acids and other compounds than the rest.

Every individual possessing a useful character in excess of the average is an improvement, and it is the object of the plant breeder to secure crops of such individuals. This he endeavours to achieve by selecting and isolating the improved form from its neighbours and cultivating it.

A glance at the literature of plant breeding reveals much confusion of thought in regard to the part which selection plays in the production of improved crops.

Beginning with a crop of a particular agricultural or horticultural variety, by the selection of certain individuals and the cultivation of these, we are ultimately able to secure crops of plants very different from those with which we commenced our experiments. It is an easy and somewhat natural step to conclude that the change of character of the crop which has occurred has been caused altogether, or in some degree, by selection, and only the most careful and continued watchfulness will prevent us from falling into this error. It is indeed difficult to speak or write about the results obtained by selection without using terms which, although apparently harmless, are responsible for much of the obscurity and perplexity which the plant breeder not unfrequently encounters. For example we find such expressions as these, "the weight of the gooseberry has been increased by systematic selection," "the doubling of flowers has often been effected by careful selection," "by selection the early maturity of peas has been hastened," "man has wrought profound changes in plants by selection." All these are inaccurate and misleading, for selection neither doubles flowers, increases the size of gooseberries, nor causes any changes, profound or otherwise, in plants.

Selection does nothing more than isolate and preserve desirable forms of plants; it plays no part whatever in their origin.

Several somewhat different methods of selection are practised, the value, respective merits and defects of which can perhaps best be appreciated by reference to a few examples of the results obtained by their application to cereals and one or two other crops.

Selection of spontaneously occurring individuals exhibiting improved characters has provided us with most of the best of our cultivated plants. The figurative description of them by the ancients, as gifts of the gods, expresses the truth, for they have been found ready made; in the production of them or their good qualities mankind has had no part. All that we have done is to select and propagate them.

As examples of famous varieties of cereals obtained in this manner, may be mentioned Fife and Squarehead wheats, Potato oats and Chevallier barley.

Fife wheat, one of the world's best varieties, which for a long time dominated the vast corn growing regions of Canada and the United States, originated from three ears, apparently the produce of a single plant, grown from a sample of Dantzig wheat sent by a friend from Glasgow about 1842 to David Fife in Ontario.

The dense-eared Squarehead variety was discovered in a field of the comparatively long lax-eared Victoria wheat about 1868; on account of its stiff straw and high yield it rapidly became the chief kind in all the wheat fields of western Europe.

The Potato oat, which has been more cultivated in Scotland and the north of England than any other, and at the present time is unsurpassed in several of its qualities, was discovered in a potato field in Cumberland in 1788.

The renowned Chevallier variety of barley had a similar casual origin, arising from an extraordinarily fine ear observed and selected by a labourer employed by the Rev. Dr Chevallier of Debenham in Suffolk.

Patrick Shirreff, who, during last century, spent about fifty years in the improvement of cereals, made a practice of closely examining growing crops just before harvest for superior ears and plants, and his best introductions were selections made in this way.

In a wheat field of moderate area, many millions of plants are grown; it is an experiment on a scale far exceeding anything we can attempt with every plant under control. The opportunity for the production of new forms which on account of our ignorance of their origin we designate sports is almost unlimited, and under such circumstances Nature is generous in providing material from which the experienced student of variation may always expect valuable gifts.

Those who are concerned with the commercial introduction of new improved forms of plants still have to depend for the majority of them upon the selection of sports—individuals chosen by the alert eye from Nature's lavish experiments—and if I were asked to suggest methods to obtain say a better variety of wheat, potato, carnation or bean, one of my recommendations would be, study thoroughly the peculiarities of your plant, grow half a million of them and search for the individuals you require.

Mass selection is another method formerly much practised and still of especial value in particular cases.

As an example of the method and the results to be expected, the production of Schlanstedt rye by Rimpau may be given. In 1867 Rimpau selected from his fields of Probstei rye a handful of the longest and best filled ears which he could find, taking care that none of them was chosen from parts of the crop obviously luxuriant through accidental local fertility of the soil.

The grains of the selected ears were rubbed out, mixed and sown. At harvest a further selection of the best ears of this crop was made, and the grain sown. Similar selection from successive crops, each the progeny of the best ears of the previous season, was continued annually and, after about twenty years, the strain obtained was so much superior in average length of ear and yield of grain, that it rapidly displaced the varieties commonly grown, not only in Germany, but in many parts of France.

Varro, and other ancient writers on agriculture, refer to the practice of selecting the best ears of cereals for seed and recommend it where the crops show signs of diminishing returns.

In all countries and in all ages this form of selection has been found to result in a gradual increase of the yield of crops, especially in those which are normally autogamous, and at the same time it has led to the production of a more uniform plant.

The process rarely succeeds in fixing the desired character, and the improvement achieved is gradually lost when selection is relaxed or discontinued.

The commonly accepted explanation of the success of this method of selection is, that the crop from which the initial selection is made consists of a mixture of several distinct, independent forms, each with its own hereditary ear-length and yielding capacity. By selecting annually the longest, best-filled ears in the crop the shorter-eared low-yielding forms are gradually eliminated, and if the process is continued long enough, there remains but one form, namely that possessing in the highest degree the characters on which the selection is based. When the selection is discontinued before the complete elimination of all the poor-yielding forms has been effected, the latter soon multiply and bring back the population to the original unimproved state. The explanation given suggests that mass selection is nothing more than a method of isolating particular forms from a mixed population, and there is, I think, no doubt that it is the true one. The explanation of course leaves unsolved the all-important question of how the mixture of forms in the unselected crop has arisen.

Screening or sifting out the larger grains from ordinarily threshed cereals for use as seed, is a form of mass selection recommended by the ancient Greek and Roman agriculturists and practised by the best farmers and seedsmen in all ages down to the present day.

Innumerable experiments have shown the value of this selective process.

Cobb in Australia investigated the yields of large, medium and small seeds of more than twenty varieties of wheat and found there was always a higher yield from the large grains, the increase being sufficient to justify the cost of special grading machinery.

Zavitz in Canada, experimenting over a period of six to eight years, found an average increase of over six bushels per acre from the use of large seed in the case of winter wheats and nearly four bushels per acre in spring wheats.

At many of the experiment stations in the United States, and in France, similar large scale trials with wheat, barley and oats have given like results in favour of large seed.

For a period of five years, I determined the yields of ten of the largest and ten of the smallest grains taken from single ears of five different varieties of wheat, and found that the large grains invariably gave stronger and more productive plants than the smaller grains, the increase obtained varying from 12 to over 60 per cent.

In all these cases the higher yields are due to the better-developed embryos and larger stores of reserves in the bigger grains rather than to any hereditary differences.

No matter what the variety, its large seeds or grains yield a greater crop than its small seeds, and mass selection of large grains by means of specially adapted machines is of very considerable value in securing the best yields from our fields.

Hallett made a special effort to send out seed of high grade in this respect, and the best seedsmen of the present day adopt the same practice.

Although it is invariably true that large seeds of all varieties of cereals give better yields than small seeds of the same line, it must be clearly recognised that varieties possessing grains of high average size are not necessarily more prolific than those whose grains are of smaller average size. Hereditary high-yielding capacity may be associated with a grain of medium size: this is perhaps more frequently the case with oats. Some high-yielding forms of this cereal produce spikelets with two or three grains, one of which is smaller than the rest. On the other hand, varieties giving poor yields very frequently produce only one or two comparatively large grains in each spikelet, the average size being greater than those of the better-yielding forms. Mass selection of large grains applied to a mixture of two such forms, by eliminating the smaller grains, would reduce the proportion of the high yielding form at each selection, and if carried on long enough the process would end in the isolation of the poorer type.

From the earliest times down to the present day the selection of superior individuals, which on account of our ignorance we designate sports, and the process of mass selection have given us most of our improved crops; these were indeed the only methods known to the ancients who accomplished so much. The modern plant breeder is inclined to discredit and discard these methods, but the results which they have given should inspire respect for them, and until we obtain far more knowledge than we at present possess of the real nature and origin of the hereditary variations of plants it is too soon to abandon them altogether.

Mass selection is now largely superseded by what is termed line or pedigree selection, followed by the progeny test. Single promising ears or plants are selected as in mass selection from ordinary crops of the cereals, but instead of mixing the grains from these before sowing, the progeny of each ear or plant is raised in separate rows or plots. The individual rows are subsequently tested and compared, and the best selected and multiplied.

First employed by Le Couteur and Shirreff in the early part of last century, line selection was extended and systematised by Vilmorin about 1860, and employed later with much success by Hays in America and Nilsson in Sweden. Johannsen applied the method to the Princess variety of Dwarf Bean (*Phaseolus vulgaris*), and by it isolated nineteen different lines, each with its own distinctive mean weight of seeds. Similar selection and testing of the progeny of individuals produced directly by hybridisation, or of individuals from the progeny of these, has given us many valuable forms of cultivated plants.

Soon after the discovery of sexuality in flowering plants in the seventeenth century, it was recognised that crossing led directly to the production of new forms, and set up variation in plants not commonly given to sporting. Instead of passively waiting for sports, there was a definite effort to create new forms of plants from which selections could be made. Hybridisation became fashionable among plant breeders; in many cases it was of haphazard nature, in others only varieties possessing valuable qualities were used for crossing.

Not content with the results of a simple cross the earlier breeders of wheat, especially in Australia and America, resorted to composite crossing, and a similar practice was adopted in other countries. Frequently, hybrids between two different pairs of parents were crossed, and in some cases the plant obtained from these was again crossed with another variety or hybrid, the final individual selected being the product of six or eight different plants.

Many of Farrer's Australian selections were hybrids between four or five varieties, and in the production of the American Early Genessee Giant, A. N. Jones utilised eight different ancestors.

Some of the crosses appear to have given rise immediately to valuable stable forms; in other instances a desirable chaos of forms appeared in the offspring from which selection of promising individuals was made. In such methods accurate knowledge of the origin and genetic constitution was impossible, but to the plant breeder, concerned only with results, this was a matter of indifference.

Through the discoveries of Mendel and the more recent investigations regarding the transmission of characters of plants from one generation to another, haphazard hybridisation is being slowly replaced by crossing with a definite object. We are now able to combine in one form some of the peculiar qualities distributed among several individuals, and the selection of individuals possessing various desirable combinations of parental characters has become, in some instances, somewhat easier and more certain. Mendelism, however, has in no way reduced the importance of selection; if anything, the knowledge of cumulative factors, inhibitive factors, complementary factors and linkages, only serve to emphasize the necessity of selection and the many difficulties attending it.

When allusion is made to selection as a means of obtaining improved crops, it is, perhaps, more especially the practice of continued selection which is generally understood. The fact that by selection repeated during many successive generations we are able to obtain crops in which there is a gradual increase in the character selected was discovered long ago and is recognised as a potent method of obtaining superior crops. Examples of the results of such continued selection are well known; two or three only need be mentioned.

The improvement of the sugar beet is apposite. Commencing with the White Silesian Fodder Beet or Mangel, with an average sugar content of about 9 per cent. or less, by repeated selection in successive generations, of roots with the highest sugar content, the average has been raised to 15 or 16 per cent., at which figure it is maintained at present by continued annual selection.

Similarly, a gradual increase in the length and grain content of the ears of rye obtained by the same method has been mentioned already when discussing the results of mass selection.

The classic experiments carried on by Hopkins and Smith at the University of Illinois for more than 25 years have given the most remarkable results. During this period repeated annual selections of Maize have been made for high and low oil content and for high and low protein content respectively.

In 1896 Hopkins selected 163 ears from an ordinary crop of Burr's White Maize, and after making an analysis of a few grains from each ear for oil and protein, divided them into four classes, viz. high and low oil, and high and low protein content.

The different classes were grown in isolated plots and the extreme individual variants of each, as determined by chemical analysis, were selected annually as parents of the next year's crop. At the commencement of the experiments the average oil content for both the high and low classes was 4·7 per cent.

After ten years' selection for high oil, the strain obtained gave an average of 6·65 per cent.: the average for the low oil strain being 2·98 per cent. at the end of the same period—a difference of 3·6 per cent.

At the end of the next ten years' selection, the average of the high oil strain had risen to 8·02 per cent., the oil content of the low strain having fallen to 2·03 per cent., a difference of 5·99 per cent. for the twenty years continued selection. The figures in 1921 for the twenty-fifth generation were: high oil content 9·94, low oil content 1·70—a difference of 8·24 per cent.

Similar though less marked results have been obtained by repeated selection for high and low protein. Starting with an average protein content of 10·92 per cent., strains have been obtained having averages of 16·6 and 9·13 per cent. respectively.

Tests have shown that the different strains maintain their distinctive chemical compositions when grown side by side.

According to Pritchard the improved crops of sugar beet with increased sugar content are due to the repeated occurrence and selection of small mutations, and the results of continued selection in the other cases mentioned may be explained in the same manner.

In connection with the practice of repeated selection reference must be made to the results obtained by Johannsen in his experiments with Dwarf Beans (*Phaseolus vulgaris*).

From a commercial sample of the Princess variety he raised separate lines from single seeds and discovered that these lines differed from each other in the mean weight of their seeds. He concluded that the sample was not a single homogeneous line but a mixture of many lines (he isolated nineteen), each characterised by a different and constant mean weight of its seeds.

Repeated selection of the heaviest beans of such a sample in successive generations and propagation from these alone, would, if continued long enough, ultimately end in securing a type possessing a higher mean weight of its seeds than that from which we began, for the line with the highest mean weight would be gradually isolated from the rest.

Now Johannsen found that continued selection when applied to a single isolated line gave a different result. By selecting the heaviest and lightest seeds of such a line for six successive generations he failed to secure a crop in which there was any appreciable difference of mean weight of seeds; the lines obtained from the heaviest seeds and those from the lighter sister seeds had practically the same mean seed weight. He concluded that the wide plus and minus variations in the weight of individual seeds from the mean are entirely confined to the generation in which they appear. Such variations, to which the term fluctuation is commonly applied, are assumed to be entirely different from the variations which are transmitted from one generation to another, and repeated selection of these is considered of no value in securing improved crops.

Johannsen's conclusions have been hastily accepted, and the practice of repeated selection within lines raised from single seeds or individuals, especially of self-fertilised plants, has been checked.

It is being taught that the single initial selection in self-fertilised plants is alone of value; once obtain your improved variety, it maintains its characters from generation to generation, and further selection is superfluous. This, I think, is to be regretted, for hereditary variations, mutations or whatever they may be called, do occur in such lines, and it is only by renewed selection that these can be discovered and isolated: even if we were quite certain that the causes which produce fluctuations do not affect the hereditary mechanism of the plant, we should be justified in the practice of repeated selection on this account.

In my opinion, Johannsen's selections were continued for much too short a period to warrant a positive decision regarding the occurrence or non-occurrence of hereditary changes in his isolated lines.

It should be remembered that the differences in the average seed weight between several of his lines were only a fraction of a centigram. I think that much more evidence is needed before we can conclude that such differences represent fundamental hereditary characters: even if we admit that they do, the annual amount of mutation needed to pass from one line to another would be exceedingly small and very difficult to detect. In fact, some of the differences in the mean seed weight observed by Johannsen, especially in the selected lines raised from the large seeds of the lines of low mean seed weight, may after all have been such significant variations rather than fluctuations.

All that can be rightly concluded from his results is that no appreciable hereditary variations in the mean seed weight occurred in the lines investigated during the time of the experiments.

We do not, of course, know anything of the causes which led to the production of so many lines differing so slightly from each other in the original Princess variety used by Johannsen; apparently there was nothing to suggest that the latter was a complex population, especially in a plant usually self-fertilised.

The causes which gave rise to this complexity are doubtless always at work, and it is only a question of time for each of the isolated lines to become similarly heterogeneous, after which repeated selection would be effective in obtaining improved crops from them.

My conception of old established agricultural and horticultural varieties of our crop plants, is that apart from accidental mixture or outcrossing with different varieties, the several distinct lines usually discovered in them originate from mutation and subsequent crossing of the mutants with the parental type and with each other. True mutations, which I consider are in all cases responses of the plant to the environment, appear to be of great rarity, but only a very few are needed to give rise to a large number of new forms by re-sorting and re-combination of their factors and those of the parental form after crossing. The time taken for a recently isolated line to become an obviously mixed population of distinct lines will depend upon the frequency of mutation, and the facility for hybridisation.

Although selection is essentially no more than the isolation and preservation of improved forms which are presented to us, it is by no means a simple operation. Really effective selection is indeed difficult, and success only comes to those who devote much time to the study of the particular crops in which they desire to secure improvements.

For selection to succeed, it is fundamental, of course, that we should be able to recognise, as soon as they appear, the variations which we wish to preserve. Sometimes this presents no difficulty, the variations which constitute the improvement being readily appreciated even by the novice; for example, extra petals, decisive changes in the colour of flowers, giant or dwarf forms among plants, are easily observed, and by chemical analysis, differences in the content of sugar, starch, protein and other compounds can be accurately determined.

On the other hand, small morphological peculiarities, slight differences in habit, earliness, and other subtle features which may be the starting points of valuable modifications can only be discerned by the experienced eye.

Yielding capacity, which is one of the most important characters of cultivated crops, is at the same time probably the most difficult of all characters to determine.

The actual yield obtained in any particular case is not only dependent upon the hereditary yielding capacity of the plant but also upon a multitude of external conditions, among which may be mentioned, size of seed, the amount sown and the time of sowing, the depth at which the seed is deposited, the geological and physical nature of the soil, manuring and previous cropping of the land, rainfall and variations in its distribution during the growing period of the crop, temperature of the air and soil, and intensity and duration of sunlight. Variations in one or more of these conditions frequently cause extensive modifications in the yield of a crop. Yields of the same line of cereals grown in different localities often differ very widely: yields even from plots of equal area side by side often differ by 20 or 30 per cent., and yields of neighbouring plants on the plots may differ by as much as 100 to 500 per cent.

That there are hereditary differences in yielding capacity which are independent of external factors must be admitted, but in the best of our cultivated varieties they are always small and completely masked by differences due to the responses of the plant to external conditions.

In no cultivated crop do we know how much of that which we see should be attributed to the response of the plant to the environment and how much to genetic constitution; for example, the roots of most varieties of turnip on well-tilled highly manured soil are always 6 to 8 inches in diameter, but on poor soil the same seed gives plants in which the hypertrophy of the root is almost absent. Accurate knowledge of the inherent differences in the yielding capacity and quantitative characters generally of our cultivated crops, probably best ascertained from sand or water cultures under standardised conditions, would be of the greatest scientific interest, and efforts should be made to determine them for all our cultivated plants. Of more immediate importance, and of equal scientific moment, however, is the thorough investigation of the fluctuations induced by the external conditions commonly prevailing on cultivated land in different localities, for it is to these, rather than to the former, that the actual yields obtained are due. Many surprises, I am certain, are in store for those who will examine these problems.

For the moment, selection for high yield should take the form of accurate determinations of the yields of varieties of the chief cereals and other farm crops at various centres throughout the country.

Another difficulty in the way of obtaining improved crops by selection is the fact that there are limits to the extent of variations, fluctuating and hereditary alike, beyond which it is impossible to advance. We can regard as absurdities beets containing 100 per cent. of sugar, potatoes all starch, cereals all grain, strawberries as large as turnips, and peas the size of walnuts; but equally impossible of realisation may be beets with 35 per cent. of sugar, potatoes with 30 per cent. of starch, and peas as large as broad beans.

We have no accurate knowledge of what the limits are, nor do we know much about the conditions which determine them. There is considerable evidence pointing to the conclusion that plants in which the variations of a character reach a certain point beyond the mean become monstrous, with a pathological physiology, and it is not improbable that most of our much improved farm and garden plants are already monsters in this sense. Sugar beets containing more than 16–18 per cent. show a decreasing vitality up to about the 25 per cent. limit, at which point they are too weak and too small to be worth growing.

How often have we met the kindly old gardener who for many years carries on selection experiments to eliminate all the green tissue from his variegated pelargoniums in the hope of one day raising a crop of pure albinos? We know enough about physiological processes to check us from repeating this, but our knowledge of them is not sufficient to prevent us from expecting to obtain by selection results equally vain, for I have little doubt that we have reached the limits of improvements in many of the characters of our cultivated plants. That selection sometimes fails to secure crops in which the desirable character is an advance on what we already possess is doubtless to be explained on this hypothesis.

The few examples to which I have referred sufficiently indicate the function and value of selection in securing improved crops; each method has its own application and it is too soon to abandon any of them. We cannot afford to neglect the selection of obvious valuable sports which are continually arising among our cultivated crops.

After the most carefully planned hybridisation, accurate selection of desirable true breeding forms from the hybrid offspring is an absolute necessity; crossing is easy enough, it is the subsequent selection which is difficult, and which alone counts: unskilled selection makes hybridisation valueless.

Selection of individual lines from old established varieties or from the progeny of hybrids, and the propagation of the best, after progeny tests, is the method most widely practised at present and gives almost immediate results; its value is obvious.

Mass selection also has its value. Applied to seeds, it is always effective in securing more prolific crops, and similar selection of ears and plants from the ordinary varieties of cereals and many other cultivated plants results in improved types. As a preliminary process, mass selection finds useful application in eliminating mediocre lines from the progeny of hybrids before the final line selections are made. It is also of great value in maintaining the high standard reached in our best varieties by preventing the deterioration brought about by crossing or by mixing with inferior forms either accidentally introduced or arising in the stock by mutation.

Finally, repeated selection in all lines should be diligently pursued. We need not be unduly depressed by the dicta that "continued selection within pure lines does not change the type," or that "selection inside a pure line cannot increase the character selected," for both statements are meaningless. Selection is not the cause of any change; it is, however, the only means we have of discovering hereditary variations which may occur at any time in any of our cultivated crops be they pure lines or mixtures.

In conclusion, I should like to suggest that the present occasion is a fitting one for the inauguration of a series of experiments designed to extend our knowledge of the nature and extent of fluctuating variations and the results obtainable by repeated selection of such variations over long periods of time, in lines derived from individual seeds or plants of self-fertilised species.

Long continued observations are also needed upon such lines with a view of determining the kind and time of first appearance of recognisable mutations within them.

Experiments should also be instituted to determine the effects of good and bad cultivation upon clearly defined varieties of cultivated plants, and the effect of cultivation upon wild species, especially those which are considered the prototypes of our cultivated crops. The particular species and varieties of the plants to be used for such experiments should only be selected after the most careful consideration and discussion, and the experiments should be laid down only at such institutions which are likely to be able to continue the work for a long unbroken period. Results from such experiments would, I doubt not, enable our descendants to settle the problems which at the present time are only controversial and provocative of fruitless polemics.

Prof. R. G. Stapledon. SELECTION WORK ON HERBAGE PLANTS

From the point of view of pure genetics, herbage plants would seem to present a problem bewildering almost to the point of hopelessness.

In the first place, amongst the all-important clovers, we have *Trifolium pratense, T. hybridum, T. repens,* and the related plants *Lotus corniculatus,* and *L. uliginosus,* not only highly cross fertile, but absolutely, or almost absolutely, self sterile.

Medicago lupulina and *Anthylis vulneraria* are, however, abundantly self fertile, but appear none the less to set slightly more seed when cross pollinated.

Medicago sativa is also self fertile, but in order to ensure fertilisation the keels have of course to be sprung, but with this species seed setting is most abundant when cross pollination takes place, and individual plants vary widely in their seed setting properties when self pollinated[1].

Of the herbage grasses of Imperial importance, I suppose *Agropyron tenurum*[2] is practically the only one which is not only abundantly self fertile, but which is as disinclined to natural cross pollination as Wheat or Oats, and possibly even more disinclined. My colleague Mr Jenkin has recently found the same to be true of the useless *Lolium temulentum*[3].

The ordinary herbage grasses are, however, by no means absolutely self sterile, but there is considerable evidence to show that the majority of species consist at all events largely of individuals which are relatively self sterile, and these grasses like *Zea Mays* afford particularly interesting material for the study of albinism, virescence and variegation[4], charac-

[1] Jørgensen, C. O. "Om Bestovings- og Befrughningsforhold hos nogle grnermarksbalgplan der med Henlilik paa deves Foraedling." *Den Kongelige Veterinaer og Landlohøjihole Aarskrift,* 1921.
[2] Malte, M. O. "Breeding Methods in Forage Plants," *Scientific Agriculture,* January, 1921.
[3] Jenkin, T. J. "The Artificial Hybridization of Grasses," *Bull.* Series H, No. 2, Welsh Plant Breeding Station.
[4] See, for instance, Kajanus, Birger: "Zur Genetik des Chlorophylls von *Festuca elatior* L.," *Botaniska Notiser,* 1921.
At Aberystwyth, albino seedlings and virescents have been met with amongst the

teristics which the agronomist does not wish to perpetuate and, therefore, unfortunately cannot ignore.

At the outset an important direction in which to pursue selection is in search of self fertile individuals of all the chief species; at Aberystwyth we have already accumulated a little evidence which perhaps affords grounds for hoping that such a search may not be entirely in vain, while, having in mind Vavilov's concept of parallel variation[1] and the selfing propensities of modern wheats, it is obvious that the search should not be lightly abandoned—the more so since the prize would be of such inestimable value. From the agronomic point of view a plant is only usefully self fertile if it will not only set an abundance of seed, but will give rise to a line of individuals of no less vigour than those resulting from cross pollination, and to a line, the individual plants of which will continue to set seed abundantly, generation after generation.

I have commenced to investigate this problem in the case of *Dactylis glomerata,* some 245 plants of which species I have so far enclosed to ensure self pollination. Unfortunately my data cannot be regarded as absolute, because it does not follow that a plant, the panicles of which set no seed under the conditions superimposed by bagging, might not have set seed had the plant in question been protected from stray pollen by isolation on some habitat at once devoid of bags and remote from all other plants of *Dactylis.*

Obviously I can only give but a brief summary of my data on the present occasion. The following statement indicates the average difference in seed setting between plants selfed and when panicles of two plants are bagged together:

Year	No. of single plants or pairs of plants under test	Selfed or two plants together	Average No. of heavy or viable seed per panicle
1921	22	Two plants together	364 heavy seed
,,	70	Selfed	36 ,, ,,
1922	20	Two plants together	85 viable ,,
,,	71	Selfed	3·3 ,, ,,
1923	7	Two plants together	78 ,, ,,
,,	104	Selfed	14·8 ,, ,,

The data for 1922 and 1923, although those years were less favourable for fertilization than 1921, are the most reliable, because, as extensive

progeny from selfed plants in the case of *Dactylis glomerata, Arrhenatherum elatius, Alopecurus pratensis, Festuca elatior, F. pratensis, Lolium perenne, Phleum pratense,* etc.

[1] Vavilov, N. I. "The Law of Homologous Series in Variation," *Journ. Genetics,* vol. XII (1), 1922.

data now being accumulated by Mr Jenkin seem to show, in the case of selfed plants the production of heavy seed is no necessary criterion of the production of viable seed[1]. The figures as a whole clearly indicate, however, that *Dactylis glomerata* sets considerably more viable seed when given the opportunity of cross pollination than when denied that opportunity[2].

It was possible in 1923 to test the seed setting propensities of plants derived from the seed of mother plants previously selfed in 1921. The following comparison stated on the assumption that any plant casually taken from any particular habitat constitutes an F_1 is of interest:

Year	No. of single plants or pairs of plants under test	Particulars	Average No. of viable seed per panicle
1923	7	Two F_1 plants together	78·0
,,	104	F_1 plants selfed	14·8
,,	16	Two F_2 plants together	51·5
,,	52	F_2 plants selfed	5·1

It would thus appear that two F_2's together set seed, not as well, but nearly as well, as two F_1's brought together, but that F_2's only allowed to self do not, on the average, set seed nearly as well as F_1's so selfed.

The above, of course, are average figures and although of interest do not take us very far. If the behaviour of single plants is considered, we find that of the plants (presumably F_1's) selfed in 1922, only five out of seventy-one failed to produce any viable seed. Of those (presumably F_1's) selfed in 1923, 91 per cent. produced at least traces of viable seed, and 25 per cent. produced over twenty viable seeds per panicle, while of the presumably F_2's selfed in 1923 only 70 per cent. produced at least traces of viable seed and only 7 per cent. gave over twenty viable seeds per panicle.

Looked at more closely, it becomes perfectly obvious, however, that considerable segregation manifests itself in respect of the ability to produce abundant seed under selfed conditions. Thus, to quote a typical example (318 *Bc*), we have five different F_2 segregates from the same mother plant, giving respectively 51, 2, ·5, 0, 0 viable seeds per panicle when selfed.

The data presented appear to indicate two things of importance relative to selection, firstly, that any particular plant selected from

[1] Cf. also the case of wide crosses, see T. J. Jenkin, *loc. cit.*
[2] It must be borne in mind that cross pollination was probably in many cases not complete, as the panicles were merely bagged together and doubtless selfing also occurred.

amongst its natural surroundings need not necessarily be an F_1, it might be an F_2 or even an F_n, and if such were sometimes the case this would be sufficient to account for the poor seeding properties (under selfing) of some of the plants removed *in toto* from their natural habitats contributing to my data; and, secondly, as I have indicated already, that the search for highly fertile pure lines can hardly be regarded as necessarily foredoomed to failure.

It is difficult to estimate accurately the vigour of the progeny of selfed plants *versus* that of their parents or *versus* that of hybrid plants; such evidence as is available appears, however, to indicate that in this respect the grasses appear to behave in a somewhat similar manner to *Zea*, in the case of which species East and Hayes[1] have been able, under selfing, to isolate types (1) on a par with the parent, (2) superior to the parent, and (3) much inferior to the parent. Mr Jenkin is collecting accurate quantitative data in respect of *Lolium*, and although it is too soon to make a definite pronouncement it would seem that on the average his hybrid plants are decidedly more vigorous than the progeny of selfed plants—there being a suggestion, however, that occasionally segregates appear which are actually more vigorous than the mother plants. In the case of *Dactylis*, and taking height as a rough measure of vigour, I have the following evidence (1924). I grew my F_2 (= progeny of selfed plants) alongside my F_1's (the original mother plants), and was therefore of course comparing plants of different ages:

Average height of 53 mother plants	= 103 cm.
,, ,, progeny of 850 plants selfed	= 98 cm.

In 36 cases the average of the progeny was less than that of the mother plant, while in 17 cases the average of the progeny was taller than that of the mother plant. Apart altogether from albinos, variegated plants and virescents, in most cases I had amongst my progeny plants which were weaklings pure and simple, and in most cases also plants which were as tall or taller, and had all the appearance of being as or more vigorous than the mother plants. In the case of hand crosses kindly made for me by Mr Jenkin the average of the progeny was about the same as that of the parents. It would appear, therefore, that on the average F_1 plants of *Dactylis* are decidedly more vigorous than F_2, but that F_2 plants undoubtedly occur which are remarkably vigorous, and thus with vigour as with the ability to set seed under selfed conditions we are dealing, at all events

---

[1] East, E. M. and Hayes, H. K. "Heterozygosis in Evolution and in Plant Breeding," *U.S. Dept. Agr. Bull.* No. 243.

in so far as potentialities are concerned, with the phenomenon of segregation; Lindhard[1] moreover having shown that at all events as far as the fourth generation it is possible to retain vigorous progeny from plants segregating for vigour in F_2.

I have dealt at some little length with this question of self-fertility, because it is obvious that it is only species which are capable of yielding vigorous pure lines in the strictest genetical sense that are amenable to those methods of improvement which have made possible recent advances in cereal breeding; whether the more important grass species can be induced to do so under systems of controlled pollination can only be ascertained after the lapse of a number of years[2].

Let us assume for the sake of argument, however, and despite the slight evidence in the contrary direction, that a robust pure line, capable of being carried to the nth generation, of the chief grasses, like a pure line of Red Clover, is out of the question, and ask ourselves what, if anything, selection can achieve despite this discouraging postulate.

In the first place, it has to be realised that improvement, like almost everything else, is relative, and everybody familiar with our grasslands and their problems will agree that the scope for improving herbage plants is tremendous, and in the second place that "purity" at all events in the agronomic sense is also relative.

Further, it is at least doubtful if the same degree of genetical purity is economically necessary in the case of a herbage plant as in the case of say wheat, or of an industrial plant like cotton or the sugar beet.

In selecting herbage plants it has to be remembered that what is wanted is something breeding true to certain major characteristics such as excessive leafiness, a propensity for early spring growth, an ability to recover quickly after cutting, and an ability to remain aggressive and persistent under adverse conditions; no matter, so long as these desirable characters are retained in full measure, how wide the variation between plant and plant in respect of intimate and easily recognisable and easily definable morphological characters (such as particular leaf markings, colour of glumes, corolla or anthers and the like) as such of no significance to the grazing animal.

---

[1] Lindhard, E. "Planteforaedling ved Tystofte," *Tidskrift for Landbrugets Planteavl,* Bd. x, 1913.

[2] It is evident, however, that from the point of view of pure genetics, grasses afford comparatively satisfactory material, for they are not entirely self sterile, and it is possible, as Mr Jenkin has shown, to achieve comparatively large numbers of F_2 and F_3 plants. This is also a fact of considerable agronomic importance, for it should make it possible to elucidate the factors controlling the inheritance of various characters of economic significance.

At Aberystwyth we have made rather a detailed study of herbage plants in relation to source of origin both in the case of ordinary commercial seed and of indigenous plants obtained from seed collected off various natural—or more strictly speaking semi-natural—habitats, and by removing plants *in toto* from such habitats. It is a remarkable and significant fact in the case of certain species in particular that an investigator devoting most of his time to the study of such species can identify the country of origin of particular samples of seed with a high degree of precision by a critical examination of the plants to which such seed gives rise. It is possible to do this with considerable assurance in the case of Red Clover[1], whilst I myself would have but little difficulty in differentiating between say French, Danish and Indigenous Cocksfoot[2].

This suggests of course that the plants which constitute different nationalities by no means show the whole potentialities of the species, but consist only of certain and more or less definite aggregate strains. The case of American Red Clover is particularly interesting in this connection. *Trifolium pratense* was introduced into America and presumably from numerous European sources, and yet to-day both the American Medium and American Mammoth of commerce constitute strains which are easily differentiated from other nationalities, and more remarkable is the fact that although individual plants identical with these strains may be met with amongst European Red Clovers my colleague, Capt. Williams, has not come upon a single lot of Red Clover seed from any European country which has given rise to a population of plants which would have justified identification as American.

It would seem evident, therefore, despite the readiness with which the clovers and chief grasses cross-pollinate, that the combined influence of different sets of climatic and soil conditions and different commercial methods of seed production none the less exercise a very real selective influence, and an influence which shows itself in the limited variability of a population of plants derived from seed produced generation after generation under identical conditions.

The effects of commercial methods of seed production are perhaps purely a matter of agronomy, and I have dealt with this question in some little detail elsewhere[3], but their selective influence must be emphasised in the present connection. Consider the difference, the aggregate strain

[1] Williams, R. D. "Importance of Strain in Red Clover," *Agricultural Progress*, 1924.
[2] Stapledon, R. G. *Bull.* Series H, 1, Welsh Plant Breeding Station, Aberystwyth, 1922.
[3] Stapledon, R. G. "Strains of Herbage Plants" in the *Year Book* of the Essex Farmers' Union for 1923.

difference, between Wild White Clover (seed harvested from old swards) and White Dutch (seed from temporary leys), between Danish and U.S.A. Cocksfoot on the one hand (seed harvested from temporary leys) and Indigenous and Banks New Zealand on the other (seed harvested from long duration swards)—between Late Flowering Red Clover and Broad Red Clover. The case of the two highly distinct strains of Red Clover is the more interesting, because both are often harvested for seed within a restricted area, and yet commercial Late Flowering and commercial Broad Red are seldom much mixed. There can be little doubt that the most important major factor responsible for the relative aggregate purity of nationalities of extreme types, and even of local strains, is the very wide range in flowering date which obtains between the different strains of one and the same species.

The following examples, taken from data obtained by my colleagues and myself at Aberystwyth, are interesting in this connection:

Cocksfoot, 1921. Of some 300 plants the first to flower exserted anthers on May 24th, and the last to flower did so on June 28th, giving a range of 35 days.

Timothy, 1923. Gave a range of approximately July 8th to July 28th.

Sweet Vernal Grass, 1922. Gave a range of May 7th to June 7th.

Red Clover, 1923. Eighty-one lots of seed representing twenty-two distinct nationalities and strains showed the earliest lot coming into flower on June 26th and the latest not until July 30th[1].

It would seem probable that even a comparatively short gap of no more than two or three days between the zenith of flowering of two plants constitutes a greater barrier to cross fertilisation than has perhaps been fully realised—Mr Jenkin has, for instance, found in the case of artificial hybridization of grasses that his chances of success are very considerably increased by pollinating several times, and in nature of course the shorter the period of contemporaneous flowering as between two divergent strains, the smaller the chances of frequent inter-pollination.

Looked at from this point of view, data obtained on Cocksfoot and Red Clover on a particular day during the flowering period of each species appear to be significant:

Cocksfoot, June 19th, 1924. On this day, which was very favourable to the shed of pollen, the flowering stage of 47 plants representing a

---

[1] For accurate comparisons date of heading can be recorded more easily than date of flowering. The range given by heading dates is rather longer than that given by flowering, since the later strains to head flower more quickly after heading than those which head earlier.

number of my types was carefully estimated, with results that may be briefly stated thus:

14 per cent. of the plants had completely finished flowering, these were all foreign types.

12 per cent. of the plants were about at their zenith of flowering.

3 per cent. of the plants had exserted about 50 per cent. of their anthers.

56 per cent. of the plants had commenced to exsert anthers, but had certainly not shed 50 per cent. of their pollen, many of them only just starting and others not having shed as much as 10 per cent.

15 per cent. of the plants had not commenced flowering, these were all typical indigenous lates.

Red Clover, July 20*th*, 1923. On this date, Capt. Williams noted the number of plants in flower in the case of his various nationalities of Red Clover, several hundred plants being under review:

Early American Broad Red had	93 % plants in flower.	
English Broad Reds averaged	87 % ,, ,,	
Swedish Broad Reds gave	43 % ,, ,,	
Ordinary English Late Flowering Reds gave	33 % ,, ,,	
Swedish Late Flowering Reds gave	4 % ,, ,,	
Extra Late Montgomery strains gave only	3 % ,, ,,	

It is important to bear in mind also that the date of flowering can be regulated, and in regard to commercial methods of seed production is regulated by the hand of man, so that natural differences can be accentuated within fairly wide limits, and are no doubt equally accentuated by pronounced local meteorological conditions.

Garner and Allard[1] have shown that different strains of Soy Beans react differently to photoperiodism; if different strains of Red Clover have also different requirements with regard to available daylight, and evidence we are obtaining strongly suggests that such is the case, nationality differences in strain would be expected on this account also— differences which would of necessity be accentuated in one direction or another by the methods of seed production adopted.

The manner in which flowering date may limit variability, particularly under commercial methods of seed production, is further suggested by the broad differences which occur between the late flowering and early

[1] Garner, W. W. and Allard, H. A. "Effect of the Relative length of day and night and other factors of the Environment on growth and reproduction in Plants," *Journ. Agr. Res.* vol. XVIII (11), 1920.

flowering strains of some of the more important species. In this case of *Dactylis glomerata* for instance, the early strains tend to be less leafy than the late, tend to have fewer tillers and a lower proportion of barren tillers, and tend to be less persistent under adverse conditions. Remarkably definite differences can also be noted in the case of *Trifolium pratense*, in both cases intermediate forms of course being met with, but to a much more limited extent than would have been expected.

A critical study of a species from the agronomic point of view involves a somewhat different procedure to that adopted by the systematist, or ideally, it involves the concurrent adoption of the two procedures, but I am bound to confess I find it difficult to be both a botanist and an agronomist. The agronomist must come to know his plants somewhat as a shepherd knows his sheep—both are primarily concerned with the organism in the aggregate. I have classified *Dactylis glomerata*, purely for my own agronomic ends, into fifteen growth form classes. I should be sorry to have to describe my classes in exact botanical language— but I at least am satisfied that they are as real biologic units as shall I say Merino, Hampshire, Suffolk or other breeds of sheep, and are to me as easily recognisable as would be the sheep to my agricultural colleague, the shepherd.

It is, however, because of and in terms of these breed classes of mine that I think I am justified in saying a few words about habitat-races, and the selection of habitat-races in relation to improvement by selection.

When studying individual plants, either from seed collected from various habitats or when the plants themselves were dug up and grown as spaced individuals, I was at the outset impressed with the fact that in some instances practically all the plants derived from a particular habitat represented but one of my breed classes, while in other cases several of these classes were represented—frequently no one class being predominant. I found that, generally speaking, plants or seed obtained from hedges were decidedly mixed, and the same was usually true of plants from thickets.

The most interesting case was that of about 30 plants which I myself dug up on the top of the cliffs at Tintagel—these were not only all true to one of my "dense pasture" classes, but all the plants were quite remarkably similar. Another lot of 200 plants grown from indigenous seed sent me from Scotland—unfortunately I do not know the precise nature of the habitat—fell into but two of my classes, both closely related, of the widely fanned and somewhat saucer type. Plants or seed taken from old permanent swards or from at least what purported to be old swards showed interesting differences—in some cases nearly all the plants from a particular field conformed to but one or two classes—usually dense pasture classes—in others several classes were represented, some of the plants having characteristics suggestive of Danish origin or of crossing with Danish plants.

In short the evidence afforded from an examination of over 5000 plants derived from some 60 habitats, despite glaring cases of diversity, showed an altogether greater uniformity per source of origin than was perhaps to be expected from a freely cross pollinating species. A uniformity that can only be explained, at least so I should imagine, by the selective influences of long continued uniform environmental influences enormously accentuated by the wide range in flowering dates of different strains previously referred to.

It is fairly obvious that if all the plants taken from a particular habitat are similar: (1) that if individual plants from such a habitat are selfed there will not be much segregation amongst the progeny and (2) if several plants are isolated together to mutually inter-pollinate, the progeny plants will conform to the characteristics of the habitat-race which the parents represented. Thus if the above postulate is correct, the agronomist has but to find (= to select) a habitat-race possessing outstanding agronomic virtue, and his plant breeding in the ordinary sense of the word in respect of a particular agronomic requirement is finished before it is begun.

The method, however, demands a considerable amount of labour, experience and a certain acumen born of that experience, for it is difficult to recognise breed classes except when plants are grown as spaced individuals.

Before I had advanced far enough with my work to attach any significance to the "habitat-race"—perhaps ecologically considered I use the term in a too restricted sense—I had begun selfing my *Dactylis* plants in a wholesale manner, and also isolating together for mutual inter-pollination, batches of two, three or four plants which I deemed to be —not genetically but agronomically or breed classly—identical.

I must be brief, and must continue to talk in terms of my breed class —rather blatantly stated my evidence to date is as follows:

Of 76 plants selfed during 1921 and 1922, 34 gave rise to progeny all of which could fairly be placed in the same breed class as the mother plant in each case, while 40 plants gave rise to progeny much of which in each case was different in breed class to the respective mother plants.

Of 30 lots of two agronomically identical plants bagged together for mutual inter-pollination—16 actually gave rise only to progeny of the same breed class as the identicals so bagged, while 14 gave rise to progeny much of which represented different breed classes to that of the two parent identicals.

It is thus obvious that a considerable proportion of the F_1 plants taken promiscuously from devious habitats breed true to their breed class—and that in many cases agronomic identicals bunched together give progeny true to class. Last year I selfed two of the plants of my Tintagel habitat-race and also encased two together—I now have a population of some 600 odd seedlings at about four months old, and they are a particularly uniform lot—and I may say that extreme breed classes reveal their differences at a very early seedling stage.

It will be apparent from the above considerations that the method of "bunch breeding" practised by Zade[1] and others has a decided application—and that apparent "identicals" bunched together may give rise to true breeding progeny (in the breed class sense) from the outset, but of course if the apparent identicals have not been taken from habitat-races, or their equivalents, true breeding will be unlikely.

Taking the evidence as a whole one can apparently formulate a fairly rational procedure for the improvement of herbage plants by selection:

(1) By selecting promising parents possessing desirable contrasting characteristics, and proceeding on lines of accurate hand hybridization and ultimately building up stocks of seed of a true breeding segregate, or by pooling apparently similar segregates in the later generations. The applicability of this method cannot yet be said to be proved or disproved in respect of a number of important species.

(2) By selecting a large number of plants agronomically identical and by bunching same together for inter-pollination. If the plants selected are known to have been taken from amongst a naturally isolated habitat-race no further precautions would seem likely to be necessary.

(3) By selecting a number of agronomically identical plants and in the first instance selfing each plant. Then only bunching those of the original plants which when selfed gave rise to a progeny agronomically similar to the mother plant, and of course, therefore, truly representative of the breed class aimed at.

(4) An alternative method would be to bunch together all the progeny plants of these mother plants originally selected which when selfed bred true to the breed class desired. This plan would be perfectly legitimate if it were known absolutely that selfing but once had no subsequent ill

[1] See Zade in *Jahrbuch der deutschen Landegesell.* Bd. xxxiii, p. 139, 1918.

effects on the vigour of the plants subsequently resulting from the inter-crossing of the plants developed from selfing.

In conclusion I would like to make passing reference to two very important points. In dealing with herbage plants the investigator has very little to guide him in respect of what constitutes the ideal to aim at, one of the most difficult problems to be solved is the correct definition of the ideal in terms of attributes of single plants that may be deemed to be inheritable characteristics. To-day I think the agronomist knows far less about the fundamentals of quality, yield, persistency and aggressiveness for instance, than the geneticist knows about the fundamentals of heredity and thus he who aspires to improve herbage plants has his own very difficult furrow to plough before he is in a position to attempt to prepare the seed bed upon which to sow the teachings of modern genetics.

With the cross fertilizing herbage plants at all stages precautions have to be taken to ensure protection from stray pollen and this applies also to the building up of stocks of seed for final distribution.

Provided the production of named and approved strains is not over-done, unique facilities are of course afforded within the confines of our vast Empire for isolation and production of abundance of seed true to type, and this question of co-ordination in the production of seed supplies of cross fertile agricultural crops is one well worthy of detailed attention, and one which would seem to call for the adoption of co-operative methods within the Empire and should therefore appeal rather strongly to the members of this Imperial Conference.

Finally I wish to express my indebtedness to my colleagues Mr Jenkin and Capt. Williams, on whose extensive and laboriously acquired data I have abundantly drawn in the preparation of this paper.

Dr W. Lawrence Balls, F.R.S. Defects in the Theory and Practice of Selection

In the Theory of Selection we seem to be in need of much clearer concepts and stiffer definition of phenomena. With the work of Johannsen, Mendel and Morgan behind us there is little excuse for the looseness that pervades selection work. To the working agricultural botanist the greatest of these three is Johannsen.

The successive stages of uncertainty in our material may be summarised thus:

(1) A pure strain, self-fertilised, reproduces itself unfailingly, except

for the very rare chance (Yule and Willis)[1] of mutation, and no selection can be made therein. It is true that in our present state of knowledge the assertion of line purity rests on negative evidence only, and a verdict of "not proven" is the best we can obtain against assertions to the contrary. Nevertheless, the author regards this as the one fixed point on which applied genetics can be pivoted.

(2) A line which is not absolutely pure may simulate mutation by very rare recombination of genes.

(3) Pure lines which have been contaminated by the presence of even a single rogue must eventually show acclimatisation or deterioration, although this change takes place very slowly at first.

(4) The mixed variety produced after a term of years in the preceding case, or immediately by carelessness, can have no clear biological reactions to its environment, since the reactions of any one year will be partially obscured in the following year through the operation of natural selection.

In the Practice of Selection the ordinary commercial seedsman's work seems disappointing in contrast with the precision obtainable with cotton —and this in spite of the fact that up to 1900 cotton was written off as one of the most "variable of all crop plants."

There would seem to be two main systems of operation in the practice of selection:

(A) The production of improved varieties as quickly as possible, without much regard to their permanence, but hoping to supplant the improvement by a further improvement later. This is the method usually followed, and is regarded as almost inevitable when "results" are demanded. While invaluable in the early stages, this method has probably reached the limits of its utility in the principal agricultural crops. Bulk selection is a case in point, where all the work of the selectors is destroyed by natural selection if—as sometimes happens for political reasons— their control is discontinued temporarily for a few years.

(B) Alternatively, we have the method of pure line formation, which is often considered too slow to be a real alternative and regarded rather as an ideal objective. The outstanding advantage here is that no work need be wasted; we have evidence that by the use of fairly cool storage we can preserve even the oil bearing cotton-seed for periods of thirty years, and probably for much longer by suitable methods. From even

---

[1] Yule, G. U., "A Mathematical Theory of Evolution, based on the Conclusions of Dr J. C. Willis, F.R.S.," *Phil. Trans. Roy. Soc.* B, vol. ccxiii, 1924.

one such seed the pure line can be re-established in bulk very much more quickly than is commonly supposed, by taking advantage of two facts:

(i) That the number of seeds produced by a single plant is (in most crops) almost directly proportional (between limits) to the spatial allowance for the plant.

(ii) That specialised sowing methods can be used to enable nearly every viable seed produced to grow into a new plant. (For cotton the best of these methods consists in covering single seeds in dibbled holes with sifted sand.)

The simultaneous control of these two variables produces striking results. Thus, the ordinary rate of seed-multiplication for the whole crop of Egyptian cotton is 10 : 1 per annum, or 1000 : 1 in three years. Results actually obtained by the author in 1912–13 gave in three separate cases 10,000,000 : 1, *i.e.* 10,000,000 seeds harvested in 1913 from one seed sown in 1911 or later. This particular piece of work was done under many handicaps, so that a figure exceeding 10,000,000 is very likely under proper conditions for the first three years of production. The following table (p. 87) gives some typical figures in round numbers.

The amount of labour and supervision required, the sacrifice of land to wide planting, the organisation of smooth routine and the provision of special appliances (such as the relatively small bee-proof cages we used for the second year's plants), all tend to make such work resemble Civil Engineering rather than Botany. But, by using such engineering methods, it is possible to carry the precision of the genetics laboratory out into the field, and not otherwise. Adequate development in this direction has probably been delayed by contrast between the especially simple and inexpensive equipment needed for the genetics laboratory and the capital cost of such large-scale "works operation."

Usually it has been considered that this intensive treatment, which aims simultaneously at economy of time and at preservation of accuracy (in pedigree control), is not worth while as compared with the cheaper procedure[1]. There is, however, one peculiarity of seed propagation to which full significance has possibly not been attached, namely that the highest annual percentage contamination (whether by zygotes or gametes) of a variety seems to happen when the area of it under cultivation is just too large for the laboratory and yet not large enough to form a self-contained field crop. If this stage can be passed in two years instead

---

[1] This alternative procedure is only cheaper in the sense that it does not demand capital outlay; its running costs are higher per seed produced. Thus, in the long run, the selfing of flowers by hand costs more than bee-proof cages.

	ACTUAL EXAMPLES WITH EGYPTIAN COTTON				ESTIMATED AVERAGE
	All Egypt, based on commercial statistics	Exceptionally good field crop, with special care on best land (figures largely hypothetical)	A field selection Strain No. 111 1911–1913	Emergency operation with one area under Strain No. 77 in 1913	Rates on average land (spacing* and germination records)
1st year. Seed sown	20 or 2 (1)	8 or 2 (1)	1 (1)	All sown seed destroyed by mole-crickets in 3rd year; only the emergency reserve of seed available for re-sowing	1
Plants grown	2	2	1		1
Seeds per plant	100	300	400		1,300
Seeds harvested	200	600	200 + 200		1,300
2nd year. Seed sown	200 or 20 (10)	600 or 150 (75)	200‡		1,300‡
Plants grown	20	150	50		1,000
Seeds per plant	100	300	600		1,300
Seeds harvested	2,000	45,000	30,000		1,300,000
3rd year. Seed sown	2,000 or 200 (100)	45,000 or 11,000 (5,605)	30,000‡	8,000§	1,300,000§
Plants grown	200	11,000	10,000	5,000	800,000
Seeds per plant	100	300	1,000	2,000	200
Seeds harvested	20,000	3,300,000	10,000,000	10,000,000	160,000,000
4th year. Seed sown	20,000 or 1,000 (1/10 kilo)	3,300,000 or 412,500 (40 kilos)	10,000,000 (1 ton)	10,000,000 or 1,250 (1 ton)	160,000,000 (16 tons)
RATIO	$\frac{1}{400}$	1	5	18	6
Crop expressed as power of annual rate	10^3	64^3	316^3	$(1{,}250^1)$	400^3

* *Phil. Trans. Roy. Soc.* B. 327.
† In small bee-proof cage.
‡ Wide-spaced crop, with belting.
§ Field crop in 60 acres, with special sowing attention but ordinary spacing, producing a paying crop of lint.

of three or more (as in the cases quoted), it is worth while to consider the cost of doing so.

We imply the reservation, of course, that in ordinary cultivation even a pure line of a close fertilised variety will require intermittent seed renewal from the original stock—unless no other kindred variety exists at all in the same country. The most striking instances of accidental mixture were disclosed by the author's work in 1912–13 through the use of various detective precautions; cotton stumps sprouted in land where cotton had not been grown for at least a year; seed-cotton was dropped on the land, presumably by birds; and even when in store the contents of seed-sacks were mixed by mice. These sources of error are, in cotton, combined with natural crossing, so that expert agriculture and ordinary reasonable safeguards are quite ineffective to keep a strain pure. One particular strain was isolated in 1906, and is still in existence in 1924; but a field crop grown from it in 1917, of which the author saw samples, already contained some 40 % of rogues, though only three generations removed from a 5 cwt. lot of pure seed handed over in 1914 and farmed with scrupulous care during the three intervening years.

The selection of Physiological Characters should give valuable results, provided that the work of selection is coupled with physiological study. Three examples of such characteristic differences in cotton may be quoted:

(*a*) The mean maturation period, from the opening of the flower to the opening of the fruit, on chequer board plots was 49 days for one pure strain cotton and 52 days for another.

(*b*) One strain shed its flowers freely in the Northern Delta of Egypt, while another held them normally; the position was reversed in the Southern Delta.

(*c*) One pure strain produced $1\frac{1}{2}''$ lint under summer cultivation in Egypt, but only $1\frac{1}{4}''$ lint under winter cultivation in the Sudan. Another pure strain produced $1\frac{1}{4}''$ lint in Egypt and $1\frac{1}{2}''$ in the Sudan.

In conclusion we would re-direct attention to the risk pointed out by Harland, that the mine of raw material provided for selection by the natural jumble of varieties is liable to be destroyed, unless plant breeders deliberately maintain a rubbish heap through which to rake for valuable material.

Mr G. O. Searle. THE VALUE OF SELECTION WORK IN THE IMPROVEMENT OF THE FLAX CROP

When considering the value of selection work in the improvement of crop plants, one must not lose sight of the fact that ultimately such work seems of very little value, unless it is capable of being developed, and in the end actually is developed, on a commercial scale. The ultimate value of selection work must be judged by the influence such work has on the national yield of any particular agricultural crop. From this point of view it is pertinent to the discussion of this question to examine some of the difficulties experienced in bridging the gap between the experimental and the commercial stage. In this connection therefore it may be of interest to describe briefly some of the problems and difficulties, which have arisen during the bulking to a commercial scale of a new variety of flax by the Linen Industry Research Association. It is hoped that a description of some of the outstanding points in such work may draw attention to fundamental difficulties met with in the exploitation of a new strain, even although experimentally such a strain has been proved fully to be of value. It is highly probable, and much to be regretted, that only very few of the new strains of agricultural plants, of which descriptions are published, are ever carried sufficiently far towards commercial bulk as to place their merit beyond question: that is to say carried to such a stage that one can decide definitely as to their commercial value, where the average yield of a whole country, or even a whole district, is concerned.

Flax, like many other agricultural crops, presents several inherent difficulties to the plant breeder. Although usually self-pollinated, cross-fertilisation takes place sufficiently freely in the field to necessitate the "bagging" of all plants which it is desired to isolate in pure lines. As a corollary to this, new strains in their early stages of bulking must be strictly isolated, if they are to continue pure. Caging against cross-fertilisation is impracticable where inclement weather prevails, and roguing the small plots is out of the question, because in flax selection one is dealing chiefly with percentage content of the fibre—an internal character unlinked, as far as we have been able to observe at present, with any definite external character. On the small scale, however, such difficulties can be fairly easily overcome, though necessitating a considerable expenditure of labour and demanding facilities for isolating small plots in different parts of the country. It is when larger acreages are being grown and the commercial scale is being approached that the chief difficulties arise, and the question of financing the work becomes one of the main problems.

As illustrating this latter phase of the problem, reference will be made to some of the difficulties encountered by the Association in bulking a new variety of flax, known as "J.W.S.," which was selected originally by Dr J. Vargas Eyre during 1911 to 1915, and which has been conclusively proved since to be a very distinct advance on any of the ordinary varieties of flax grown for fibre purposes.

Having a seed of proved merit to develop one is confronted by serious limitations. Northern Ireland is the largest of the flax growing areas within the Empire, but, although suited to the production of good fibre, the climate there is unsuitable for saving flax seed, at any rate with any certainty, and in consequence the practice of saving seed is very seldom followed. To attempt to bulk a new strain of flax in Ireland consequently is largely a waste of effort. Furthermore, the ideal plant, from the flax grower's point of view, is one with but few branches and seed capsules, and, taking the "J.W.S." as an example, it seldom yields more than a four-fold return of seed as compared with the eight-fold yield of ordinary varieties: a fact which necessitates extremely rigorous precautions being taken against contamination of the strain during bulking. As a further consequence of this poor seed yield, the cost of raising the "J.W.S." variety for its seed alone becomes prohibitive, and makes it necessary to win the fibre from the straw in order to defray the cost of growing. This difficulty becomes more and more acute as the bulk of seed increases and inevitably means the undertaking of factory operations on a large scale.

It is now generally recognised that the future of flax growing not only in the British Isles, but in other parts of the Empire, is mainly dependent upon an improved sowing seed coming into general use. Although the Linen Industry Research Association has been able to bulk the "J.W.S." seed, which is the most advanced new strain at present available, up to a quantity of over 10 tons, sufficient to sow more than 200 acres of crop, it is clear from a consideration of the points mentioned above that an entirely different organisation is necessary to carry this and other new strains on to the point when they may become of material benefit to the community.

Presumably other crops, where seed is not the product of primary importance, present similar difficulties; but it is doubtful whether with any other crop such a high cost of manufacture and such a very low seed yield are combined. Amongst other textiles for instance, cotton can be bulked at a very much greater rate, the picture given by Dr Balls showing a ton of seed derived from a single cotton seed in two years naturally excites the envy of the flax breeder. Where the conditions of climate are suitable, as in America, it is possible by alternating the flax growing districts to obtain two crops in the year and thus greatly increase the rate of bulking. It was thought at one time that some such scheme might possibly be brought into operation by using England and N. Africa as the alternate cropping areas, but the small time available between the seasons and the difficulties of transport caused us to abandon the idea.

Owing to these difficulties attempts are now being made to develop a Flax Seed Bulking Organisation for the benefit of all flax growing districts within the Empire and to encourage the opening up of new areas to this crop. The necessity for this step will be appreciated from a brief account of the efforts made to accomplish the work without such state assistance.

In 1922 the situation with regard to the "J.W.S." flax had become critical. There was just over two tons of this seed available, and it had taken over ten years to reach that amount. For the reasons already given, it was realised that to distribute this seed to growers in Ireland would undoubtedly result in its disappearance from cultivation within a very short time. The probable produce from the crop, say some 80 to 100 tons of straw, was more than the Association could deal with in its own experimental plant. Our experience of sending parcels of seed to Canada and France respectively to be bulked had resulted in a complete loss in each instance. Turning to possibilities nearer home, negotiations were entered into with the National Institute of Agricultural Botany, which it is understood is expressly for the purpose of filling the gap between the plant breeder and the commercial user; but whether through a lack of interest in the crop, or a prevision of the trouble involved, that Institution expressed itself quite unable to undertake the further bulking of the two tons of this seed. As there was no other organisation which would undertake the bulking in the British Isles, there was no alternative but for the Research Association to make private arrangements with a flax factory to carry on the work. An arrangement was made, with the only flax company existing in the south of England, on the following terms. The flax company were provided with the seed free, and undertook to grow the flax in a district apart from their other crops. They guaranteed to take rigorous precautions in deseeding the crop in special machinery away from other flax, and they promised to hold the resultant seed at the disposal of the Research Association. The fibre to be won from the straw was their source of profit. This was found to be a perfectly satisfactory arrangement to both parties, and the scheme would have allowed of considerable extension. Unfortunately the area devoted to the "J.W.S." flax was quite small compared with their main crop, and, although the former proved to be a source of very reasonable profit to the company, they went into liquidation at the end of the first year's working of the agreement. Only one other flax factory was available at that time, and it was situated in Scotland and was therefore not so suitable for flax seed saving. Another of the Association's new varieties however was entrusted to them on a similar agreement, over a ton of seed being provided. This company also went into liquidation during the general slump in trade, and in this instance, owing to the technicalities of Scottish law, the whole bulk of seed was lost.

After considerable trouble an arrangement was made with two farmers in Somerset to carry on the bulking of the "J.W.S." seed in 1923 under a somewhat similar arrangement. These two farmers are probably the only two private scutch mill owners in England. This scheme worked very satisfactorily, and showed such a good profit to the farmers concerned that they have both undertaken the maximum acreage possible in 1924, but of course are unable to deal with the whole quantity of seed, which would sow some 300 acres.

As a final solution of the problem, the Association is endeavouring to put on foot a scheme whereby the bulking operations will be carried on by a state-aided separate organisation. In the meantime in order to tide over this year, the Association has taken over from the Government two factories in the Yeovil area, where about 200 acres have been sown, although owing to the delay it was impossible to obtain the best land and give the seed a good chance. In this way it is hoped that the "J.W.S." flax, and other new varieties as they are produced, can be grown on an increasingly large scale each year, until such a quantity of this seed can be distributed to growers in Northern Ireland, and other flax-producing parts of the Empire, that it will have a material influence on the whole flax industry.

This short sketch of some of the difficulties encountered, when endeavouring to carry the results of flax selection to a successful issue, is put forward as an illustration of one of the broader aspects of the work, which must be borne in mind when considering the value of plant breeding. We have in this instance a concrete example of a new variety of flax, which has proved to be superior to any of the ordinary varieties.

It has been asked for very widely, applications in fact for quantities of this seed have been received from twenty-seven different countries. It has been bulked up to a considerable amount, and is available for further exploitation. Yet, so far it has not been possible to find any organisation willing to relieve the Association of the further work and responsibility, or to get relief from the financial burden, which is now beyond its means. It can only be suggested that the cost and financial risk of an undertaking of this nature should, when it has been carried to such an advanced stage, be made an Imperial matter.

Mr W. N. Sands. SELECTION OF HIGH YIELDING VARIETIES OF RICE IN MALAYA

The total area of British Malaya devoted annually to rice cultivation averages 669,000 acres; this represents about 2 % of the total area of the country. The annual local production of rice is 3,578,700 pikuls (pikul = $133\frac{1}{3}$ lbs.) whereas the normal annual imports total on the average 5,462,000 pikuls. The average yearly cost of the imported rice is 34,000,000 dollars (dollar = 2s. 4d.). The local production of rice is estimated to be sufficient to feed only 1,142,326 persons of all ages, out of a total population which in 1921 amounted to 3,332,603.

It was this undesirable state of affairs which led Government to offer exceptional facilities and encouragement for the production of rice in different parts of the Peninsula, where suitable conditions existed, and several large schemes for opening up new areas and irrigating them are receiving attention.

The Department of Agriculture has been, and is, intimately concerned in the examination of areas on which rice may be grown; the improvement of cultural and milling methods; the improvement of the yield of local varieties; the importation and acclimatization of new varieties, and in many other ways.

Experimental research work connected with the improvement of rice has been conducted on an extensive scale by the Botanical Division of the Department. Since the year 1916, the Economic Botanist, Mr H. W. Jack, has been in charge of the investigations, and for the past four years the writer has been associated with him in the work.

The Department's chief Rice Breeding and Testing Station is situated at Titi Serong in the Krian District of Perak. In this district there are 55,000 acres of land under irrigation. The area consists of low-lying alluvial plain, a few feet only above mean sea-level. The soil is, for the most part, heavy clay rich in plant nutrients, and capable of producing 500 gallons of padi (unhusked rice) per acre in average seasons.

The area of this Experiment Station is 21 acres and all of it is used annually for tests of local and imported varieties, pure line selection and breeding, manurial and other experiments.

Method of cultivating Wet Rice.

A brief account of the method of cultivating rice in the irrigated area of the Krian District may perhaps be usefully included here, as the operations have been closely followed where practicable in the Experiment Station.

At the commencement of the planting season fresh water from the distributaries of the main canal is let in until the land is flooded to a depth of six inches or more. The luxuriant growth which consists mainly of grasses and sedges is then cut down to the level of the soil by means of the "tajak"—this is a heavy iron-bladed scythe about 18 inches long with a tapering blade attached to a stout wooden handle which is set almost at right angles to the blade. In cutting down the vegetation, the "tajak" is raised well above the shoulder and swung sharply downwards so that the edge of the blade meets the plants just at or below the level of the soft soil. No tillage of the soil is practised and the cut vegetation is simply allowed to rot in the water until the fields are required for planting some two to three months later, when the undecayed portions are raked to the "batas" or raised divisions between the three to six acre fields. It will be readily understood that the soil near these divisions is richer in plant nutrients than in other places. Under local conditions it has been found that there is no "batas" or border effect on the rice plants if they are planted in the experiment plots at a distance of not less than 20 ft. from the divisions.

The nursery beds are prepared at the same time as "tajaking" is started. In the construction of these the cut grass and weeds are piled into a long strip, 3 to 4 ft. wide, and raised about an inch above water level. On this foundation of vegetable matter sufficient mud is plastered to form a level surface. A thin layer of mud rich in organic matter is placed on this to complete the work. Prior to sowing, the seed is steeped in water for two days and two nights; then taken out and kept in a moist condition for a further two days. At the end of this time the grain which has started to germinate is sown by broadcasting it thickly and evenly over the surface of the prepared bed. In the case of pure lines, the ears with the grain attached are laid on the beds with their zinc labels. The nursery beds are covered with grass, banana or palm leaves to afford shade as well as protection from birds. The beds are watered morning and evening if the weather is dry. The nurseries of the Experiment Station are protected from birds and rats by wire-netting of $\frac{1}{2}$-inch mesh over which leaves are placed to give the necessary shade. As the seedlings grow, the shade is gradually reduced; when they are 4–5 inches high the beds are lowered by pulling out some of the grass foundation from underneath until the surface of the bed is level with the water.

When the seedlings are 10 to 12 inches high or about 14 days from the time of sowing the grain, they are taken up in clumps of 150–200 with the nursery soil adhering to their roots and placed in a convenient place in the field where the depth of water is about 4 or 5 inches. After an interval of 8 to 10 days the clumps are re-divided into five or six smaller ones and again planted in water. This second transplantation, however, is sometimes omitted. After a further period of three weeks the clumps of plants are divided again into masses of 12–20 plants, each group occupying about 2 square feet. There are, therefore, two or three transplantations before the final one. In the earlier stages great care and watchfulness has to be exercised to prevent damage by excess of water and damage by pests.

When ready for the final planting the plants, which are then from 2–3 ft. high according to the variety, are pruned back and lifted up by cutting around the roots. They are then transported to the field and divided into small bundles. In the Krian District all planting is done with the aid of a short iron tool with a forked end, known as the "kuku kambing" or goat's hoof. In planting, the planter takes up a bunch of plants in the left hand; pulls out the number of plants to be planted in the hill, which may be one to four; places the plants between the prongs of the fork of the "kuku kambing" at the collar and deftly inserts the plants into the soft soil in an erect position. After planting, the rice needs little attention beyond an occasional weeding in the earlier stages and the regulation of the water supply and drainage, until harvest three to five months later.

Under the usual Krian conditions which apply to the Experiment Station, high tillering is constant and it is usual to allow an average of $1\frac{3}{4}$ square feet (15″ × 17″) per hill, each hill usually containing three seedlings, except in the line cultures, where only one is planted. In areas where the depth of water is constantly over 6 inches and in others where the land is less fertile, a somewhat closer planting distance is followed. It has been demonstrated that the number of tillers within a variety depends on the fertility of the soil, distance of planting and depth of water. For example, a good 8-months Seraup variety may produce as many as 50 tillers per plant under the most favourable conditions, 18–20 under average conditions and 2–3 under poor conditions. In breeding work, tillers can be used to supply vacancies in plots in definite areas. This is a most useful method of propagation in experimental work because it causes no damage to the stock plant and at the same time affords a ready method of increasing a particular strain.

The first indication of the appearance of the flowers is a distinct swelling of the stem of the leading shoot. Shortly after, the panicle thrusts itself through the leaf sheaths. The time taken to reach this stage varies, of course, with the maturation period of the different varieties, but it is a signal for the gradual withdrawal of the water from the fields. The grain is usually ready for harvest some six weeks later.

About ten days from the emergence of the panicle the flowers commence to open, starting from the top. At the end of six days all the flowers on a panicle have matured.

The flower opens about 10 a.m. in fine weather; remains open for some 20 minutes only and then closes again. It would appear that the flowers of the Seraup and Radin groups with which experiments have been made, are self-pollinated just before the opening of the flower. The anthers when they emerge are practically empty of pollen, which can be seen thickly coating the inner surfaces of the glumes, and the stigmas are freely dusted with it. Up to date our efforts to make crosses between the varieties along the usual lines have failed. Judging from the ease with which it has been possible to maintain the purity of the lines, it is probable that natural cross-fertilization, if it occurs, must take place at rare intervals. As opportunity occurs, it is hoped to undertake further researches in the hybridization of the local varieties.

The ripe ears are harvested either by means of a small sickle or with a peculiar knife called a "pisau menuai." The sickle enables a handful of ears with about 2 ft. of straw attached to be reaped at one operation, whereas with the small knife the ears are harvested singly. In the experimental work, the former tool is used in reaping multiplication plots, and the latter for the line cultures.

The grain is threshed either by striking the ears with the grain attached against a short ladder placed in a large tub, the tub being backed by a high screen of matting; or, in the case of the ears harvested with the knife, by treading the ears under foot.

The threshed and uncleaned "padi" is then sun-dried and winnowed.

Experimental lots are usually sun and air-dried for at least 14 days after harvest before they are measured and weighed, because the padi loses its moisture with fairly uniform rapidity up to the 14th day, but after this period the variation in weight is dependent on weather conditions.

Over 1700 so-called varieties of rice have been collected in different districts. Although they have not been critically examined and classified, it would appear that there are not more than 300–400 distinct varieties.

Varieties of rice differ in size, shape, colour, and hairiness of grain; in the length of the maturation period; in yield of grain per unit area; in length, strength and thickness of straw; in tillering capacity; in culinary and milling qualities; in chemical composition, and in many other characters.

The local varieties of hard rice, that is those which have a vitreous fracture, can be grouped broadly into the following classes according to the shape and size of the grain:

(a) Seraup type, which has a distinct shoulder on the anterior extremity of the grain.

(b) Radin type, which is very symmetrical in outline and of medium length.

(c) Siam type, which is always very long and frequently slightly curved.

(d) Rangoon type, which is markedly broad and thick in proportion to its medium length.

All the rices so far found in Malaya can be placed in one of these categories, but by far the most important classes are the Seraup and Radin types, which probably produce 80 % of the total output.

It is with these types principally that investigations, having for their object the isolation of high yielding rices, have been made.

The two Seraup varieties which are popular amongst the millers because of their weight and good milling qualities, namely, Seraup Kechil and Seraup Besar, are chiefly confined to the large flat coastal areas of Krian and Province Wellesley and to the heavier clay soils around Kuala Kangsar, and these probably represent 25 % of the annual crop.

They are commonly known as "berat" or heavy paddies because of their high yielding characters and their long maturation periods, which average 235 days, though strains can be bred to shorten or lengthen this period by 14 days. Strains of these varieties in very fertile soil are capable of producing crops up to 850 gallons of padi per acre in exceptionally good seasons and in first-class land. Under average conditions the crop is rarely less than 500 gallons per acre.

The Radin type is more popular because of its excellent culinary qualities and for this reason is common in all the smaller padi areas where the crop is grown for home consumption only, though it is also found in the larger areas in Kelantan, Kedah, Krian and other places. This padi is not as popular with millers because of its lightness compared with "seraup" and because of the smaller size of grain. These factors have adverse effects on its milling qualities by raising the ratio of husk to kernel as compared with "seraup." Probably 50 % of the annual crop is derived from this type.

Besides the hard rices there are numerous varieties of glutinous rice; these are of little economic importance, being used chiefly on ceremonial occasions. They never have a purely vitreous appearance when fractured, and when boiled they form a soft gelatinous mass.

Several methods may be adopted to improve the rice crop of any particular locality; these may be grouped under five headings:

(1) Amelioration of environment by improved agricultural methods and irrigation.

(2) Importation and acclimatization of varieties, both local and foreign.

(3) The selection and propagation of desirable mutant types.

(4) Hybridization of varieties selected for particular characters.

(5) The pure line method of isolation and selection.

In this paper it is intended to give an account of some of the results of investigations undertaken under the last-named head only, as it is with the isolation, selection and testing of pure lines that most progress has been made to date; besides, the pure line method is one of the most important methods of crop improvement.

As mentioned previously, there are between 300–400 distinct varieties grown in Malaya. About 300 of these were chosen from a collection of some 2000, and cultivated as pure lines for five seasons. Each successive season, grain from a single plant of each variety was sown and the resulting seedlings planted out in lines of 100 plants. Records of the botanical and agricultural characters were also registered and checked each season.

Particularly promising types, from the point of view of yield, were multiplied and tested against selected "seraup" and "radin" strains, these tests still being in progress, as all tests must cover at least three seasons so as to give an insight into average yielding capacity under the seasonal variations which occur from year to year. Thus these tests raise the study of varieties from the level of a mere piece of academic research to a work of practical utility and at the same time furnish a check to observations made while the particular variety was grown in line culture.

Seasonal variations, with particular reference to rainfall, complicate the testing of varieties or strains considerably, for it is frequently found that a variety or strain under even slightly different climatic conditions on the same soil may give very diverse yields in two successive seasons.

Hence, in testing work, it is necessary to strike averages over a period of several seasons in order to make sure that a given variety or strain is really more valuable than others, and of course the averages become the more reliable in proportion to the number of seasonal tests from which they are derived.

The agricultural characters of the leading Seraup and Radin varieties, which produce 80 % of the total local production, vary considerably when grown under similar conditions, as shown below:

Seraup Varieties

Maturation period	167 to 246 days
Height	48 ,, 65 inches
Tillers per plant	15 ,, 23
Yield per plant	44 ,, 94 grammes

Radin Varieties

Maturation period	151 to 240 days
Height	33 ,, 60 inches
Tillers per plant	8 ,, 28
Yield per plant	38 ,, 91 grammes

In every variety of rice as in other self-fertilised plants there are distinct and constant forms termed biotypes or elementary species. From among the best producing Seraup and Radin varieties grown in the Krian district, 300 plants which appeared to possess desirable characters were selected for testing in 1915–16, and 1000 more in the following season. These 1300 individuals have formed the basis of the work with pure lines to-date.

Even with this comparatively small number the various operations, particularly those connected with the planting, harvesting, and recording of results, have proved most onerous and exacting under local conditions. In certain seasons wet weather at harvest time has added considerably to the difficulties encountered, so that it has not been practicable to deal with a wider selection of strains.

In the first year a single ear of each selected strain was sown in the nursery and the resulting seedlings planted out in a line varying in length according to the number of seedlings obtained from each parent ear. In finally transplanting the seedlings only one plant was set in each hill 15 inches apart, with a space of $3\frac{1}{2}$ ft. between each line. In each succeeding season an ear with its attached grain from a single plant was used as seed. Any strains which proved to be mixed were eliminated.

All the strains were grown as pure lines in triplicate for two seasons

Strain No.		1919	1920	1921	1922	1923	Av. of 5 years per plant in grms.
\multicolumn{8}{c}{Variety: Seraup Kechil}
36		95	108	80	71	117	94
52		93	113	74	70	115	93
48		85	102	69	85	120	92
68		87	111	68	77	111	91
20		86	113	73	64	109	89
257		88	119	67	82	87	89
371		87	101	73	75	107	89
146		86	103	76	79	96	88
6		85	95	71	76	107	87
8		79	97	65	79	104	85
151		83	98	68	76	99	85
517		80	95	67	69	98	82
\multicolumn{8}{c}{Variety: Radin—best six strains}
7	Radin Kuning	84	86	91	83	83	85
2	,, Merah	72	81	92	86	75	81
13	,, Puteh	81	79	91	69	80	80
1	,, Merah	77	78	96	71	69	78
16	,, ,,	75	79	85	74	75	78
4	,, ,,	72	76	81	64	85	76
\multicolumn{8}{c}{Variety: Seraup Besar}
15		82	106	76	58	101	85
509		75	93	58	82	93	81
3		76	97	64	54	98	78
590		70	97	74	61	82	77
\multicolumn{8}{c}{Variety: Padi Pahit (Seraup type)}
1		75	86	78	78	101	84
4		72	84	81	80	91	82
8		69	83	92	75	83	80
7		71	78	73	63	85	74
Average yield from un- selected seed		49	51	45	43	49	47
Average yield per plant from 629 strains under selection		66	79	54	64	84	69

before any of the low-yielding ones were discarded. Since the 1918 season, about half of the annual plantings have been eliminated each year. The method adopted was to obtain the mean of all the strains under test in respect of weight of grain per plant and to reject those strains which had an average weight per plant below this mean. In the 1923 season there were only 77 strains representing five varieties grown. Each of these was planted in quadruplicate plots of 100 plants per plot.

For the past five seasons the yield of plants grown from unselected seed in adjacent fields has been taken and compared with that of the pure lines.

The table on p. 100 gives some of the yields of the best selected strains of the chief varieties compared with the yield of local unselected varieties grown in the neighbouring fields.

The yields per plant in each season were derived by weighing the grain of 400 plants of each strain. The 400 plants were grown in four separate lines of 100 plants each every season, each line being planted in a different part of the experimental area. The yields produced by unselected seed were obtained by harvesting 400 plants in 10 straight lines of 40 plants each in adjacent fields planted by Malays. Whilst the yields obtained from plants of unselected varieties are not strictly comparable with those of the pure lines on account of the wider row or lateral spacing—$3\frac{1}{2}$ ft. as against an average of 17 inches for the former— yet experiments have indicated that in the fields where the pure lines are grown the optimum planting distance is 15 inches by 17 inches, which spacing is the average for native grown rice on first class land in the neighbourhood of the Experiment Station where the yields of the native varieties were obtained.

The number of plants per hill is also different, namely, one in the pure lines and one to four in the native holdings; but here again experiments have shown that as far as yield is concerned it appears to make little difference whether one or four plants are used, although the number of tillers per hill is higher by an average of nearly four when four plants are set to a hill. The actual results for one of the Radin varieties at the optimum planting distance ($15'' \times 17''$) were as under:

No. of plants per hill	No. of tillers per hill (av. of 100 hills)	Av. yield of padi per 100 hills	Av. yield of padi per hill in grms.	Yield per acre in pounds
1	13·3	5,460	54·6	2,964
2	14·5	5,370	53·7	2,917
3	15·5	5,480	55·0	2,976
4	17·1	5,270	52·7	2,862
5	17·0	5,110	51·1	2,775

The yields per plant of the best pure lines are so much higher that even allowing for a lower yield per plant when grown in large plots and by native cultivators, increased yields from the pure lines are anticipated where similar conditions exist. As however the best yielding strains have been multiplied in plots of ½ acre to 4 acres each and the seed distributed for numerous large scale tests in different parts of the Peninsula, and native cultivators have grown and compared them with their own varieties, it has been possible to show that where local conditions and methods of cultivation are suitable, increased yields of from 5 to 25 % can be obtained by growing the pedigree seed.

The yields of the pure strains which are given in the table above (p. 100) were adversely affected by the wet harvesting seasons of 1921 and 1922; on the other hand good weather was experienced during the 1920 and 1923 harvests, with the result that the yields were satisfactory.

Before dealing with the tests of the best yielding strains in large plots in different districts it should be mentioned that a good deal of attention has been devoted to the estimation of the Probable or Experimental Error of the results obtained in connection with the pure line selection work for increased yield of grain in the Krian Station. The application of biometrical methods for comparing the yielding abilities of a large number of different pure strains is most important because they enable selections to be made with the minimum cost from small plots, always provided that the calculations are applied only to the land on which the experiments were conducted, or land in the same neighbourhood where the climate, soil, water supply, cultural and harvesting operations are similar.

In a normal harvest it was found that the Experimental Error was ± 3 %, whereas in a wet one it worked out at ± 10·7 %. Under Krian conditions an Experimental Error of 10 % is considered a safe margin on which to base calculations except when a wet harvest occurs, when it is raised to 12 %.

Details of the experiments performed by Jack which enabled these results to be obtained will be found in *Bulletins* Nos. 32 and 35 of the Department of Agriculture, Federated Malay States and Straits Settlements.

Experience has shown that a variety or strain of rice which thrives in one locality may not do so in another, so that it is necessary (a) to isolate a wide range of strains and to work with more than one variety, (b) to have large comparative plots in different districts, in order to ascertain which strain or strains are superior to those grown locally.

Although the strains isolated in Krian are suitable for a wide range of conditions, yet it is possible that not one may be superior to the local forms, with the result that in a particular locality it may be desirable to have a local experiment station where strains of local varieties may be isolated. Test stations where selected strains are grown in large plots have been started in several places during the past two seasons, and placed under the supervision of local agricultural officers, who keep records of the yields obtained by native cultivators who have been supplied with selected seed and the yields obtained from unselected seed for purposes of comparison. Most encouraging and informative results have been obtained in this way.

The question of the effect of the change of environment has also to be considered, and there is some reason for assuming that a pure strain introduced into a particular area may not at first give satisfactory yields, but after being grown there for several consecutive years it may prove to give yields superior to the local forms.

That certain of the pure strains have become popular in different districts is instanced by the fact that the native planters in the Krian district alone had, last season, not less than 2000 acres under pedigree strains. The even growth, uniformity of ripening, full ears, heavy grains and strong straw of the rices were points quickly noticed by the native grower and found favour with him, with the result that large quantities of seed of the strains have been re-distributed. Again, millers like the padi because it gives good milling results. It is expected that in a few years' time practically the whole of Krian irrigated area will be planted with seed originally obtained from the Experiment Station.

In other districts progress has been slower, but good reports of certain strains have been received from Province Wellesley, Perak North, Negri Sembilan, Kedah, Perlis, Selangor and Malacca, and a large extension of the area to be planted with selected seed will be made next season. In some parts of Perak, Province Wellesley and Malacca, selected strains did not thrive as well as the local varieties, and it is in these places that local breeding work may be necessary.

It is worthy of note that the demand for pure seed of the heaviest yielding strains exceeds the supply, and it is estimated that 5000 gallons will be insufficient for the coming season's requirements, notwithstanding that a large re-distribution of seed is taking place among the native cultivators themselves in most districts.

I have drawn freely on the writings of my colleague—Mr H. W. Jack— in this paper, and I am indebted to him for reading the proofs of it.

PLANT PATHOLOGY AND MYCOLOGY

THE RELATION OF PLANT PATHOLOGY TO GENETICS

(CHAIRMAN: PROF. J. PERCIVAL)

Mr F. T. Brooks

All plants which have been cultivated for any considerable period of time exist in innumerable forms or varieties. One of the most striking differences between the varieties of any one crop plant lies in their varying susceptibility to specific diseases; one variety may be extremely liable to a certain malady and another may be completely immune to it. On the other hand the same variety of cultivated plant may differ in susceptibility to a specific disease from season to season or place to place, so that in such a variety the genetical factors determining susceptibility or resistance must be modifiable in expression by changes in environmental conditions. Susceptibility or resistance to disease may be due to a single genetical factor or to multiple factors. It has been shown with many crops that resistance to disease may be deliberately combined with other desirable characters such as high yield and good quality, although it is sometimes difficult to break asunder the linkages often existing between resistance to disease and certain undesirable characters.

One of the most interesting results of attempts to elucidate the inheritance of resistance and susceptibility to disease is the indication that in the hybrid derivatives there are sometimes forms which are more resistant and also more susceptible than the original parents. On the University Farm at Cambridge some of the F_2 derivatives of a cross between "Rivet" and a "Vulgare" form of wheat are extremely susceptible to ergot, whereas the "Vulgare" form is practically immune and "Rivet" only moderately susceptible to it.

It is clear therefore that one of the most effective means of combating plant diseases is by breeding disease-resistant varieties. Prevention is better than cure, and one ideal in crop production is to possess varieties which are not liable to disease of a serious nature.

With plants, resistance to disease is not usually of a clear-cut nature, the degree of resistance being modifiable by environmental factors, although, with wart disease of potatoes, an immune variety appears to be immune under all conditions. It is important to realise the amount of modification in disease-resistance that can be induced by changes in environment whether of soil or weather. In fact the increase in susceptibility is sometimes so considerable that it almost seems as if the genetical factors for resistance had been obliterated. Thus last season the variety of apple, Bramley Seedling, which is usually free from attacks of the scab fungus, was badly infested by this disease. Although the genetical factors for resistance may thus be exceptionally masked they still exist, and over a period of years will confer a weighty advantage upon varieties possessing them over varieties in which these factors are absent.

A valuable line of enquiry is to try to determine in what way changes of environment modify susceptibility to disease. The physiology of the crop plant may be considerably influenced by alterations in environment and a change of this nature may assist the development of the parasite, but the precise alterations may be difficult to elucidate, especially where the influence of soil condition is paramount. The physiology of the parasite also may be influenced in the direction of increasing its ability to attack the host. A part of the increased susceptibility under changed conditions may be due merely to greater physical opportunity for the parasite to cause infection. For instance, with leaf parasites it is essential that comparatively moist conditions should prevail in order to allow of infection. Again, with many parasitic fungi certain conditions of temperature within the range tolerated by the host may either greatly facilitate infection or entirely prevent it.

It is admitted that the genetical constitution both of the host and of the parasite may alter with the lapse of time, but mutations probably occur only rarely. The plant breeder can undoubtedly work faster in producing new desirable types of crop plants than genetical changes in either host or parasite are likely to occur. One of the most striking features of the work of Stakman and his colleagues in America in elucidating the various biologic forms of *Puccinia graminis* is the remarkable constancy of these types in infective capacity, so that fears of sudden changes in violence and range of infection by such parasites are probably unfounded.

Crop plants resistant to disease are usually obtained either by the deliberate combinations of the plant breeder or by the selection of resistant individuals in a mixed population. Both methods have been adopted with success, especially in dealing with annual crops such as cereals and cotton. To be economically profitable these new types must have been subjected to rigorous tests as regards disease-resistance over a period of years, so that the average measure of resistance can be gauged. It is of the utmost importance in these tests that abundant opportunity for infection should be provided. For instance, in testing the resistance of barley to *Helminthosporium graminum* every opportunity should be afforded by copious provision of spores for all derivatives of the cross to be exposed to infection, otherwise plants which may remain free from attack may be looked upon mistakenly as immune or resistant when in reality they have only escaped infection through lack of infectious material.

It is generally recognised that resistance to one disease does not imply resistance to other diseases, although occasionally varieties of crop plants arise which remain free from attack by more than one serious disease. Again, owing to the existence of various biological strains of parasitic organisms and to profound environmental differences it is now realised that the problem of obtaining disease-resistant varieties is one which is specific almost to each country. It is a vain hope that in plant-breeding institutions in this country types of crop plants may be produced which will be available for commercial use in other parts of the Empire. For example, it is almost useless to attempt to obtain varieties of wheat in this country resistant to *Puccinia graminis*, which is the chief curse of wheat growing in Canada, Australia and East Africa, simply because the strain of the fungus which occurs on wheat is exceedingly rare in this country, so that an adequate test of resistance cannot be provided. It is important that a decision should be come to in each country as to what are the most important diseases of crop plants, and for the attention of the plant breeders to be concentrated upon these.

With perennial plants the problem of creating resistant types is more difficult, largely because of the longer time taken to achieve results. On the other hand the problem is of even greater importance than with annual crops because of the long period over which a perennial plant is productive. The devastation of the Weymouth pine by *Cronartium ribicola* and of the American chestnut by *Endothia parasitica* need only be recalled to realise the destruction that can be occasioned by fungi which attack trees. With the former disease there does not seem to be any hope of amelioration by the plant breeder, as not only is the Weymouth pine but other five-needled pines are attacked. With *Endothia parasitica* a closely related species of chestnut has been found to be markedly resistant to the disease. It is, however, in connection with fruit culture that results of great importance in the near future may be expected by the combination of the characters of disease-resistance and other desirable qualities. If for instance the resistance to silver-leaf disease of the Pershore plum could be combined with the quality of the Victoria or Czar, fruit growers would be relieved from a constant menace. With apples too there is an immense field for the plant breeder. Some of the best varieties as regards productivity and quality are seriously attacked by canker and scab, both of which diseases are difficult to control by the usual methods of plant pathology, and it is to be hoped that in the process of time these troubles will be practically eliminated by the creation of new varieties. In this connection Salmon's work on hops leading to the creation of commercial types resistant to mildew is of great value.

It is with cereals that the most important work has been done hitherto in the production of disease-resistant varieties, following upon the demonstration by Biffen that susceptibility and resistance to yellow rust (*Puccinia glumarum*) of wheat were inherited according to Mendel's law. By following up this discovery an increase in cropping capacity has been obtained by the creation of such varieties as Little Joss and Yeoman. The only other wheat rust widely prevalent in this country is *Puccinia triticina*, the brown rust, but as this usually only occurs here in quantity during the later part of the growth of the wheat plant it probably has no material influence upon the yield of grain. In the Argentine, however, brown rust of wheat does serious damage, and it is of interest to know that Backhouse has recently found that a Chinese variety of *Triticum vulgare* is extremely resistant to this rust, thus giving hope that new combinations can be effected which will greatly surpass the old forms in use in that country. In countries where *Puccinia graminis*, the black rust, is prevalent the position is much more serious. As is well known, this fungus in attacking chiefly the stem has a much more serious influence upon the crop than the brown and yellow rusts, and it is by no means unknown for more than 50 % of the crop in North America and other wheat-producing countries to be lost by its ravages. In North America the production of resistant types is complicated, as Stakman has shown, by the presence of many different biologic forms of *Puccinia graminis*. Some of these biologic forms occur in different areas, so to this extent the problem of breeding resistant types will be specific to those areas, unless varieties can be gradually built up which are resistant to most of the forms of black rust. It has been shown that the factor for resistance in some varieties is the same for several different forms of the rust. For instance, the variety Kanred is practically immune to at least eleven different forms of *Puccinia graminis*. Little is yet known about the resistance of varieties of cereals to smut fungi, although the losses occasioned by them in some countries are not less severe than those caused by rusts. Bunt is a very serious disease of wheat in many parts of the world, especially in winter-hardy varieties, and it is greatly to be desired that commercial varieties should be introduced which are more resistant than the present types. Steeping or sprinkling the grain with a fungicide is an efficient protection against this disease in most countries, but in some parts, *e.g.* the Pacific States of North America, infection is liable to occur through the soil. Gaines[1] has shown that resistance to bunt is a segregating character so that there is hope of achievement along genetical lines as regards this disease. Perhaps the most serious disease of barley is *Helminthosporium graminum*, the cause of heavy mortality of young plants and blindness of the ears, but little is yet known as to the comparative susceptibility of varieties. Work is in progress at Cambridge to try to determine this and to ascertain whether resistant types can be built up along Mendelian lines.

With other herbaceous crop plants work is proceeding in many parts of the world to discover or to create types which are resistant to some of the most troublesome diseases. In connection with this work important results are being obtained as to the nature of disease-resistance and as to the manner in which the factors for resistance are coupled with other characters. For instance, Fromme and Wingard[2] have shown that with *Phaseolus vulgaris* all varieties with red or mottled seed are resistant to attacks by the rust fungus *Uromyces appendiculatus*, while white varieties are specially susceptible. Another biologic strain of this fungus attacks the cow pea (*Vigna lutea*) both in America and in Egypt, and in the latter country[3] a type of this plant has recently been isolated by selection and propagated on a large scale, which seems to be almost completely immune. Another interesting correlation between the presence of pigment and resistance to disease has recently been demonstrated by Walker[4] in the case of onions in America. He finds that with the disease of onion called "smudge," caused by *Colletotrichum circinans*, coloured varieties are highly resistant and white varieties are susceptible. The pigment in the coloured varieties is chiefly present in the outer

[1] Gaines, E. F. "Genetics of Bunt Resistance in Wheat," *Journ. Agric. Res.* p. 445, 1923.

[2] Fromme, F. D. and Wingard, S. A. "Varietal Susceptibility of Beans to Rust," *Journ. Agric. Res.* p. 385, 1921.

[3] Hefnawi, M. T. "On the selection of a rust-immune variety of *Vigna sinensis*," *Mem. of Agric.*, Hortic. Sect., Egypt, 1919.

[4] Walker, J. C. "Disease resistance to onion smudge," *Journ. Agric. Res.* p. 1019, 1923.

scales, and when these are removed the inner fleshy scales of these varieties are found to be susceptible. Furthermore, Walker has shown that a substance associated or identical with the pigment has a very detrimental influence on the growth of the fungus and is unquestionably bound up with the resistance of these varieties.

Perhaps the most difficult class of diseases to be dealt with by the ordinary methods of plant pathology are those which enter through the root systems and which frequently cause wilting. Soil sterilisation is employed on a considerable scale in greenhouse cultivation to deal with diseases of this class, but it is obvious that this is out of the question on a field scale, and it is equally clear that the production of resistant varieties would be a far better and cheaper means of control. Much work has been done in America in introducing varieties of herbaceous crop plants which are resistant to wilt diseases. In California for instance some varieties of tomatoes remarkably resistant to the wilt caused by *Fusarium lycopersici* are stated by Shapovalov and Lesley[1] to have been produced by breeding. At present in this country considerable damage is being done by a wilt disease of carnations where these are grown on a large scale under glass. This disease, caused by *Fusarium dianthi*, is particularly troublesome to deal with, but as with most cultivated plants, there are great varietal differences in susceptibility. These differences are of such a character as to afford a sufficient basis on which to build up by breeding a wide range of resistant types.

It may be argued, from the considerations put forward above, that in the process of time the measures devised by plant pathologists for the control of disease will give place entirely to the ameliorative measures of geneticists. To abolish disease by the elimination of susceptible varieties is, however, a counsel of perfection, and is never likely to be achieved in practice. As has already been pointed out, changes in environmental conditions sometimes exercise a profound influence upon the incidence of disease in some classes of plants ordinarily resistant, and again the pathogenicity of micro-organisms is probably liable to modification over long periods of time. The best outlook for the future is for the plant pathologist and geneticist to work in co-operation with each other in the production of disease-resistant types, and for the plant pathologist to perfect his methods for dealing with disease in susceptible varieties by means of plant sanitation, soil sterilisation, the application of fungicides, and, above all, by a careful study of the environmental conditions

---

[1] Shapovalov and Lesley, J. W. "Behaviour of certain varieties of tomatoes towards *Fusarium*-wilt infection in California," *Phytopathology*, p. 188, 1924.

under which crop plants are least liable to disease. Frontal attacks upon fungus diseases of crop plants are usually futile, and in general these troublesome organisms can usually be defeated only by the application of subtler methods.

Dr R. N. Salaman

(Abstract)

Congenital infection by Virus Disease of the Potato has been thought to be very rare. In a previous communication the writer estimated that it might occur about once in a thousand seedlings.

There is some reason to believe that the observations in this respect are misleading. The seedlings of a selfed Curly Dwarfed plant raised in this year, when 1–2 inches high, showed unmistakable evidence of Mosaic infection in 75 % of their number. In three weeks' time scarcely a trace of this was to be observed.

The warm moist atmosphere of the hot-house may cause the symptoms of the seedling to disappear.

Seedlings apparently in perfect health, planted in 1923 in a garden which had been used continuously for potatoes for 20 years, from which all tuber raised plants were removed, developed abundant evidence of Mosaic and Leaf Roll although they had been sprayed twice a week since the sowing of the seed.

Similar seedlings planted out on virgin soil and at distances of 300 yards from the nearest potato showed no signs of Mosaic in their first year (1923).

A further batch of the same seedlings planted in the midst of Mosaic-infected tuber raised plants were uniformly infected, the symptoms varying from extreme Curly Dwarf to light Mosaic mottling.

Infection by insects may be an incomplete explanation for the spread of Mosaic Disease and Roll.

Whether Mosaic and Leaf Roll are invariably conveyed by insects or not, it is certain that seedling plants can acquire the most advanced form of the disease in their first year of growth.

Experiments designed to show whether *Solanum nigrum* is a carrier of Mosaic to the potato in the field proved negative, in the year 1923.

Evidence has been found of susceptibility to Mosaic Disease being an inherited character; the same is true for the inheritance of Leaf Roll.

Susceptibility is not necessarily linked with a high mortality, but lack of vigour allows of an intensification of the effects of the lesion after infection.

There would appear to be a relation between the earliness of maturity and a high mortality amongst seedlings.

Dr William B. Brierley

In the present paper I do not so much wish to bring forward matters of new fact, as to emphasize points of view. To carry out experiments in one's garden or laboratory, quickly fit the new results into the old framework and pass on to new experiments, is so entertaining that one rarely questions the capacity of the framework to hold the facts. In mycological research, for example, I cannot help thinking that we are fitting our facts into a creaking framework of theory borrowed largely from bacteriology and in many respects incompatible with our mycological needs. Again, where our studies have a genetic bearing we are thrusting our results somewhat blindly into the framework of genetics, perhaps a little unappreciative of the fact that owing to incommensurable organisms and different criteria, the results of our work with microorganisms cannot be equated with the results of apparently parallel work with Primulas, Fruit-flies and so forth. It is obviously impossible here to treat of these questions in detail, but certain aspects relevant to our discussion may be considered briefly.

The relation of plant pathology to genetics is an extremely complex one and care must be taken not to falsify the perspective by over-simplification. The situation is of course comparatively straightforward where a state of disease is due to the development of a host under unfavourable conditions in its physical environment. Here the problem largely resolves itself into a study of the values of climate, soil and other factors in relation to the genetic qualities of a given variety of plant. As the limiting factors are physical, they can frequently be isolated and investigated quantitatively by standardised methods. Such problems come, in fact, almost more within the range of applied physics and chemistry in their relationships with genetics, and it is only by intimate co-operation with physicists and chemists that plant pathologists will throw any fundamental light on these problems.

In a disease relation where the state of ill-health is due to invasion by parasitic organisms, the situation is greatly more complex, and it is not easy to overstate the difficulties of its analysis. The principal factors involved in such a condition may be represented schematically in the following diagram:

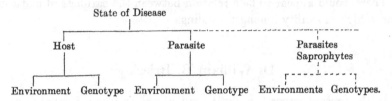

At its simplest, such a state of disease is the equilibration of the vital activities of a specific invasive organism and a host-plant. The host and the parasite, severally, are the resultants of congeries of specific germinal potentialities expressing themselves under the moulding influence of particular environmental conditions. Assuming a constant environment, it will be evident that any change in the genetic qualities of host or parasite will alter the balance of disease. On the other hand, assuming constancy of genotype, any alteration in the environmental conditions may modify the equilibrium. It will thus be evident that the complete understanding of a case of ill-health involves the genetic and physiological analysis of both host and parasite, and the physical and chemical analysis of the conditions under which both organisms have developed and at present exist. In actual practice one can only adopt a Baconian technique and rejoice if any single factor can be isolated and examined. Such investigations necessarily can only give very partial views, and a real comprehension of any particular case, considering the disease as an individual entity, requires that such partial views be synthesised as separate photographs are merged in a cinematograph film. An investigation of a disease, is, as it were, a cross-section of a process that is essentially a continuum; and a serious danger in plant pathology to-day is the tendency to accept isolated studies of a disease at any one moment as a true picture of the whole.

What has been said holds for even the simplest disease relationship; but the picture in nature is often almost infinitely complicated by the additional presence of one or another unfavourable condition in the physical environment, or by the intervention of parasites and saprophytes having relationships subsidiary to the primary disease initiating organism. Such, for example, would be the case of a potato plant growing in the present (1924) wet season, infected with one or more kinds of virus, with Phytophthora blight on its foliage, and perhaps Oospora scab and Wart Disease on its tubers, together with the multitude of saprophytic organisms entering the lesions. In the investigation of a disease relationship, however, one must as far as possible return to experimental conditions of minimal complexity; and it will be obvious that the relation of any particular host plant growing under such controlled conditions to an invading organism of pedigreed ancestry is a subject for genetical analysis of the most critical nature.

Considerable attention has been given to this problem by geneticists and plant breeders; but the emphasis has been laid almost entirely upon the host, and the fungus complement has been sadly neglected. The earlier studies appeared to show that resistance or susceptibility to any particular disease was due to single factors behaving as simple Dominants or Recessives. In a very few more recent cases this has been confirmed; but the general weight of evidence makes it clear, I think, that in the majority of disease relationships, even when considered from the host standpoint only, the genetical situation is not a simple allelomorphic one but extremely complex. In his practical work the plant breeder must assume a very high degree of stability if not entire constancy of genotype in the pathogen just as he assumes constancy of genotype in his homozygous F_2 and later segregates of the crop plant. Were this not the case, breeding for disease resistance would not be the feasible undertaking that results have shown it to be. The assumption of germinal stability is fundamental to the theory and practice of breeding for disease resistance; but germinal stability is by no means generally accepted by mycologists.

Leaving for the moment the point of view of the plant breeder let us turn to that of the mycologist or plant pathologist, *i.e.* the view-point of, as it were, the pathogen *quâ* pathogen. The first group that the investigator works with is the "species" of the systematist, *i.e.* the several individuals having in common certain specified morphological features. The moment however that critical experiments are made, the student finds, what is now so well known, that almost any common species of *Botrytis, Septoria, Penicillium, Fusarium, Helminthosporium, Peronospora, Puccinia* and so forth almost endlessly, consists of a greater or smaller number of discrete physiological strains or races. In perhaps a majority of cases these strains cannot be classified by morphological criteria, and so, beginning with the systematists' identification of individuals which look alike, the pathologist further divides the species into races or strains containing individuals which behave alike. Each strain is characterised by certain distinctive qualities and, speaking generally for the moment, these genetic qualities are constant. This stability of genotype is well illustrated by certain of the better known rust fungi where the exact relationships between particular strains of *Puccinia* etc., and particular pure lines of Wheat, Barley and so forth form a classical case. In essence, the exact disease relationship means, that under relatively standardised conditions, the degree of susceptibility or immunity of, for example, any pure line of Wheat is specifically directed only to a particular strain of *Puccinia graminis*. This implies, what Stakman and his colleagues have shown to be a fact, that a pure line of Wheat can be used to differentiate rust-fungus strains; or a "pure-line strain" of rust fungus may be used to differentiate wheat strains. This "host strain—fungus strain" relationship, so well known in rust diseases of cereals, is now being found true of many other disease relationships, the systematists' species of both host and parasite being shown to consist of discrete genetic strains which are complementary in finite and distinctive physiological relationships which we term diseases. What I wish to emphasise is that every disease is a two-faced relationship, one face turned towards the host and the other to the parasite; that the germinal qualities of the host and the germinal qualities of the parasite are complementary in the physiological summation which is the particular disease, and that the latter can only be truly understood in the light of a genetical analysis of both complements.

A finite relationship of "pure-line host" with "pure-line parasite" does not, however, exist aloof from external conditions, and, as already noted, a change in these external conditions may sway or entirely alter the physiological equilibration of "host-parasite." In certain of these cases the physiological relationship appears to be unaltered fundamentally, but owing, as it were, to morphological or anatomical interference, the disease summation is modified. The work of L. R. Jones and his colleagues at Wisconsin has demonstrated how critical a limiting factor soil temperature may be, whilst soil aeration, soil and air moisture, soil acidity, lack or excess of food-stuffs, have all been found to have very definite and in many cases accurately measurable effects on specific disease relationships. Fundamentally, the genetic relationship of "host strain—parasite strain" appears to remain unaltered, but the phenotypic expression of this relationship is in many cases susceptible of extreme modification.

The experimental control of biological and physical conditions, and the isolation of particular factors which are vital in genetical and pathological research, have no real existence as individual entities in nature. The conditions in nature are infinitely complex, and the application of laboratory and plot results and, still more, of interpretations derived therefrom to a field or commercial scale is one of the most difficult and dangerous steps possible. Certain of the difficulties are immediately relevant to our consideration, but time will not permit of more than the mention of one illustration.

Plant cultivation in its widest sense is very rapidly becoming the commercial selection of, relatively, a very few pure lines of economic plants. The greater the genetic purity of commercial seed the more valuable it is, and vast world areas are increasingly being planted to "pure-line" crops of Maize, Wheat, Potato, Cotton, Sugar-cane and so forth. Of primary importance in the estimation of the value of seed are its disease-resisting qualities. Now it has been pointed out that the disease relationship is very sharply defined and that a particular variety of crop plant resistant to one strain of a pathogenic organism may be susceptible to another strain. In nature or under commercial conditions, and to a certain extent in addition to and independent of climatic and other physical determinants, the limiting factor in the relation of plant pathology to genetics is the geographical distribution of particular strains of fungi, a subject of which our ignorance is complete.

Even from the foregoing discussion the complexity of the relation of plant pathology to genetics will be evident. There are at least three primary groups of factors which determine the appearance of immunity or relative susceptibility to disease in nature: (*a*) a fundamental genetic host-quality which, as in the varieties of potatoes immune to Wart Disease, supersedes all environmental conditions; (*b*) environmental conditions which, as in most cases, determine the phenotypic expression of the fundamental genetic relationship; (*c*) the comparative geographic distribution of host and parasite.

Incidentally a fact may be noted here which is very insufficiently recognised in disease research, and which has considerable bearing on our discussion. It is almost axiomatic in veterinary and human pathology that a strong and vigorous subject is less liable to infectious disease than a weakling. This anthropomorphic conception of a see-saw balance between an attacking parasite and a defending host has permeated plant pathology, and it is a commonplace that if the welfare and hygiene of plants be attended to disease will be absent. Now in certain cases this is true but in other cases it is certainly not true. A striking instance is seen in many rust diseases of cereals where the conditions favouring the development of the host are those which favour the development of the parasite and the literature of plant pathology abounds in other examples.

Perhaps I may mention three widely different cases from my own experience. The tubers and shoots of weakling potato plants are quite commonly clean whilst the vigorous plants adjoining are heavy with warts caused by *Synchytrium endobioticum*. I have little doubt that in the case of Wart Disease the incidence of disease is in almost direct ratio to the "health" and vigour of the host. Under experimental conditions certain strains of *Botrytis cinerea* only infect sickly or injured plants; whilst other strains regularly pick out the strongest and most vigorously growing hosts. Occasionally the mangolds on certain plots at Rothamsted are black with a bacterial disease, whilst certain immediately adjoining plots are perfectly clean. The plot distribution of the disease entirely cuts across the hygienic relations or vigour and strength of the mangolds, which are determined by the particular manurial compositions, and is, apparently, directly related to the presence or absence of certain chemical elements in the mixtures used.

I think there can be no doubt that the anthropomorphic conception of disease in plants as a simple conflict between invader and defender needs fundamental revision. The conception of disease presenting itself to me, is one of a summation of the physiological activities of two (or more) organisms—not the resultant of one organism *versus* another, but of one organism *plus* another; the factors involved being the germinal qualities of both host and parasite and the organic and inorganic determinants of the environment.

Let me now turn to a closely related but slightly different aspect of our problem. I have already emphasised the well-known fact that the morphological species of the systematist often consists of a greater or smaller number of stable physiological strains. The larger morphological species as in the genera *Aspergillus, Penicillium, Fusarium* etc. may often be divided into smaller groups or sub-species of morphological value, and these into strains of physiological value; and I would again suggest that the terminology put forward by Lotsy—"Linneon" "Jordanon" "Species"—could very usefully be adopted in more critically defining these grades of specific values in mycological literature.

These morphological and physiological groups are well known and accepted; but it has been very largely overlooked, that two or more discrete physiological strains may co-exist either in culture or more commonly in single lesions on diseased plants. Very often indeed in my experience what is apparently a pure growth of a parasitic fungus on a host contains two or more physiological strains which can be differentiated in culture and which thereafter remain distinct. Thus I have found, over and over again, that what is morphologically a pure growth of *Botrytis cinerea* on a plant, may contain two or more quite distinct strains of that species (or "Linneon") and the diseased condition, local or general, may thus be not a simple physiological state but a very complex resultant. Now a stable difference in physiological properties implies a difference in genetic qualities, and it therefore follows that what is morphologically a unit, may be genetically a population. There is further implied the possibility, that certain phenomena which apparently represent true genetic change, may more correctly be ascribed to the selection of strains within a genetic population. It was, for example, the non-realisation of the possible genetic complexity of a morphological unit with the implications I have noted, that underlay the earlier apparently successful work on the "bridging species problem" in rust diseases of cereals. The more recent critical work of Stakman and his colleagues has shown the truer explanation to be one of strain selection, and not one of genetic change. I think that no one who has followed the Minnesota researches carefully, and still more who has had the privilege of seeing them in progress, can accept the hypothesis of genetic instability in the rust fungi which is implicit in the bridging species interpretation. The co-existence of physiological strains in morphological units throws light on much previous work which seemed to demonstrate almost conclusively the occurrence of extreme variability and genetic instability in micro-organisms and it cannot be too clearly recognised in plant pathological research that a morphological entity may be a genetic population.

One may, however, carry these considerations much further. It has become almost a fashion in mycological laboratories to make hyphal-tip or single-spore cultures of any particular fungus, sub-culture extensively on various plated media and study the "sector" phenomenon. The preliminary technique is to ensure that only one fungus strain is present in the initial culture. Subsequent operations are based on the assumption that the initial culture and its sub-cultures are genetically pure; *i.e.* the genetic purity of a single-spore culture is equated with that of a homozygous F_2 segregate in, for example, *Drosophila*. If any changes occur in the sub-cultures, and if, when isolated, these sector variants are found to be more or less stable they are then described as mutations and equated genetically with *Drosophila* mutations. Now granted the premisses and the logical nature of the subsequent steps the interpretation of the phenomena as one of fundamental genetic change in a pure line, or "mutation," is the obvious and simple one; and it seems to many students wilful blindness and sheer love of casuistry that compel one to remain unconvinced by their interpretations and to seek some apparently more recondite hypothesis.

If however one examines critically the astonishing and often mutually contradictory phenomena recorded in experimental mycology which, on the above premisses, must all be interpreted as "mutations," and further if one follows to their conclusions the implications which logically are involved, I think that even the most empirical of mycologists might feel a little alarmed.

I will say at once that I believe the initial premiss to be false and that I see not the slightest reason for the acceptance of a single-spore or hyphal-tip culture as genetically pure or in any way comparable with a pedigreed homozygous organism. All the available evidence seems to me to negative this assumption and to support the hypothesis of the occurrence of very considerable genetic complexity and consequent genetic segregation in fungi.

Both plant pathology and genetics are aspects of biology which have developed so enormously in the last twenty years, and which during this period have lived so remotely from each other, that it is the rarest thing for any mycologist to have more than the slightest acquaintance with genetical phenomena, theory and criteria, or to possess any trace of a "genetical perspective." Yet, in the last few years, mycologists have blundered like little children into the genetical aspects of their problems, with childlike assumptions and interpretations that are, in many cases, almost ludicrous. They have crammed their results into the framework of genetical theory regardless and mostly very ignorant of the fact, that in the great majority of cases these results are quite incommensurable with the facts out of which the framework has been constructed.

No consistent attempt has yet been made to co-ordinate the results of experimental work on the genetics of fungi as a contribution to the study of evolution. The facts and interpretations of such mycological research are, however, almost an exact replica of those of bacteriology; and the methods and criteria, points of view, premisses, and logical implications are held in common. Now Adami has synthesised a good deal of the bacteriological work in his *Medical Contributions to the Study of Evolution* and this, I think, very fairly expresses the viewpoint of mycologists toward the genetic aspects of their problems. If there is any difference, it is that Adami adopts perhaps the more cautious and conservative attitude. Now if one critically analyses Adami's use of genetic terms, his conception of the meaning of genetic purity, his interpretations of facts, and the assumptions underlying his interpretations together with the logical implications of these assumptions; and then concept for concept, criterion for criterion, assumption for assumption, and implication for implication compares Adami's volume with Morgan's *Physical Basis of Heredity* or *The Mechanism of Mendelian Heredity* one will, I think, gain some idea of the utter incommensurability of genetical research on bacteria and fungi with the genetical research which serves as the basis for genetical theory. Most of the concepts and terms of genetics simply cannot be applied in mycological research, and such a term, for example, as "mutation" when used of fungi, is sheer obscurantism having about as much value philosophically as the term "magic" in former days.

I think it is evident that during the next few years the attention of mycologists and plant pathologists will increasingly be turned towards the genetical aspects of their problems. Already a considerable number of phenomena have been recorded and the volume of literature is steadily increasing. We are, as it were, laying the foundations of a new aspect of our subject, an aspect that I, personally, feel may be of fundamental importance in genetical philosophy, and, because of its relation to disease in plants, and the theory and practice of plant breeding for disease-resistance, may be of tremendous import to the world's agriculture and so to the welfare of human kind. In the laying of these foundations we can be content with our present uncritical premisses, our empirical methods, gratuitous assumptions, analogic processes of thought and our childlike interpretations, but if we are, and we continue as at present, I very much doubt whether our work will have the slightest interest for geneticists or anyone else. If however we can raise our concepts and criteria to the plane even of present day genetical theory and practice, we may reap to the full all the possibilities that our subject holds.

I expect that I shall be told that I preach a counsel of perfection; but I am very sure that it depends upon our attitude now, whether the mycological researches of the next decade build a nebulous and inchoate *Oenothera*-like edifice or an exact and quantitative *Drosophila*-like structure.

Dr W. F. Bewley

(*Abstract*)

In spite of improved methods for disease control under glasshouse conditions two diseases stand out in which the breeding of resistant varieties is at once the simplest, cheapest and most efficient means of control. These diseases are mildew of the tomato (*Cladosporium fulvum*) and Mosaic disease of the tomato and cucumber.

Cladosporium fulvum can be partially controlled by a combination of spraying and thorough ventilation of the glasshouses. It is a comparatively simple matter to control this disease under laboratory and experimental house conditions, but those who are familiar with the glasshouse industry must at once realise the extreme difficulty of spraying under commercial conditions.

Our own experience has provided the following figures:

Diseased crop sprayed, 38 tons per acre.

Diseased crop unsprayed, 31 tons per acre.

Healthy crop, 43 tons per acre.

These figures confirm the general experience of benefits derived from spraying and ventilation, but show that the present methods cannot produce a yield equal to that from a healthy crop.

It has been shown that an important factor in the setting of tomato fruits is related to atmospheric humidity, and daily overhead damping invariably results in an excellent setting of the four lowest trusses. Towards the end of May, damping is discontinued because of the dread of mildew, and the fifth and sixth trusses frequently "set" badly. It would appear from our experiments that if damping could be continued the "setting" of the top trusses should be almost as good as that of the lower ones.

The fear of mildew is the main argument against overhead damping during June, July and August, and when a mildew resistant variety is obtainable it will be possible to continue this process with excellent results throughout the season. Selection methods are giving promising results at the Cheshunt Station.

Perhaps the most striking example of the need for resistant varieties is that of mosaic disease of the tomato and cucumber.

Five main lines of control suggest themselves:

(1) The determination and elimination of infection centres.

(2) The determination and elimination of agents by which the disease is spread.

(3) The determination of environmental conditions which will induce a high resistance in the host plant.

(4) Spraying, dusting and soil sterilisation.

(5) The breeding of immune varieties.

(1) Carrier plants show no outward symptoms and under practical conditions it is impossible to eliminate them.

(2) It is impossible to eliminate all transmitting insects.

(3) In spite of considerable investigation spread over a number of

years, it is not possible to increase resistance in the host by manurial treatment or any adjustment of the environmental conditions.

(4) No method of spraying, dusting or soil sterilisation is effective.

(5) The breeding of resistant or immune varieties seems to be the sole instrument whereby control can be effected.

An organised search for individuals showing resistant characters is being made in the tomato and cucumber nurseries in the Lea Valley. Some 24 promising cases have been tested but all proved susceptible on copious inoculation.

Verticillium Wilt of the Tomato. One variety—Manx Marvel—has shown high resistance to this disease. It is not satisfactory commercially, for while setting well, and producing an abundance of fruit, the fruits are small, and ripen some 14 days later than other varieties like Kondine Red and Ailsa Craig. Manx Marvel has been crossed with Ailsa Craig (a good commercial variety) to combine the resistance of the former with the good fruiting qualities of the latter.

The first generation resembled Manx Marvel in external characters but were less resistant to the disease than this variety. The F_2 generation yielded one group resembling Ailsa Craig in external character, but more resistant to *Verticillium* wilt than this variety although less resistant than Manx Marvel. Further work is in progress.

Mr T. S. Venkatraman

The sugar-cane offers perhaps one of the best instances of the value of breeding in fighting diseases in field crops. It is becoming increasingly realised that vigour and health of plants are assets of considerable importance in fighting diseases.

In the case of the cane, Java was perhaps the first country to take advantage of breeding in this manner. Crossed seedling canes, raised between the Javanese canes and the Indian kinds, proved resistant to the then little known disease "Sereh" and helped that country to tide over a critical situation in the eighties of the last century.

The Indian canes are generally very hardy and comparatively more resistant to diseases. It is not, therefore, impossible that some of them might prove useful in fighting new and little understood diseases like "Mosaic." Already at least one Indian cane, the *Uba*, has been found to be resistant to "Mosaic" and the station at Cuba is already trying to produce mosaic-resistant canes by crossing with Uba. In India there are many indigenous canes as hardy or hardier than Uba. Three Indian seedling canes sent to Antigua last year have displayed a remarkable vigour of growth in that country.

PLANT PATHOLOGY AND MYCOLOGY

OBSCURE PLANT DISEASES OF WIDESPREAD OCCURRENCE

(CHAIRMAN: DR E. J. BUTLER, C.I.E.)

Prof. S. F. Ashby. SUGAR-CANE MOSAIC

It may be said at the outset that sugar-cane is propagated by means of pieces of the stalk each carrying one or more buds or eyes. These cuttings or setts are planted and the buds grow up into primary shoots which in turn give rise to secondary shoots and these to others of a higher order forming a stool or clump. This first crop from the sett is the plant cane. When mature it is cut down to the ground, and the dormant underground buds spring and give rise to a first ratoon crop which in a similar way may be followed by a succession of ratoon crops. In some countries, however, only plant cane is grown or, at most, one or two ratoon crops are taken before replanting. Sugar cane is also raised from true seed in the nursery but this is undertaken only when new varieties are desired; canes of such known origin termed "seedlings" are propagated by cuttings in the usual way.

Distribution. It may be of interest if an account of Mosaic disease is preceded by a brief history of its introduction into most cane-growing territories during the last thirty years. This will be understood better if I say now that the disease is regularly transmitted in the cutting and is infectious. Maize, sorghums and a number of annual weed-grasses become naturally infected in or near diseased sugar-cane cultivations, but in no case, hitherto, has mosaic been seen to be transmitted by the seed. The evidence points to sugar-cane as the prime source of infection.

By the end of the third quarter of the last century the modern industry had come to be based mainly on the cultivation of the colour-variants of four kinds of thick cane—Otaheite from the Society Islands, the Batavian from Java and Tanna and Cavengerie from New Caledonia(1). The distribution had been mostly from collections in Mauritius. There is no evidence that any of these kinds or any others in cultivation up to that time were infected by mosaic and Mauritius[1] is one of the few territories still free from it. In 1882 a new disease "sereh" was detected in Java which spread steadily during the eighties through the main cane area(2). The standard Black Cheribon proved to be susceptible

---

[1] Just found, June 1924.

and sensitive and the stability of the industry seemed to be threatened. In the hope of finding varieties resistant to "sereh" many native kinds were introduced during the eighties from the neighbouring islands and some also from New Guinea, and these were distributed freely among the plantations (3). The conditions were, therefore, favourable for the introduction and extension of another pseudo-hereditary and infectious disease. Van Muschenbroek was the first to notice mosaic in 1890 and he published an account of it in 1892 (4) under the name Gelestrepenziekte (yellow-stripe disease) by which it is still known in Java. He observed it to be regularly transmitted by cuttings. In 1893 it was found to be already rather widely distributed. The disease was present in Queensland at the beginning of the present century, but could not have come from Java as there is no record of cane introductions from that island between the seventies and 1912. The Department of Agriculture, however, had introduced some 70 varieties of native canes from New Guinea in 1896 (5). Some years later mosaic was seen on native canes in the interior of New Guinea (6). It is probable, therefore, that the disease entered Java and Queensland independently but from the same source.

From these two infected territories mosaic was introduced either directly or indirectly into many other cane-growing countries during the present century. The Planters' Association in the Hawaiian Islands introduced seedlings from Queensland in 1900 and many more in 1902 (7). Mosaic was noticed first in 1908 and a few years later good evidence of its infectious character was obtained (6). From Hawaii it reached the Philippines by 1910 (8), while in 1909 it was detected in Egypt on Java seedlings introduced some years earlier (9) and was taken to the Argentine in Java seedlings in 1911 (10). With the possible exception of a private station in Cuba mosaic was evidently not present in the West Indian Region prior to 1910. An experiment station in Porto Rico introduced Java seedlings direct from Java and also by way of Egypt, about 1912 (11) (the precise dates have not been published) and infection was found to be widely distributed over a restricted area adjacent to this station in 1916 (12), being regarded at first as a new disease. Most of the island was infected three years later—in parts very heavily (12). It was brought into the Experiment Station at New Orleans in or shortly before 1912, and from 1914 (13) until it was definitely recognised five years later this Station was distributing it in cuttings to the plantations in Louisiana, to a number of other Gulf States and to some of the West Indian islands. The experiment station in Jamaica introduced two Java seedlings in 1912 (14) but mosaic was not definitely recognised until 1920 (15) when, as might be expected, it was already rather widely distributed. It was recognised in the same year in Trinidad(16) and Barbados but some of the other British islands are still free from it. Mosaic has not been recorded from India with the exception of the experimental farm at Pusa(17) where it was recognised in 1921 on a seedling introduced from the United States and on a native variety to which it had spread. It was recognised last year in Natal on Java seedlings introduced some years earlier, apparently by way of Egypt and the Argentine. The disease is also undoubtedly present in parts of Brazil and in Formosa and has been extending steadily in Cuba within the last few years.

Symptoms. In regard to symptoms it is necessary to distinguish between primary and secondary infection. If top-cuttings from diseased stalks are planted the buds are already infected and the shoots therefore are diseased from the start; the same condition applies to ratoon shoots arising from fully infected stools; these are cases of primary infection. In secondary cases the shoots become infected during growth. The progressive development of the leaf symptoms can be seen in these secondarily infected shoots. The first visibly infected leaf-blade shows, as it unfolds from the bud or spindle, a few isolated light green spots usually near its base. The spots are narrow lenticular in form with the longer axis parallel to the veins and from a fraction of an inch to an inch or more in length. Each succeeding blade as it unfolds shows an increase in the number of pallid green spots which may unite into irregular longitudinal bands. In the fully developed pattern the general effect is that of a striping, streaking or mottling, varying somewhat in the same variety of cane but to a marked degree in different varieties. Every gradation can be found between a mosaic of well defined and isolated light green stripes on a predominantly dark green ground and a predominantly light green or yellowish ground, with the normal dark green hue in more or less isolated stripes or irregular blotches. The effect is visible on both sides of the blade and the under side of the midrib and also, but less distinctly, on the sheath. It appears fully developed on the leaf blades within the spindle as soon as they begin to show colour, the spots undergoing no change of shape nor increase in number subsequently. The contrast is always greater on the unfolding or newly unfolded leaves, as the tendency in the older leaves is towards a regeneration of the normal hue in the pale areas. In a number of varieties, a striping is present also on the internodes or joints; on green or faintly coloured stalks narrow pointed longitudinal red stripes may be developed and on red or purple canes light green or light yellow stripes. If the infection is primary the leaves of many thick cane varieties in the West Indian region show opaque white streaks in addition to the usual mottling, and this secondary symptom is frequently accompanied by a white striping of the internodes in addition to the colour striping. The longitudinal white streaks on the leaves, which may be several inches in length, are confined to the light green parts and may occupy as much as a quarter of the total leaf area. Although dried out and functionless they usually undergo no change of tint in the older leaves. They can be detected on the rolled leaves in the spindle as soon as these begin to colour. The white stripes on the stalks, which result from the drying up of the outer cortex, are already present as somewhat depressed areas on the young internodes. They vary in length without reaching the nodes and may occupy much of the joint surface by lateral fusion. Distinct from this early necrosis is the development of red or rusty spots on the fully expanded leaves of a few seedlings in Trinidad[6] and Hawaii[16]. It is of interest to note that the white striping of the leaves and joints so frequent in cases of primary infection in the West Indies has not been observed in Java and Hawaii, even on the same canes or their colour-variants.

A condition in which the white striping of the joints is associated with longitudinal shallow fissures has been referred to as "canker" in Porto Rico[12]. Mosaic causes a stunting of growth in thick cane varieties, the stalks being shorter and narrower than those of healthy canes, and this symptom is always more marked and often substantially so when infection is primary, so that the disease is, in some measure, cumulative. The economic loss[18] is expressed mainly in a reduction of tonnage owing to the stunting of growth and the lesser number of stalks which grow up to maturity. Where infection is primary this loss may range from 10 to 60 % in fully diseased fields depending on the variety cultivated. The amount and composition of the juice is little if at all affected except when the stalks are cankered severely[12, 19].

Varietal resistance. In this connection it is convenient to distinguish between susceptibility or resistance to infection and sensitiveness or tolerance when infected.

The thick cane seedlings 247 B and 100 P. O. J.[18] formerly for many years the standard canes in Java show a marked difference. 247 B is quite susceptible to infection and also very sensitive, being severely stunted by mosaic; the other cane is very resistant to infection but quite sensitive. Otaheite and Cavengerie are both susceptible to infection and sensitive, while the best known Tanna variant—yellow

Caledonia—is resistant to infection but very sensitive, being killed by mosaic in Porto Rico [19]. The colour variants of the Batavian cane formerly so widely cultivated (Crystalina or White Transparent is still the standard cane of Cuba) are the most tolerant of the old kinds. It is safe to say, however, that no thick cane is sufficiently tolerant to primary infection to be cultivated with profit under conditions of heavy infection. The hybrid seedlings raised by Kobus at the East-Java Station between 1897 and 1910, by crossing the thin Indian Chunnee variety as male parent with the thick Batavian cane in the hope of securing seedlings highly resistant to sereh disease, are very susceptible to infection by mosaic but so tolerant that they can be cultivated when every stool is diseased [3]. Two of them, although fully infected, are now the standard canes in the Argentine [20] and one of them in Formosa [21]. These hybrids frequently show the leaf symptoms so obscurely or doubtfully that it is difficult to free them from mosaic by selection, and the disease has been introduced or reintroduced into a number of territories on them.

Some semi-thin Indian varieties, among which Uba, the standard cane of Natal, is now widely known, show no symptoms of mosaic even when exposed to very active secondary infection: they have been regarded, therefore, as immune [11, 19]. It has not been experimentally determined if this immunity is complete or whether these varieties may, for a time at least, following exposure to infection, behave as symptomless carriers. In a recent trial by Brandes [22] with Uba and similar canes from India, China and Japan including a wild cane (*Saccharum narenga*) exposed to infection in a greenhouse, several kinds showed mottling in three weeks including the wild species, but Uba showed no symptoms; in those varieties which became infected the symptoms were slight and growth was not perceptibly influenced. Cases in which the unfolding young leaves ceased to show mosaic have been observed in maize and weed-grasses by Brandes [13] and in maize by Williams [16] in Trinidad, while Lyon [6] in Hawaii has recorded an interesting case of such apparent recovery from secondary infection in sugar-cane. He planted a row of healthy cuttings beside a row of plants already showing disease. When the healthy plants were from two to four months old a number of shoots became infected. Each of these shoots was tagged as soon as mottling was detected. At the end of nine months there were 35 of these marked stalks all showing characteristic symptoms on the young leaves as they unfolded from the spindle. During the following two months seven of these stalks gradually threw off the mottling, the young leaves as they unfolded showing no signs of disease, and this condition continued until the canes were cut four months later. Cuttings from four of the stalks showing this loss of leaf symptoms were planted. All the top cuttings produced normal shoots but cuttings from the lower joints which had earlier borne visibly infected leaves gave rise to mottled shoots. It is not known whether such "recovered" shoots retain the power to transmit infection to healthy plants.

When stalks become infected after a number of well developed joints have been formed, cuttings from the lower parts frequently give rise to symptomless shoots indicating that the virus moves downwards rather slowly. Single stools often carry both infected and healthy stalks up to maturity and cuttings from the latter give rise to healthy shoots.

Transmission. Although mosaic has been under observation in Java for over 30 years and its very regular transmission by the cutting was perceived at the outset, it was not recognised by investigators as infectious until 1922. In Hawaii[6] on the other hand, evidence of transmission from diseased to healthy canes was obtained at an early date although direct inoculations yielded negative results and the natural vector remained undetected. The behaviour of the disease in Porto Rico and the Gulf States, pointing strongly to transmission by some aerial agency and its resemblance to the mosaics of other crop-plants known to be transmitted by insects, led to renewed investigation. In 1919 Earle[19] in Porto Rico injected juice, pressed out of infected stalks under mineral oil to prevent oxidation, into the spindles of healthy stalks and after a month found that five out of ten had developed mosaic. The experiment was made in a field where natural secondary infection was occurring. In 1920 Brandes[24] repeated it at Washington under controlled conditions, and found that eight stalks out of the ten inoculated showed mottling after a month. He observed also that when healthy cane plants were brought into a greenhouse in which sorghum plants infested by the corn aphis (*Aphis maydis* Fitch, *A. adusta* Zehntner) and mottled cane plants were growing, the healthy cane plants began to show mosaic in 23 days. In another experiment in the same house five healthy cane plants were left exposed and ten were placed in insect-proof cages. The exposed plants showed mosaic in 19–30 days while the protected plants were still healthy after three months. Some of the sorghum plants in the same house also became infected and so did three species among the weed-grasses present. When corn aphis from the mosaic sorghum was transferred to healthy cane in insect-proof cages it showed mottling after 20 days while controls remained healthy. A leaf-hopper and a mealy-bug taken from infected plants failed to transmit the disease.

On two later occasions(22, 24) in the same year juice pressed out of infected young cane stalks under mineral oil was injected into the spindles of healthy shoots and 11 out of 12 so treated showed mosaic in 14–24 days. Controls injected with juice from healthy stalks remained healthy. Mosaic was also transmitted in this way from sugar-cane to sorghum, pearl millet (*Pennisetum glaucum*) and the weed-grass, *Panicum sanguinale*, but failed with maize. Using the artificially infected sugar-cane and sorghum plants the disease was transmitted by means of the corn aphis to sugar-cane, sorghum, maize, seven annual weed-grasses and one perennial grass—the ornamental *Miscanthus sinensis*. Wheat, barley, oats, rye and rice did not become infected. Two species of leaf-hoppers failed to transmit infection in all attempts. Mosaic was also found in maize, pearl millet and four of the susceptible annual weed-grasses under field conditions in the Gulf States but only in proximity to diseased sugar-cane. Bruner(25) working with twenty insects infesting cane in Cuba obtained transmission by the corn aphis alone. Chardon and Veve(26) using insect-proof cages in the field in Porto Rico secured transmission in sugar-cane by means of the same insect. Kunkel(27) in Hawaii transmitted mosaic by means of that insect from maize to sugar-cane, and showed also that a leaf-hopper (*Peregrinus maydis*) could be a vector in maize but not between maize and sugar-cane. In 1921–22 Wilbrink(28) in Java showed conclusively that the corn aphis could transmit mosaic from cane to cane, and that the common yellow cane aphis (*A. sacchari* Zehntner) which infests only the older leaves was not a vector. Infection was found also in maize, sorghum and two annual weed-grasses but only in and near cane cultivations. Mosaic has also been transmitted from cane to cane by the injection of juice by Kunkel(27) in Hawaii, Fawcett(10) in Argentina and Dastur(17) in India.

The corn aphis has been looked upon as a rare insect on sugar-cane, and economic entomologists were at first very sceptical regarding its importance as a vector of mosaic in the field. Wilbrink(28) in Java sought it on sugar-cane for a year in vain, but then found it frequently on cane growing near infested maize, sorghum and weed-grasses. It is difficult to detect on cane as it infests only the rolled leaves in the spindle and prefers other hosts when they are available. It may colonise sufficiently on the leaves in the spindle to cause a stagnation of growth and even a yellowing and withering of the young leaves. Visits of ants to the rolled leaves are a good indicator of its presence. In addition to maize and sorghums it infests a number of annual and perennial weed-grasses some of which are common in and near cane-fields in both hemispheres. Three of the annual weed-grasses of wide distribution, the *Panicum colonum*, *Panicum sanguinale* and *Eleusine indica* of Linnaeus, also become infected by mosaic in the field. Mosaic[22] has also been found on a number of annual weed-grasses in the Southern States of more restricted distribution, but hitherto on no perennial weed. That these hosts of the corn aphis exercise an important influence on the spread of mosaic in sugarcane is now certain[26, 28]. The rapid increase of secondary infection following the weeding of infected young cane fields is due to the migration of the aphis from the preferred weed-hosts to the cane shoots and its movement from plant to plant. Such active infection is most noticeable during rainy weather which favours growth of weeds and aphis and necessitates more frequent weeding. After the cane covers in and restricts weed growth secondary infection is usually much lessened.

A very light infestation of the cane is evidently enough to bring about a serious spread of mosaic. The stay of the aphis on young cane is frequently of short duration so that by the time that new infections become noticeable, usually two to three weeks after each weeding, it may no longer be possible to detect the insect. The intensity of secondary infection varies much from season to season and may be regularly very different in two localities of the same island. In Java it is an old experience that mosaic spreads less on heavy land than on loams or light soils. Wilbrink states that the weed-grass flora of heavy land contains few individuals of species which are hosts of the corn aphis. In Java[28] there is a continuous succession of winged and wingless viviparous forms of the aphis, while in temperate Northern India[29] a sexual generation occurs.

Brandes[22] failed to get transmission of mosaic by the seed in maize, sorghum and weed-grasses and Wilbrink[28] obtained negative results with sorghum and maize. Earlier observations in Java indicated that the disease was not transmitted by sugar-cane seed[4, 18].

Mosaic infection is not retained in the soil and there is no positive evidence of transmission through the soil from stool to stool.

Observations made hitherto indicate that infection occurs only on young parts such as the rolled leaves in the terminal bud or the youngest joints near the growing point. Presumably the virus passes to the growing point as all leaves and buds developed subsequently are infected except possibly in the infrequent cases of apparent recovery. Attempts to infect the expanded leaves artificially have failed and the aphides which infest such leaves do not transmit the disease[27, 28, 25].

Kunkel[30] in Hawaii has made a histological study of mosaic, par-

130

ticularly in maize which is often severely stunted and rendered sterile. The cells in longitudinal pockets adjacent to vascular bundles in the stalks become enlarged and break down so that internal cavities result. So-called "foreign bodies" are present in these hypertrophied cells. A similar condition occurs in the pale stripes on the veins of the maize leaf; at first the affected parts are thicker than the normal leaf tissue but later become thinner owing to collapse of the cells.

The "foreign bodies" are admitted to be unlike the intracellular bodies in the tissue of the vein-galls in Fiji disease of sugar-cane which McWhorter[31] in the Philippines has found to be a true protozoan.

Control of the disease has been effected by sett selection, roguing out infected plants from the young cane, and by replacing susceptible varieties with kinds resistant to infection, preferably such as are also fairly tolerant. The presumed immune semi-thin Uba cane is now being planted extensively where infection is heavy in Jamaica and Porto Rico.

The observations on the vector, its hosts and other members of the grass family susceptible to infection are too recent for their application in control to have been fully worked out. They point clearly enough to the need for close attention to sett selection and the early and thorough roguing of young cultivations of susceptible and sensitive varieties. Basing its action on field observation alone, Jamaica now prohibits the planting of maize in and near cane fields.

In recent years mosaic has been one of the important factors furthering a tendency to seek hardier varieties for cultivation within the tropics than the known thick canes can furnish. It is becoming clear that this need can be satisfied best by combining the hardy characters of the thin extra-tropical canes with the high tonnage and sugar content of thick tropical varieties. Some of the Java seedlings of Chunnee blood meet these requirements, but their susceptibility to mosaic in spite of high tolerance is a serious drawback and, although capable of solving the problem of sereh disease in Java, they have attained no extended culture there. It is probable that a high degree of resistance to infection by mosaic will be essential in any crosses between thick and thin canes suitable for wide cultivation within the tropics. Attempts are now being made in the West Indies and Hawaii to raise such seedlings by crossing the apparently immune Uba or similar varieties with thick canes and in Java by crossing immune wild species with thick varieties. In the meantime Uba is proving valuable as a means of cleaning up heavily infected localities in the West Indies and if nothing better is forthcoming its cultivation may be largely extended in Cuba within the next few years.

Reference may be made here to the series of Co seedlings raised by Barber at Coimbatore in South India. Some of these are on trial at several points in the West Indies and it may well be that the mosaic resistant tropical canes of the future will be found among them.

Literature cited

(1) DEERR, NOEL. *Cane Sugar.* Rev. ed. 1921.

(2) WAKKER and WENT. *De ziekten van het Suikerreit op Java* (1898).

(3) JESWIET, J. *Med. Proefstn. Java Skind*, No. 5 (1918).

(4) WILBRINK, G. and LEDEBOER, F. "Bijdrage tot de kennis der Gelestrepenziekte." *Med. Proefstn. Java Skind*, 39 (1910) and references.

(5) EASTERBY, H. T. "Varieties of sugar-cane in Queensland." *Bur. Sug. Exp. Stns. Bull.* 2 (1915).

(6) LYON, H. L. "Three Major Cane Diseases: Mosaic Sereh and Fiji Disease." *Hawaiian S.P.A. Exp. Stn. Bull.*, Bot. Series, vol. III, Pt. I (1921).

(7) ECKART, C. F. *Hawaiian S.P.A. Exp. Stn. Reps.* (1901–04).

(8) HINES, C. W. "History of Cane varieties in the Philippines." *Philip. Agr. Rev.* VIII, No. 3 (1915).

(9) KOBUS, J. D. *Rep. Proefstn. Java Skind* (1909).

(10) FAWCETT, G. L. *Rev. Indust. Agric. Tucuman*, 13 (1922).

(11) EARLE, F. S. "Sugar-cane varieties of Porto Rico. II." *Journ. Dept. Agric. P.R.* v, No. 3, pp. 35, 109, 110 (1921).

(12) STEVENSON, J. A. *Journ. Dept. Agric. P.R.* III, No. 3 (1919).

(13) BRANDES, E. W. "The Mosaic Disease of Sugar Cane and other Grasses." *U.S. Dept. Agric. Bull.* 829 (1919).

(14) —— *Bull. Dept. Agric. Jamaica*, New Ser. vol. II, p. 239.

(15) ASHBY, S. F. "Mosaic Mottling or Yellow-stripe Disease of Cane." *Pampl. Dept. Agric. Jamaica* (1920).

(16) WILLIAMS, C. B. *Bull. Dept. Agric. Trinidad*, XIX, 1 and 3 (1921).

(17) DASTUR, J. F. "The Mosaic Disease of Sugar Cane in India." *Agric. Journ. India*, XVIII, Pt. V (1923).

(18) KOBUS, J. D. "Vergelijk Culturproev. om't Gelestrepenziekte." *Med. Proefstn. Java Skind*, No. 12 (1908); also (6), (12), (13), (15).

(19) EARLE, F. S. "Yellow-stripe Disease investigations." *Journ. Dept. Agric. P.R.* III, No. 4 (1919).

(20) CROSS, W. E. *Louisiana Plant*, 66, No. 12 (1921).

(21) AGEE, H. P. *Louisiana Plant*, 74, No. 4 (1924).

(22) BRANDES, E. W. "Cultivated and wild hosts of sugar-cane mosaic." *Journ. Agric. Res.* 24, No. 3 (1923).

(23) —— "Mosaic Disease of Corn." *Journ. Agric. Res.* 19, No. 10 (1920).

(24) —— "Artificial and Insect transmission of sugar-cane mosaic." *Journ. Agric. Res.* 19, No. 3 (1920), also (2) and (23).

(25) BRUNER, S. C. *Rev. Agric. Com. y Treb.* (Cuba), 5, No. 1 (1922).

(26) CHARDON, C. E. and VEVE, R. A. "The transmission of sugar-cane mosaic by *Aphis maydis* under field conditions in Porto Rico." *Phytopath.* 13, No. 1 (1923).

(27) KUNKEL, L. O. "Insect transmission of yellow-stripe disease." *Hawaiian Plant. Rec.* 26 (1922).

(28) WILBRINK, G. *Med. Proefstn. Java Skind*, No. 10 (1922) and references.

(29) DAS, B. "The Aphididae of Lahore." *Mem. Ind. Mus.* 6, No. 4.

(30) KUNKEL, L. O. "A possible causative agent for the mosaic disease of corn." *Hawaiian S.P.A. Exp. Sta. Bull.*, Bot. Series, vol. III, Pt. I (1921).

(31) McWHORTER, F. P. "The nature of the organism found in the Fiji Galls of Sugar-cane." *Philipp. Agriculturist*, 11 Nov. 1922.

Mr H. H. Storey. Streak Disease, an Infectious
Chlorosis of Sugar-cane, not identical
with Mosaic Disease

The existence in Natal of sugar cane mosaic disease was announced
by the present writer in 1923 (10). On the scientific side, the problem in
this country appeared to present no unusual features, and local experience has in general confirmed observations of the many workers in
this field in other cane-growing countries. There can be no question of
the identity of the South African disease with the mosaic disease of
Java, and the New World; the agreement of symptoms is absolute,
while there is even good evidence to suggest that the disease was introduced to this country from Argentina in setts of the self-same Java
seedling canes (P.O.J. 36, 213 etc.), which are believed to have carried
mosaic to most of the cane-growing countries of the New World. But
regarded as a practical problem in disease control, the situation in Natal
is probably unique, 99·8 % of all the sugar-growing area being planted
to a cane which other countries have found to be immune to mosaic,
the variety Uba.

A condition of this Uba cane however, which was at first believed to
be mosaic disease, has been held under observation for the past eighteen
months. The institution of control measures for mosaic disease caused
the determination of the exact status of this condition in the Uba to be
of the utmost importance: for a recommendation had been made to
destroy the small balance of susceptible varieties in an attempt to
eradicate mosaic disease, a policy which would clearly be frustrated
should the Uba be susceptible itself.

The evidence upon which the writer has based an opinion that this
condition in Uba is not mosaic disease, but a separate and distinct disease
of a similar type, will be brought out in this paper. One weakness in the
chain of evidence, of which the writer is fully sensible, lies in the absence
of evidence of transmission of the disease under controlled experimental
conditions. Had a successful technique been available, there would have
been opportunity of clear proof of conclusions, which are now based,
perhaps somewhat insecurely, upon laborious field observations. It was
his intention to defer publication of an account of the disease until clear
evidence upon this point could be offered; but the necessities of the
practical situation in the Natal sugar industry were such as to require
early action and it was therefore considered inadvisable to delay publication further.

This condition has been observed in ten varieties of cane in addition to Uba, all of which ten varieties are susceptible to and may exhibit the typical symptoms of mosaic disease. In maize there exists a diseased condition which bears a relation to mosaic disease in that plant similar to that which is exemplified in the case of cane. Finally, similar symptoms have been discovered in a number of species of wild grasses. To this condition the term "Streak Disease" has been applied, in an attempt to obtain a name sufficiently descriptive to enter into popular use, while avoiding the confusion, involved in the word "stripe," with mosaic, *Sclerospora*-disease, and ribbon varieties of cane. The term is used in this paper in a generalised sense, and is not to be understood to imply the identity of the disease occurring in different hosts. While much evidence suggests that this implication is in fact a true one, the separation must be maintained until proof of transfer can be offered.

History and present position

No description of this disease in sugar cane in South Africa has previously appeared. In 1920 Wuthrich(11) published an article upon the occurrence in Natal of "Yellow Stripe Disease," without descriptions or definite data, but from the context it must be assumed that this term referred in this particular instance to this disease of Uba. This observer was impressed by the rapid spread of the disease. No reference has been found in the literature of other countries to a disease agreeing with that under consideration, and it would appear to be unlikely that its existence can have been overlooked in view of the present great interest in mosaic disease. Brandes and Klaphaak(3) however draw attention to the possibility that more than one disease of the mosaic type may exist within the family Gramineae. Definite evidence of the existence of this disease in 1914 occurs on the files of the Department of Agriculture. Its present distribution is that of the sugar cane; no areas exist where it is absent, and over large tracts more than half the cane is affected.

Streak disease in maize has been known in Natal for many years under the names of "Blight," "Variegation," "Yellows" or "Leaf Striping." Fuller(7) in 1901 described it accurately, directing attention to the reduction in yield caused, and recording that, while seed from diseased plants might produce a healthy crop, seed from an unquestionably healthy source might equally produce a diseased crop. He was of the opinion that the disease was not contagious, but a result of soil conditions. He noted a similar disorder in a number of other monocotyledonous hosts. A correspondence in the *Natal Agricultural Journal* at this time showed that the disease had existed at least 20 years prior to the beginning of this century. At the present time it occurs throughout the coastal belt of Natal and the Eastern Province of the Cape, up to an altitude of about 3000 ft., and rarely up to 4500 ft.

Symptoms

The symptoms of streak disease are confined to the leaves, the stem showing no distinctive characters which may distinguish this disease from the results of unfavourable conditions in general. The leaf of an affected sugar-cane plant will show over its whole surface a pattern produced by broken, narrow, pale stripes, running in the direction of the veins of the leaf. The stripes may be more or less crowded, but their frequency will not vary upon different parts of the leaf blade. As with mosaic disease, diagnosis is most readily carried out by examination of the youngest leaves. The central, as yet unrolled, leaf will show the colourless stripes as plainly as those more mature, thus affording a means of differentiation from what may be termed the accidental spotting and marking which frequently simulates streak upon the older leaves.

The leaf symptoms in mosaic disease have been frequently described. But for purpose of comparison, it is advisable now to consider them in somewhat more than usual detail. Considered simply as a colour pattern, the typical mosaic leaf shows dark green areas upon a ground of pale green, such areas being irregular, diffuse-edged, frequently elongated in the direction of the leaf axis, but apparently unrelated to particular veins, and always spreading unevenly over several adjacent veins. In the streaked Uba leaf we have a pattern formed of light areas upon a dark ground. These areas are narrow, elongated and of uniform width, and each bears a clear relation to a particular leaf vein. Furthermore they are almost transparent, and in this respect show a pronounced difference from the normal mosaic leaf, when viewed by transmitted light. By way of illustration, one might say that while the streak pattern might well be produced with a drawing pen, mosaic would require brush work.

With only slight modifications this distinction applies throughout the plants which show this condition. In maize the streak pattern assumes a most distinctive character. The analogy previously used applies with equal force, if one may allow that the ink from the pen has run somewhat, causing the streaks to coalesce in places. Each pale stripe is seen to be disposed along a particular vein, causing an appearance suggesting that some substance had diffused out from the vein, inhibiting the production of green pigment. Generally in the streak pattern the light areas are of a uniform shade of very pale green, but in certain cases some of the streaks may be quite colourless, sometimes even shrunken and showing slight browning of the tissues. The mosaic symptoms produced in maize by experimental transfer from sugar cane, agree with those described and figured by Brandes(2). As noticed by the writer, the leaves in ageing tend to lose the mosaic pattern and to offer difficulty in diagnosis; in streak, on the other hand, the pattern becomes more pronounced in the mature leaf, the contrast between light and dark areas being increased by the deepening of the green in the latter, while the former lose what little colour they possessed. It is difficult however entirely to reconcile the figures and descriptions of Kunkel(8) with the symptoms of maize mosaic observed by the present writer, certain of his figures being even somewhat suggestive of streak disease.

A list of the species of wild grasses which have been found exhibiting streak and mosaic diseases will be found in a later section of this paper. The leaf patterns both of streak and mosaic diseases conform closely in all species to those in *Digitaria horizontalis* Willd. The mosaic diseased leaves photographed are from a plant to which the disease has been experimentally transferred from sugar cane. The difference is as pronounced as in maize, and the symptoms of each disease are of the general type described.

The difference between the symptoms of the two diseases is a sharp and clear cut one. There is no question of any continuous variation between the two types. This statement is based upon the extended observation of canes of all stages of growth, seven varieties having been studied for more than a year growing alongside in an experimental plot. Mosaic disease in the grasses, maize and the majority of sugar canes conforms to the general description given above and admits of no confusion with streak, showing always an irregular pattern in shades of green, without transparent areas. In certain varieties of cane, however, areas may occur upon the mosaic-diseased leaf which are almost devoid of pigment—the "secondary symptoms" of Brandes(1)—but these areas are never linear, but of irregular outline. The nearest approach to streak which has been observed is that exhibited by the mosaic-diseased leaves of a cane, possibly a seedling of Uba raised in Mauritius.

But again the light areas, which may be nearly transparent, are of irregular outline. On the other hand, all the varieties of cane which have been observed to be streaked have conformed to the general type with remarkable closeness. The same applies to all maize examined, both flint and dent varieties exhibiting identical symptoms; and in general also to all species of wild grasses which have been observed to be affected. But certain aberrant cases call for notice. The stripes in the leaves of a young cane shoot arising from a streak-infected sett may be so far broadened as to coalesce, and produce a pattern simulating in some measure mosaic disease. But at a later stage in its growth, this plant will produce leaves exhibiting a normal streak pattern. In a similar way maize and grasses infected at an early age may show an extreme widening of the streaks; indeed not infrequently affected plants of *Digitaria* spp. may develop leaves almost completely pure white. Even in this condition however the symptoms in no way resemble mosaic disease.

It is believed that any doubt as to the non-identity of the two diseases will be removed by the occurrence in the plainest manner of the two diseases in one plant. Within a few yards have been observed individuals of a single variety showing mosaic disease, streak disease, and the two diseases clearly superimposed one upon the other. There is no possibility here of a transition stage between the two conditions. No suggestion of any such transition has been detected in the many plants which have been kept under close observation. An example of *Digitaria horizontalis* Willd. has been found showing the same condition. It should be noted that the plant in this case has only recently become affected with streak disease, the striping consequently being restricted to the lower parts of the leaves.

The range of affected plants

The plants mentioned in the following list have been observed showing the condition now called streak disease. In addition the plants marked with a star (*) have been observed showing the typical symptoms of mosaic disease. A double star (**) indicates that mosaic disease has been experimentally transferred to the plant, and it is by comparison with the latter that the disease has been diagnosed in the remainder. Support to this diagnosis in the case of wild grasses is given by the situations in which such diseased grasses occur, none having been found except in the immediate vicinity of mosaic-diseased cane.

CULTIVATED

Sugar cane, *Saccharum officinarum* L.

Varieties

Uba of Natal.

*Agaul (Pansahi group of North India).

*Green Natal (Rose Bamboo, Cheribon series).

*Louzier (Otaheite series).

**Port Mackay (Cavengerie).

**Mauritius (possibly a seedling of Uba raised in Mauritius).

*Badila.

*H.Q. 694 (Queensland seedling).

*Black Innes (? Mauritius seedling, grown under this name in Queensland).

**Egyptian No. 3 (probably identical with a cane imported under the name "Java 147").

*Queensland No. 3 (identity unknown).

Maize, *Zea Mays* L.

White Dents: **Hickory King.
             **Silver King.
             **Ladysmith Pearl.

Yellow Dents: **Golden Beauty.
             **Sahara Yellow.

Yellow Flint: *Yellow Cango.

Teosinte, *Euchlaena mexicana* Schrad.

WILD GRASSES[1]

Digitaria eriantha Steud.

**   ,,   *horizontalis* Willd.

*   ,,   *marginata* Link.

       ,,   *ternata* Stapf.

       ,,   sp.? *Smutsii* Stent.

Paspalum scrobiculatum L.

**Urochloa helopus* Stapf.

**Eragrostis* spp.

(Four species of *Eragrostis* have been found showing streak disease, and one of them showing mosaic disease. This genus is in great

---

[1] The nomenclature is that of Stapf (*Flora of Tropical Africa*, vol. IX). I am indebted to Miss S. M. Stent and Miss H. M. Forbes, of the Division of Botany of the Union Department of Agriculture, for the determination of these grasses.

confusion and in need of revision, and specific determinations are therefore not given.)

Eleusine indica Gaert.

Fuller(7) recorded streak disease in sorghum, oats, cannas and *Setaria verticillata* Beauv. in addition to teosinte, *Digitaria horizontalis* Willd., and *Eleusine indica* Gaert. now confirmed by the present writer. It will be observed that, with the exception of the genus *Eragrostis*, all the grasses mentioned are nearly related to or identical with those cited by Brandes and Klaphaak(3) and Chardon and Veve(5).

On the other side, experience with cane varieties in this country, with one exception, has been too limited to allow of conclusions as to resistance to streak disease. There is however reason to believe that the Chunnee-Cheribon crosses, P.O.J. 36 and 213, are immune, no single case of this disease having been observed in many plots growing alongside severely streaked Uba cane.

The effects of the disease

The variety Uba exhibits a high measure of tolerance to streak disease and no case has been observed where this disease was causing damage to cane comparable with that in a normal mosaic infestation. But, apart from the variety Uba, only odd affected plants have been observed and it is possible to form an idea of the damage only in the case of that variety. This tolerance, however, in the case of Uba is almost certainly not absolute. At present this statement can be supported by little beyond general field observations, which are admittedly unreliable. But in all localities healthy plants in the young cane stand out as markedly superior to the diseased. An attempt to determine the amount of the difference at harvest time was made by reaping and weighing in commercial plantations the whole of a streaked stool and an adjacent healthy one. By this means a figure was obtained for the average weight of the diseased and healthy individual cane. This process was repeated in four localities upon 13 diseased and an equal number of healthy stools, involving 300 canes in all. The result of this work was to give, upon the assumption that healthy and diseased plants gave an equal number of shoots per acre, a figure of 10 % for the loss of possible yield in an area fully diseased. There is no means of determining the measure of justification for this assumption however, and this line of investigation was not carried further. The results of plot trials now in progress are not yet available. But it is contended that, even if the loss be less than the figure stated, the existence of this disease already in a large proportion of all the cane growing in the country, and the probability of its early spread to the remainder, is not a matter for equanimity within the sugar industry.

The effect of streak disease on maize, on the other hand, is of a most serious nature. Plants affected at an early stage may never exceed a foot in height, and, though reaching the tasselling stage, may form no ears. In this plot two plants were growing from the majority of holes, and, where not diseased, were both standing over six feet. Late infection has a less serious effect, but causes nevertheless a marked reduction in yield, in respect of malformed, undersized and partially sterile ears. Severe stunting of grasses produced by streak disease is frequently to be seen, the ill-effects being much more severe than those of mosaic upon those particular hosts.

Infection phenomena

The fact of sett transfer of streak disease in sugar cane has been fully established. The production of a diseased plant by a sett cut from a fully diseased stool has been invariable under all conditions of soil, moisture and temperature. No exception to this rule has been observed, nor has any case of recovery been detected. But to account for the present large proportion of affected cane in Natal—upon many estates exceeding 75 %—some secondary method of infection must be postulated. In this there is clearly a close parallel with mosaic disease. One piece of positive evidence only will be quoted to show that the present wide distribution is due to secondary infections, and not to the original propagation of a stock of diseased material. A field manager detected this disease upon his estate for the first time in 1916. The condition was so far novel to him at the time that he reported the matter to the Department of Agriculture. Upon his return after four years' absence on service, he early observed the disease to have spread to every part of the estate, much of which had not been replanted in the interval.

Definite cases of secondary infection have been held under observation. Two plants of Uba cane growing in the experimental ground, after remaining healthy for six months, developed streak disease. In March, 1923, certain cane varieties newly imported from Queensland were planted in a plot in the vicinity of which was Uba cane with a proportion of plants streak-diseased. These plants were kept under careful individual observation and all developed healthily. In July streaked plants began to appear in the variety H.Q. 694, and during subsequent months many new cases appeared in this variety and in Nadila and in Black Innes, although on discovery affected plants were at once removed. Numerous cases of recent infections have been observed in the field, in which the leaf pattern was visible only upon one or a few of the shoots arising from a stool. Such plants show clearly the history of the development of the disease; the lowest leaves have no markings whatever, later formed ones bear the streaks at the base only, while the youngest are fully streaked from tip to base. Attempts have been made to eliminate streak by rogueing through young plant cane. At a subsequent inspection new cases of the disease have been found, the majority showing, by their position adjacent to spaces whence diseased plants had been removed, that they were secondary infections still latent at the time of the original rogueing. On the other hand, it is clear that there are long periods when secondary infection is at a standstill. In the experimental ground at the Natal Herbarium many varieties of cane, healthy and diseased, have been growing alongside, but the only cases of secondary infection observed are the two Uba plants previously mentioned. It is common to find in the field stools which are half diseased and half healthy. All such, where not due to recent secondary infections as already described, have proved on examination to have arisen from two setts, healthy and diseased, the practice in Natal being to plant a double row. Such a case was observed in May, 1923, and now a year later, the cane having been cut in the interval, the ratoons show the same condition.

A clear demonstration of secondary infection is seen in the case of streak disease of the annuals, maize and wild grasses. For primary infection is here non-existent, in the absence of transfer of the disease by true seed. That the seed is incapable of carrying the disease is supported in the first place by the fact that several hundred maize plants, of several varieties and obtained from different sources, have been raised in a greenhouse without a single case of streak appearing. The same seed exposed to infection in the open has produced plants which have in a large proportion developed the disease. Direct evidence upon this point has been obtained. Seed of *Digitaria horizontalis* Willd. was collected from streaked and healthy plants. Several hundred plants from both sources showed no traces of disease. It was thought that the drying out of the seed might account for the disappearance of the infectious principle. Eight maize cobs were therefore taken in an immature state from streak-diseased maize plants, the seed at this time being well swollen but still soft. This seed was planted at once, and 200 plants were raised in the greenhouse, but all remained healthy until discarded after two months.

Among the maize crop in particular secondary infection is very active. Few areas exist along the Natal coast where late planted maize escapes universal infection. But it appears that it is lateness in relation to the earliest plantings in the vicinity, rather than any seasonal cause, which is the determining factor. The plots of the Indian cultivator, who plants a succession of crops to supply the cobs green for the local market, generally show a high proportion of streak even before the main crop is planted elsewhere. The following case is typical of an extensive series of field observations carried out during the season 1923–24. A garden at Pinetown, 12 miles from the coast, was kept under observation from December 14th, 1923. On that date two plots of maize existed; one, about four months old and starting to flower, showed 40 % streak, and a second of half that age showed 5 %. A fortnight later there had been a slight increase of disease in the first plot, which was then starting to ripen off, while the second plot had become diseased to the extent of 30 %. On January 15th, 1924, this last plot was 50 % streaked, and ripening off. Late in February a third plot was planted, but before reaching a foot in height every plant became diseased. The behaviour of the grasses in these plots is interesting. At the first inspection an odd case of the disease in *Eleusine indica* Gaert. was found. Subsequently a rapid increase occurred, so that by January 15th hardly a healthy plant of this species could be found on the site of the original maize plot, now reaped. This observation is typical of the behaviour of the weed grasses in general, a marked rise in the percentage of diseased individuals being frequently noticeable as the maize ripens off. The significance of this fact in regard to the identity of the disease in the several hosts is discussed later.

Results of a similar type have been experimentally obtained. Maize of five dent varieties was planted in mid-November; 25 % became streaked. A second planting alongside at the beginning of February produced a crop which was 90 % infected within a fortnight from planting. A third planting at the end of March grew more slowly at this season, but not a single plant was healthy after a month from planting. Meanwhile, as has been recorded, plants raised in the greenhouse alongside developed no single case of disease. Twelve such plants, of ages varying between one and six weeks, were removed to the open at the end of February. Streak made its first appearance in 13 days, and subsequently developed in 75 % of the plants.

An experiment was set up early in March, in which maize plants were raised in a graded series of wire-gauze cages. The cages were in the form of cylinders, closed at the upper end, the lower open end being pressed into the soil, and were made with gauze of a mesh of 64, 32, 16 and 6 holes to the inch, two of each grade. Maize was raised in each from seed and the cage was never lifted from the day of planting. After two months all the surrounding exposed control plants were streaked. Disease had appeared upon the plants in all the cages of the two larger meshes. But the plants protected by gauze of 32 and 64 mesh were healthy.

The evidence appears to indicate that the infective principle passes down into the plant from the leaves. Several canes showing new infection in the field, as described previously, have been halved and planted. In each case the upper half produced streaked foliage at once, while shoots from the butt were mostly at first healthy but subsequently gradually became diseased. This observation admits of several interpretations, but in the following experiment it was found possible to separate the two halves of the cane before infection had passed into the lower. Six newly infected shoots of the variety H.Q. 694 were graded according to the extent of the leaf symptoms exhibited, and cut off below the node bearing the oldest living leaf. It should be said that these examples were detected at a very early stage in the development of the disease, the earliest showing only a few of the typical streaks towards the base of the youngest unfolded leaf. The butts on planting produced, in the order of grading, three plants definitely streaked, one doubtful and two healthy. Subsequently the doubtful case became fully streaked, but the two remained healthy, and are still so six months after planting.

During the period of these investigations numerous experiments designed to obtain the transfer of the disease under controlled conditions have been carried out. Attention has been particularly given to the suctorial insects considered likely to act as vectors. The result of this work is as yet inconclusive, and it is proposed to defer an account of it until a later date.

Discussion

The purpose of this paper has been to present evidence for the view that in the first place the condition described as streak disease is actually an infectious chlorosis of a type similar to mosaic; and secondly that it is not identical with the latter disease. The former of these conclusions will probably not be disputed. In the majority of its characters streak affords a close parallel with mosaic disease: in the general type of symptoms, in the fact of sett transfer and the failure of transfer by true seed, and in the phenomena of secondary spread. Support would be added by the similarity of host-range if the identity of the streak disease found upon the various hosts were established. At present the probability of this identity rests upon similarity of symptoms and contiguity of occurrence in the field. But the latter does in fact give support to the former, and no streaked grasses have been found except in close proximity to streaked cane or maize. The evidence already quoted of the marked increase in the proportion of diseased grasses as a maize crop ripens off is very suggestive that some insect carrier has been driven to seek new food-plants, upon the drying out of the maize. The fact that certain species of grasses have been found showing identical symptoms both in cane and maize fields serves in a measure to link the disease as occurring upon these two hosts.

The conception of a number of virus diseases affecting a single host plant has become familiar through the work of Quanjer(9) upon the potato. The writer may therefore be accused of having stressed the difference between streak and mosaic diseases to a point even tedious. But he has felt that the separation of two diseases upon minute detail of symptoms alone requires the fullest justification. Even, however, if his criteria for the separation be regarded as insufficient, there remains the reconciliation of the fact of the susceptibility of Natal Uba to streak, and the immunity to mosaic of the same cane (imported within recent years from Natal) in Cuba(4), Jamaica(6) and apparently also in the Washington experimental houses(3). Wherein lies the difference is a matter merely for conjecture in the absence of exact knowledge of a causative agent, and there are a number of possibilities, such as that each type of symptoms is merely an expression of the transfer of a single causative agent by a particular carrier. But as a working hypothesis in the present state of our knowledge, it is believed that the present separation is justified.

DESCRIPTION OF COLOURED PLATES

PLATE I. THE LEAF SYMPTOMS OF STREAK DISEASE

FIG. 1. Healthy leaf of the variety Uba.

FIG. 2. Leaf of the same variety from a plant affected with streak disease.

FIG. 3. Leaf from streak-diseased plant of the variety known in Natal as Egyptian No. 3.

FIG. 4. Leaf from a young diseased plant of Uba, showing the broadening of the chlorotic areas, simulating mosaic disease. Such a plant will subsequently produce typically diseased leaves as shown in Fig. 2.

FIG. 5. Leaf from a young diseased plant of the variety Port Mackay (Cavengerie).

(All approximately natural size.)

PLATE II. THE COMPARISON OF STREAK DISEASE WITH MOSAIC DISEASE

FIG. 1. Streak Disease. Portion of a leaf of the variety Uba, showing the linear nature of the chlorotic areas, and the almost complete lack of colouring matter therefrom.

FIG. 2. Mosaic Disease. Portion of a leaf of the Java seedling variety P.O.J. 36. The chlorotic area here forms the greater part of the leaf surface, but the contrast between it and the small fully green areas is slight.

FIG. 3. Mosaic Disease. Portion of a leaf of an unnamed variety, possibly a seedling raised in Mauritius. The pattern here approaches most nearly to streak disease of all varieties seen, in respect of the almost complete absence of pigment from the chlorotic areas. Their irregular shape is however in marked contrast to the pattern of streak disease.

(All enlarged by about 3 diameters.)

Literature cited

(1) BRANDES, E. W. (1919). "The Mosaic Disease of Sugar Cane and other Grasses." *U.S. Dept. of Agriculture, Bull.* No. 829.

(2) —— (1920). "Mosaic Disease of Corn." *Journ. Agric. Res.* XIX, 10, p. 517.

(3) —— and KLAPHAAK, P. J. (1923). "Cultivated and Wild Hosts of Sugar-Cane or Grass Mosaic." *Journ. Agric. Res.* XXIV, 3, p. 247.

(4) BRUNER, S. C. (1923). "Mosaic and other Cane Diseases and Pests in Cuba." *The Planter*, LXX, 22.

(5) CHARDON, C. E. and VEVE, R. A. (1923). "The Transmission of Sugar-Cane Mosaic by *Aphis maydis* under Field Conditions in Porto Rico." *Phytopathology*, XIII, 1, p. 24.

(6) COUSINS, H. H. (1922). *Ann. Rep. Dept. of Agric. Jamaica* for year 1921.

(7) FULLER, C. (1901). "Mealie variegation." *First Rep., Govt. Entomologist, Natal*, 1899–1900, p. 17.

(8) KUNKEL, L. O. (1921). "A Possible Causative Agent for the Mosaic Disease of Corn." *Bull. Ex. Sta. of Hawaiian Sugar Planters' Assoc.* III, 1, p. 44.

(9) QUANJER, H. M. (1923). "General Remarks upon Potato Diseases of the Curl Type." *Rep. Int. Conference of Phytopath. and Econ. Entomology*, Holland, p. 23.

(10) STOREY, H. H. (1923). "The Major Cane Diseases." *S. African Sugar Journal, Congress Number*, p. 54.

(11) WUTHRICH, E. (1920). "Yellow Stripe Disease in Natal." *S. African Sugar Journal Annual*, 1920–21, p. 157.

PLATE I

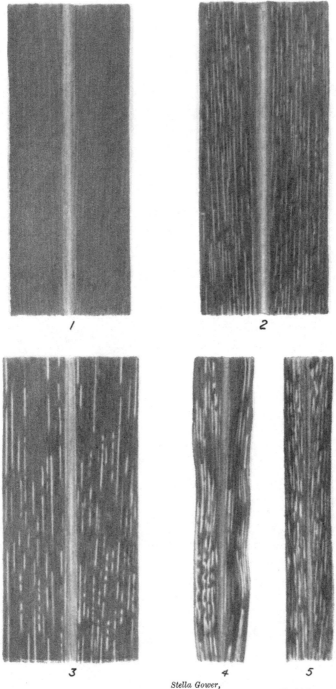

Stella Gower,
Division of Botany, Union of S. Africa.

STOREY. STREAK DISEASE OF SUGAR CANE.

PLATE II

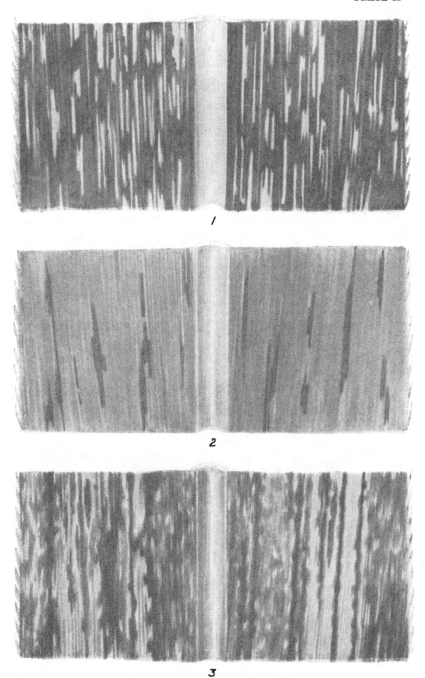

1

2

3

Stella Gower,
Division of Botany, Union of S. Africa.

STOREY. STREAK DISEASE OF SUGAR CANE.

Dr E. J. Butler, C.I.E. BUD-ROT OF COCONUT AND OTHER PALMS

A disease of palms, which was stated to have many points of similarity to that known as bud-rot in the West Indies, was described by the writer in 1906 and attributed by him to a species of *Pythium*, which was subsequently, in February 1907, named *P. palmivorum* n.sp. This was the first *named* cause of bud-rot which subsequent work has substantiated. The species is one of those forming a connecting link between the genera *Pythium* and *Phytophthora*, and in 1919, when knowledge of the latter genus had considerably advanced, the palm fungus was re-named by the writer *Phytophthora palmivora*.

In the West Indies the bud-rot of coconut palms seems to have been known for about a century, and was ascribed to various causes such as insects, bacteria, soil, and climate. Davalos believed that *Bacillus amylobacter*, which he isolated from the rotted tissues in 1886, was the cause in Cuba. Fawcett attributed it to "an organized ferment" in Jamaica in 1894. Earle also, in 1902, believed it to be a bacterial disease. In 1904 Dr E. F. Smith definitely stated, as a result of his investigations on the spot, that the disease in Cuba was a bacterial soft-rot of the terminal bud and its wrappings, the causal organism probably entering through wounds, though no experimental evidence of these statements was given. Petch stated, in 1906, that bud-rot in Ceylon was due to bacteria. Smith induced his assistant, J. R. Johnston, to undertake the detailed study of the disease, and the results of four years' work in Cuba and elsewhere were published by Johnston in 1912. Successful inoculations were claimed with several varieties of bacteria. Only a few of these inoculations were done on the spot, the rest being on seedling coconuts under glass in Washington. The greater number of the inoculations were made with *Bacillus coli* or an organism indistinguishable from it, while *B. coli* from animals was found capable of setting up a soft-rot similar to that caused by the strains from coconut, when inoculated on greenhouse seedlings. A preliminary account of these inoculations was published in *Phytopathology* in 1911. It may be mentioned, in passing, that the isolations were made by the poured plate method, which would be unlikely to reveal the presence of such a fungus as *Phytophthora*, and that the bulk of the inoculations were made with a gouge or some such instrument penetrating at least 30 cm. into the bud. But in 1911 Rorer agreed with Johnston's findings and stated that pure cultures of a bacillus (which he isolated but did not identify) could cause bud-rot if poured into the crown without wounding.

In 1914 Shaw and Sundararaman published certain criticisms of Johnston's work and produced detailed evidence, founded on inoculations from pure cultures, that *Pythium palmivorum (Phytophthora palmivora)* caused coconut bud-rot in India. Further inoculation experiments, to meet objections raised by Sharples and Lambourne in Malaya, have been carried out by McRae and by Sundararaman in India, with the result that the earlier work was again confirmed. There is now not the slightest doubt that the epidemic outbreak of bud-rot in palms on the east coast of India was due to this fungus and only to this fungus, which is a true parasite capable of penetrating the uninjured tissues almost as readily as *P. infestans* does those of the potato.

In 1918 Ashby stated that there were two forms of bud-rot in Jamaica, one of which was due to the same *Phytophthora* as he had earlier found causing a leaf disease of coconuts. In 1920 this was identified as *Phytophthora palmivora*, the cause of the disease in India, and proof of its ability to penetrate the unwounded tissues was given. In the second type of bud-rot only bacteria were found, but some doubt is expressed as to their causal connection with the disease. In 1923, however, Nowell was still of opinion that after excluding the forms of bud-rot due to fungi there may still remain a large residue of bud-rot of bacterial origin in the West Indies.

In the Philippine Islands the study of coconut bud-rot was taken up by Reinking who, in 1918, believed it to be a bacterial disease. This view he soon abandoned and, in 1919, he stated that a *Phytophthora*, which he identified as *P. faberi*, was the cause, the bacteria which are always present in severe cases being secondary agents. In 1923 further studies were published by Reinking which seem to leave no room for doubt that this fungus is a true parasite and is capable of infecting uninjured coconut palms, whether the organism is isolated from coconut or from cacao (the original host of *P. faberi*).

This at once raises the question whether *P. faberi* is distinct from the earlier described *P. palmivora*. Reinking answers in the affirmative, but his argument is based on a misconception. No oospores were found in the Philippine fungus, whereas he states that they occur in that in India. It is true that in my 1907 paper I stated that oospores were formed, but these were subsequently found to be chlamydospores or resting conidia, of a type previously unknown in *Phytophthora* but subsequently found to be not uncommon. From a careful study of Reinking's figures and description I am of opinion that his fungus is morphologically identical with my *P. palmivora*. Thus both in the west and in the east it has been established that this fungus is at least a chief cause of bud-rot.

As to Reinking's identification of this species with *P. faberi*, widely known as a parasite of cacao and rubber, a few words more are required. Ashby admits that *P. palmivora* and *P. faberi* are morphologically indistinguishable but, unlike Reinking, he finds biological differences. In his last paper on the subject, in 1922, he reports the discovery of oospores in mixed cultures of the cacao fungus with other morphologically similar strains from coconut and cotton and also with the distinct species *P. parasitica*. In view of the biological differences and the absence of oospores from all but the cacao strain he maintains *P. faberi* and *P. palmivora* as distinct species. But it is clear from Reinking's work that the biological differences are not preserved in the Philippines and it is equally clear that oospore-formation is rare and difficult to induce in the cacao fungus, so that I prefer to regard the two as at most strains of the one species, less highly differentiated indeed than many of the strains of the rusts and Erysiphaceae.

Whether there is any such thing as a primary bacterial bud-rot of palms seems to me to be doubtful. Most of the successful inoculations have been carried out through wounds in the tender tissues of the interior of the bud, and with such methods not one but several distinct bacteria have been found capable of setting up a soft rot. Sharples and Lambourne have indeed shown that various common moulds, such as *Mucor*, can cause such a rot in wound inoculations. The same is well known to be the case with ripe fruit and young seeds, always only after injury. I do not regard the fungus as a primary parasite in such cases, and it is at least in a very different category from the parasitism of such organisms as *Phytophthora infestans* or *P. palmivora*. *P. palmivora* remains the only vegetable parasite that has been proved to cause a destructive bud-rot of palms, capable of attacking perfectly healthy trees and inducing severe epidemics of disease.

[A short account of the campaign against bud-rot of palms on the east coast of India was then given.]

Mr A. Sharples. OBSERVATIONS ON BUD-ROT OF PALMS

The world-wide distribution of *Cocos Nucifera* and its economic importance ensure interest in the pathological disturbances commonly met with in plantations. An epidemic disease in this important crop demands serious attention, and considerable time and energy over many years has been devoted to the question of the unmistakable rotting of the cabbage or bud, owing to the intrusion of various organisms causing decay, which in its most severe form results in the death of large numbers of trees wherever coconut plantations exist. At the present time, the usual conception of "bud-rot" takes the form of the possibility of large losses owing to epidemics caused by fungi belonging to the genus *Phytophthora*. Recently, a more reasoned statement has been issued respecting a more general conception.

"Bud-rot" in the West Indies

Nowell, in his recently published book entitled *Diseases of Crop Plants in the Lesser Antilles* devotes considerable space to this subject and his observations are of much importance. His observations parallel those of the writer in Malaya, but as "bud-rot," associated with various *Phytophthora spp.* appears a well-marked feature in the West Indies, this author adopts a cautious attitude. In Malaya, the writer has seen much "bud-rot" both in coconut and African oil-palms, which might have been considered epidemic and therefore caused by a species of *Phytophthora*, but in all cases a consideration of other factors suggested a more probable explanation, and the possibility that further spread of the disease need not be anticipated.

A few extracts from Nowell's book under the heading of "'Bud-rot' in General" might be given here:

Nematode infestation has been recently found to be responsible for a large amount of so-called "bud-rot" occurring in British Honduras and may be expected with confidence to account for a good deal more elsewhere. The writer (Nowell) has found that typical "bud-rot" of this kind, indistinguishable in appearance from the infectious forms, follows rapidly on the death of healthy trees from poisoning or severance of the stem. The fact is, that a characteristic stinking rot, dominated by bacterial putrefaction, forms the natural process of decay of the large amount of tender tissues, deeply enclosed by the successive sheathing bases of the leaves....It is now becoming apparent that the true infectious "bud-rot" existing in the West Indies is of more than one kind.

"Bud-rot" in Malaya

The writer was brought into contact with the "bud-rot" problem owing to an attempt made to establish plantations of African oil-palms (*Elaeis Guinensis*) in Malaya during the last few years. These palms appear to suffer more severely from "bud-rot" than coconut palms. Young oil-palms, 2–3 years old, show a 10 % infection of a disease of the central leaves, the earlier stages of which might well be considered an incipient form of "bud-rot," while the older palms show numerous cases of typical "bud-rot."

The symptoms on the young oil-palms are well-marked; the third or fourth leaf from the centre of the crown is attacked towards the base and the leaf tissue at this spot becomes water-logged, soft and transparent looking, the leaf quickly falling over. The outer leaves, in succession, are attacked and fall over. The appearance is then a series of collapsed outer leaves, with the light-green, tender looking young central leaves standing erect in the middle. In all cases, recovery of the young palms has been complete, the central meristematic parts of the bud remaining untouched. This feature is of interest, for the lines followed in nature practically parallel those demonstrated by inoculation experiments with *Phytophthora sp.*, *i.e.* that rotting of the bud does not necessarily follow even a severe infection of the young, tender heart-leaves.

Little information could be found in the literature of palm diseases to help in the interpretation of the symptoms observed in the young oil-palms, but the writer's experience of coconut "bud-rot" led to the inception of experiments to test the significance of conclusions based on "stab inoculations" in the various researches published. Much importance has been assigned to the success of "stab inoculations" practically in all published work on "bud-rot" of palms, chiefly owing to the difficulty of obtaining definite results by other methods. The results of the experiments proved beyond doubt that numerous widely separated organisms, other than members of the genus *Phytophthora*, were capable of producing typical "bud-rot" symptoms when stab-inoculated directly into the cabbage or bud. The results provide strong indications that only carefully controlled inoculations conducted without injury to the bud could be of significance in interpreting experiments towards elucidating the true value of any organism as the causal agent in "bud-rot."

Phytophthora spp. *associated with "bud-rot" of palms*

Phytophthora omnivora var. *Arecae* has been shown to be associated with a disease of the crown of the areca palm (*Areca Catechu*), though principally a nut disease, in India.

Phytophthora palmivora (Butler) = *Pythium palmivorum* (Butler) has been connected with "bud-rot" of palmyra palms (*Borassus flabillifera*) in India. Successful experimental inoculations of the young tender heart-leaves were recorded. The same fungus was shown by Shaw and Sundararaman to be associated with coconut "bud-rot" in India. Similar inoculation results to Butler's were recorded, but under conditions of extreme humidity, the falling over of the central leaves was brought about in a seedling plant grown under a bell-jar.

Phytophthora faberi has been associated with coconut "bud-rot" in the Philippines by Prof. Reinking. His conclusions were based largely on stab inoculations.

Phytophthora palmivora has been shown to be associated with coconut "bud-rot" in the West Indies. Similar inoculation results to those obtained in India were recorded. Ashby specifically remarks that his artificially inoculated plants were not affected in the heart tissues.

Phytophthora parasitica has been shown to attack coconut palms in the West Indies but has not been associated with "bud-rot."

Factors associated with "bud-rot" in Malaya

There are no records of *Phytophthora sp.* being associated with "bud-rot" in Malaya. Among planters, this disease naturally attracts attention in its later stages even if only a single tree is killed, without any sign of attack on neighbouring trees. As with *Hemileia vastatrix*, the cause of the Coffee Blight in Ceylon, it serves as a standard instance of the ravages caused by epidemics of plant disease.

Factors definitely associated with severe attacks of "bud-rot" of coconut palms in Malaya are: (a) injury caused by attacks of Black Beetle (*Oryctes Rhinosceros*); (b) entry of Red-stripe Weevil (*Rhyncophorus Schach*) into the presumably previously injured portions of the cabbage; (c) tidal floodings in fields adjacent to rivers where sufficient protection has not been given to prevent the entry of water.

The above factors could not be confused with those which would have suggested epidemic "bud-rot" caused by a *Phytophthora sp.*, but further field investigations have brought to light factors which lead the writer to emphasise the quotations from Nowell's book given above. A disease of coconuts has been under observation in Malaya over the last two years which parallels the red-ring disease caused by nematode infection and described by Nowell. The symptoms in the last stages of decay of the crown are typical of "bud-rot" and the disease has been reported as such from all parts of the country where coconuts are grown. In a few cases, it was noticed that the central leaves were the last to collapse, a feature directly opposed to the usual "bud-rot" symptoms, in which the central leaves collapse before the outer leaves. Careful cutting of the whole length of the stem of such cases shows the base of the stem with the internal tissues discoloured, a beautiful salmon-pink with patches of yellowish discoloured tissue. The bud is quite healthy in these specimens. Apart from a narrow external ring of healthy tissue, the whole stem in transverse section is discoloured. In Nowell's red-ring, the discoloured tissue forms a circular zone enclosing healthy tissue. The absence of nematodes has been carefully checked and no sign of nematode infection has been observed. The discoloured tissue is found passing to varying heights in the stem but usually there is some healthy stem tissue intervening between the upper limit of discoloration and the crown, even when the whole of the leaves have collapsed and the bud completely rotted.

The facts suggest a gradual cutting down of the water supplies, owing to the entry of an organism through the roots which gradually passes up the stem. The symptoms have been well established in a severe outbreak, in which every tree in a two acre block of 22-years-old trees had to be cut out. Little has been done beyond field observations, but anatomical examinations of diseased tissue show a frequent occurrence of a coccus bacillus, while cells filled with packets of spores, suggesting algae rather than fungi, are fairly common. The symptoms described are invariably labelled "bud-rot" by the planter and until the occasional standing of the central leaves with healthy bud tissue had been observed, it would be difficult to contradict the suggestion.

Recently another variation of the above was discovered. The writer had successfully demonstrated the symptoms associated with the disease described above to a group of planters, when, walking from the main infection, a typical case of the erect central leaves with collapsed outer leaves was observed. The tree was cut down but there was no sign of discoloured tissue at the base of the stem. This was split from the base to the crown and for about 12 feet upwards the stem tissue was perfectly sound. About this height a typical bacterial soft-rot of the stem tissues became apparent and continued for 8 to 10 feet to within 9 inches of the crown; at this point there was a sharp line of demarcation between diseased and healthy tissue, and the stem tissue below the bud was still healthy. The bud showed no signs of attack. There is little doubt that if this case had not been examined until a few days later the bud would have been involved, and the verdict would have been typical "bud-rot," the organism passing downwards from the diseased bud into the stem. Bacteria were present in the affected tissues in enormous numbers, but proved difficult to culture. Morphological work shows a rod-like bacterium present in the cells of the ground tissue and in the intercellular spaces in large numbers, while there is no suggestion of fungus mycelium.

The facts elicited, coupled with Nowell's work in the West Indies, afforded strong reasons for a change of policy respecting the drastic treatment of cutting out and burning all palms suffering, or suspected to be suffering, from "bud-rot." Coconut palms are difficult to treat because of their height, but African oil-palms can be got at more easily by cutting away some of the outer leaves. A group of mature oil-palms was cut out after being badly attacked and replaced in the centre of the rows of healthy palms. Treatment consisted in scraping out all the diseased bud tissue as completely as possible, followed by placing a few handsful of salt in the cavity. Oil-palms readily re-root in presence of plentiful supplies of water, but the crux of the matter was whether new leaves would appear; presumably the whole of the meristematic tissue comprised in the bud had been destroyed owing to the attack and it is usually taken for granted that with the death of the bud, no more leaves can be produced. However, in every case new leaves were produced and the palms recovered, a feature observed in many cases in the coconut palms artificially infected in the experiments mentioned above, though oil-palms recover comparatively rapidly. This method of treatment of oil-palms has been carried on since, and only one case, out of many dozens, has succumbed, the rot travelling down and completely killing the stem.

From the applied standpoint, it appears to the writer that too much stress is placed upon the epidemic form of "bud-rot" associated with *Phytophthora spp.*, though during the last two or three years there has been more attention directed towards inoculation experiments conducted without injury to the bud. It is doubtful if a sharp distinction can be drawn between epidemic "bud-rot" associated with *Phytophthora spp.* and "bud-rot" due to other causes, for it is easy to imagine that with some appropriate change of conditions, the latter might cause much damage over large areas, instead of occurring sporadically as at present. Each addition to the list of agencies which cause symptoms simulating those described for epidemic "bud-rot" tends towards a decrease in confidence in the belief that a definite cause, or a cause which should have superior claims over others, has been established. Until exhaustive inoculation experiments without injury to the bud have been successfully carried out to prove undoubted rotting of the bud tissues, there appears little reason why we should accept the position usually taken up in connection with this problem. Much of the significance for the present attitude might be discounted did the fungi connected with the problem belong to a genus other than that of *Phytophthora*, but knowing the potentialities of the members of this genus, it is "playing safe" to exaggerate the possible effects, in the absence of exact knowledge. The writer is of the opinion that it would be advisable to consider rotting of the bud as a purely secondary phenomenon, following on attacks from widely different organisms, or resulting from changes in physiological conditions, small in themselves but which may release large potentialities. The acceptance of this view would lead to a careful investigation of points which might be otherwise neglected and which might be of much significance in a correct understanding of the present complicated position, besides being of much importance in the future development of industries dependent on successful palm cultivation.

Careful comparative inoculation experiments with *Phytophthora faberi* obtained from Prof. Reinking and *Phytophthora palmivora* obtained from Dr Butler have been carried out in Malaya, but no positive results have been obtained.

Prof. S. F. Ashby. Bud-rots of the Coconut Palm in the West Indies

In a number of palms all the leaf-stalks pass downward into thin closed sheaths forming together a rigid cylindrical column several feet long. This is the case in the palmyra and the West Indian royal and cabbage palms (*Oreodoxa regia* and *O. oleracea*). In the coconut palm the column is confined to a restricted number of leaf-stalks in the centre of the crown, the closed sheathing portion being a thin fibrous lateral extension of the stout stalks above their enlarged bases. As the fibrous sheath or strainer ruptures the leaf-stalks bend away from the vertical and cease to be in close contact. Low down in the middle of the central column or central shoot is the bud consisting of the massive meristematic tender tissue at the apex of the stem surmounted by the primordial leaves and known commonly as the cabbage. In the West Indies the term "bud-rot" is taken to mean a rot of the tender non-fibrous tissue at the apex of the stem; the growing-point being destroyed, further growth ceases and the tree dies. In order that the well-protected bud may be reached by a parasite it must either penetrate the central shoot laterally through the enveloping leaf-stalks or vertically downward by way of the folded central leaves or work upwards through the fibrous tissue of the stem. A specific primary bud-rot cannot therefore be recognised and it would appear equally justifiable to describe red-ring as bud-rot as the so-called bacterial disease in both of which marked symptoms are present in the outer crown before the bud is destroyed. From the standpoint of field observation the nearest approach to a specific bud-rot may be seen in the *Phytophthora* rot present in Jamaica and undoubtedly elsewhere in the West Indian region. Usually the first symptom noticed is the dead broken end of the youngest visible leaf; this leaf can be drawn out and shows a soft malodorous rot where the stalk has ruptured. If the tree is cut down and the central shoot split longitudinally, the bud will be found fully destroyed forming a pasty or semi-liquid mass. As a rule even when the youngest leaf or leaves are wilting and turning pale the bud will be found fully soft-rotted. On a few occasions, however, the bud has been found in a wilted but not disintegrated condition, its tissue being fully invaded by the mycelium of a *Phytophthora* with a soft-rot beginning at a few points on the surface.

In most instances where the bud was already fully soft-rotted spots were present going through the thick and thin parts of the successive leaf-stalks in the central shoot: the internal tissue of the spots was penetrated by a *Phytophthora* and felts of its mycelium were frequently developed on their faces; it was isolated (1) from the internal tissues and got into pure culture. Less frequently infection occurred also on the folded leaves within the shoot, resulting in the appearance of rows of spots on the pinnae when the leaves unfolded; the same fungus was isolated from webs on the inner faces of these spots. Observation showed that during the rainy season this symptom of leaf-spotting was often followed by bud-rot. In some instances the spots ceased to appear and the trees showed no further effects; this was the case when zoospore suspensions from pure cultures were poured into the hearts of young vigorous trees about four years of age; three or four successive leaves emerging from the shoot showed rows of spots, but later leaves were free from them and the trees were still healthy a year later. An inoculation experiment showed that mycelium from pure culture could penetrate an uninjured young leaf-stalk, and when a zoospore suspension (2) was introduced into the shoot of a seedling the bud was found rotted by the fungus at the end of a month.

The disease appears to be endemic in Jamaica in districts where there are considerable areas in coconuts, as most large plantations lose a few trees from it every year; during periods when rainfall exceeds the average, cases increase. In situations subject to flooding and difficult to drain, protracted rains have been followed after an interval of one to two months by serious losses in trees from two to twenty years of age. It is not believed that *Phytophthora* bud-rot is restricted in the West Indies to Jamaica; a mycelium of similar character has been found in tissues of specimens from cases of bud-rot in St Lucia, the Virgin Islands and Trinidad, and a *Phytophthora* has been isolated recently from bud-rot in Porto Rico. Reinking's[3] observations in the Philippines show how easily the fungus infection can be masked by secondary bacterial soft-rot, in producing which species closely related to or identical with *B. coli* appear to be active.

The symptoms of this bud-rot in Jamaica are in agreement with the *Phytophthora* disease of coconuts in India[4,5] and the Philippines[3], and the fungus seems to be the same species. Although[2] the *Phytophthora* of pod-rot and canker of cacao in the West Indies appears to be the same species as that on the coconut, it is not believed to be the same strain, as the form from the palm has not been found able to rot cacao pods, and oospores[6] hitherto have been developed only by the strain from cacao.

Some years ago a disease showing different initial symptoms but terminating in a soft-rot of the bud received much attention in Cuba. The first indication was the shedding of young nuts followed by dark discoloration of successive flower-spikes as they emerged from the swords. The outer leaves turned yellow, died back and gave way at the base; finally the youngest visible leaves withered and growth ceased. Examination showed a wet rot near the bases of the opened and unopened swords, a decay at the bases of the leaves, a wet rot of the leaf-stalks in the central shoot and a soft-rot of the bud.

This succession of symptoms was seen mostly in large trees with trunks up to 60–80 ft. in length; in younger trees the failure of the heart-leaf was usually the first visible symptom. In the Baracoa district, a coastal strip about 30 miles long in N.E. Cuba, the losses were especially heavy during a period of about ten years. In 1912, J. R. Johnston[7] recorded his observations based mainly on four years' study of the disease in this district, but also on visits to other parts of the West Indies. He found *B. coli* and related bacteria regularly in the rotted tissues and succeeded in setting up a soft-rot by boring holes through the central shoot into the bud or adjacent leaf-stalks and injecting suspensions of pure cultures, anticipating Reinking's similar methods with similar bacteria on the bud-rot in the Philippines, which he recognised later to be due in the first instance to a *Phytophthora*[1]. After 1912 the author in Jamaica saw large trees dying off with symptoms like those

---

[1] Johnston also found *Diplodia*, *Pestalozzia* and an unnamed fungus, evidently *Thielaviopsis*, frequently present in the central shoot, and Fredholm attributed a disease with similar symptoms in Trinidad to *Diplodia*.

of the Cuban bud-rot. There were occasional cases in the same spots
where trees had been recorded as dying ten years earlier. They could
be interpreted as cases of premature senescence under unfavourable
external conditions. Coconut palms are now known to die off with quite
similar symptoms in red-ring disease, trunk rot, one or more ill-defined
root troubles and severe drought.

It is difficult to estimate the significance of the external factors in the
Cuban disease as Johnston's report is very bare of ecological data and
his tendency to regard any condition of disease terminating in a soft-rot
of the bud as due to a specific bacterium increases the difficulty. Some
plantations in the affected district, stated to be well cared for, were
practically free from the disease, but the usual condition was one of
neglect. According to Horne[8] the disease progressed during a long
period of severe drought which killed off numbers of the resistant royal
palms.

The etiology of this disease must be regarded as still obscure; if a
specific bacterial rot of the coconut palm exists it seems doubtful if the
causative agent has been found. A bacterial bud-rot of canna has been
described by Miss M. K. Bryan[9]; the organism was able to infect uninjured rolled leaves in the shoot though the stomata and its cultural
characters showed it to be unrelated to *B. coli*. Two clearly recognisable
diseases of the central shoot which do not as a rule terminate in a rot
of the bud, occur in the West Indies. These are the *Thielaviopsis*
leaf-rot[10] and the little-leaf disease[10].

Thielaviopsis paradoxa is the cause of large wedge-shaped spots with
broad dark margins which run across the folded segments of the young
leaves in the shoot before they are pushed up into the light: as the leaves
open out the mid-ribs of the pinnae break at the diseased places and the
terminal parts wither or tear away, giving the crown a ragged appearance. The fungus is invariably present in the rotted spots and sporulates
freely on and in the tissue. The disease was epidemic in Jamaica for
some years on young bearing trees from 10–15 years of age in a coastal
belt about 40 miles long, having a high to moderate rainfall. Natural
recovery occurred, but many cases were cured by cutting back the
central shoot below the spotted region and applying a dry mixture of
lime and sulphate of copper to the severed surface. The strain of
Thielaviopsis present was the same as that causing the rot of cane cuttings,
pineapples and banana suckers; it differs from the saprophytic strain
common on recently cut surfaces of coconut trees in not developing
coremia. The disease has been recorded also from Porto Rico[11].

In little-leaf disease the central shoot sends out a succession of crumpled and distorted leaves which become progressively smaller until they are completely dwarfed; the upper surfaces of the leaf-stalks show raised brown stripes and patches which become woody and cracked. The rigidity imparted to the central shoot by these indurated leaf-stalks is apparently the cause of the crumpling and, in part, of the distortion of the young leaves. It is not clear how it can be the cause of the dwarfing. Spontaneous recovery not infrequently occurs, or occasionally the bud may wither up or soft-rot. Recovery is much facilitated by slitting the strainers of the central leaves and pouring into the heart a diluted Jeyes fluid or tobacco extract. The cause of this peculiar condition is obscure. F. S. Earle[12], who studied little-leaf in Cuba, concluded (in 1912) that it was caused by an attack of aphis (*Certaphis latania* Bdv. Per.) on the very young leaves in the shoot. The aphis was tended by ants which probably were responsible for transporting it. The dwarfing of the leaves would on this view be a case of stigmonose; it would also account for the efficacy of contact insecticides if they are an essential part of the treatment. Little-leaf occurs in Cuba, Jamaica, Trinidad, Tobago and Grenada and doubtless in other parts of the West Indian region.

A complication in which this disease was associated with beetle attack resulting in considerable rotting of the central shoot has been observed in Grenada. Some trees which were recovering naturally showed the development of a shoot having peculiarities resembling those described by Sharples in cases of beetle attack followed by central shoot rot in Malaya[13].

In addition to the bud-rot caused by *Phytophthora palmivora*, another species[1] of that genus has been found in the crowns of coconut palms in Jamaica. In this disease infection seems to start on the thick bases of leaf-stalks commencing to separate from the central shoot. Penetration progresses slowly so that the affected leaf has reached an intermediate position between the shoot and the outer fringe of leaves before it begins to turn yellow at the tip and gradually dry up. Large spots up to 3 inches in diameter or elongated patches are present on the stalk several inches above its attachment to the stem. The *Phytophthora* can be isolated at once in pure culture from the interior of these spots as the tissue is usually free from bacteria. Several leaves may become infected apparently independently as neither sporangia nor mycelium has been found on the surface of the spots. Bunches of fruit in the axils of these leaves are liable to drop the nuts before they are full grown owing to loss of support. This attack has not been found to result in bud-rot as the central shoot remains healthy. The fungus forms oospores in culture and agrees closely with *P. parasitica* Dastur. It was found also in Jamaica in a leaf-base rot of pineapple plants, a foot and stem-rot of tobacco, and fruit rots of egg plant and tomato, and causes in the Lesser Antilles a boll-rot of cotton. It was present in leaf-stalk rot of coconut palms with trunks over 20 ft. long.

In Jamaica it has been customary for many years to cut down trees which have died or appeared to be dying with symptoms resembling those of the Cuban disease and to burn, at least, the central shoots. Since 1911 the treatment has been compulsory, and was later made to cover cases of *Phytophthora* bud-rot. In the earlier years there can be no doubt that many trees so treated were affected by the curable *Thielaviopsis* leaf-rot and little-leaf diseases as at that time almost any symptom of disease showing up in the crown was attributed to an infectious bud-rot. The procedure seems justified in cases of *Phytophthora* bud-rot in view of the evidence as to its infectious character obtained in India and should hold good for the complete destruction by fire of red-ring cases where that disease exists. In cases where the failure is of the Cuban type the presence of an infectious disease has been assumed without good grounds and what evidence there is as to the effect of this treatment on control seems to be of a negative character (14).

Literature cited

(1) ASHBY, S. F. "Notes on two diseases of the Coconut Palm in Jamaica caused by fungi of the genus *Phytophthora*." *West Indian Bulletin*, XVIII, 1 and 2.
(2) —— *Agr. News* (Barbados), XX.
(3) REINKING, O. A. "Philippine Economic Plant Diseases." *Philip. Journ. Science*, XVIII, Sect. A, 1918; also *Philip. Journ. Science*, XIV, No. 1, 1919; also *Journ. Agric. Research*, XXV, No. 6, 1923.
(4) BUTLER, E. J. "The Bud rot of Palms in India." *Mem. Dept. Agric. India*, Bot. Sect. III, No. 5, 1910.
(5) McRAE, W. *Mem. Dept. Agric. India*, Bot. Sect. XII, No. 11, 1923.
(6) ASHBY, S. F. "Oospores in cultures of *Phytophthora faberi*." *Kew Bull.*
(7) JOHNSTON, J. R. "History and cause of the Coconut Budrot." *U.S. Dept. Agric. Bur. Plant Ind. Bull.* 228.
(8) HORNE, W. T. *Est. Agronom. Cuba, Bull.* 15, 1908.
(9) BRYAN, M. K. "A bacterial budrot of cannas." *Journ. Agric. Research*, XXI, No. 3, 1921.
(10) NOWELL, W. *Diseases of Crop Plants in the Lesser Antilles*, London, 1923.
(11) FAWCETT, W. S. *Ann. Rep. Porto Rico Agr. Exp. Sta.* 1910.
(12) EARLE, F. S. *Cuba Magazine*, III, No. 10, 1912; also (8).
(13) SHARPLES, A. and LAMBOURNE, J. "Observations in Malaya on Budrot of Coconuts.' *Ann. Bot.* XXXVI, Jan. 1922.
(14) STAHEL, G. *Dept. Agr. Surinam Verslag*, 1922–23.

Mr W. J. Dowson. Some Observations on the Bud-rot Disease of Coconut Palms on the East Coast of Africa.

This brief paper embodies some observations made and recorded during the years 1913 to 1920 in the vicinity of Mombasa on plantations of coconut palms laid out under European supervision. Many of these were examined on account of the presence of bud-rot disease, but two, in particular, were visited a considerable number of times. One of these lay 30 miles to the north of Mombasa, about half-a-mile from the sea, and the other some 30 miles to the south, and about the same distance from the sea. The first contained palms mostly of local origin, but a few acres had been laid out with palms raised from nuts imported from Ceylon. The second had been planted up with local African and with Zanzibar nuts.

The cultivation of both plantations can only be described as poor, as the palms were growing in rank grass and weeds, chief of which was the doum palm (*Hyphaene thebaica*). Occasionally these were cleared and burnt as labour became available, but cultivating was hardly ever done regularly. Under such conditions palms generally commenced to bear nuts in their seventh year, and it was just at this period that the disease made its appearance with devastating results. Older palms in full bearing, in either European or native owned plantations, were not attacked so far as we know. An early and striking symptom of the disease, obvious at a distance, was the bending downwards and yellowing of the outermost (and therefore the oldest) leaves. The appearance of the rest of the foliage was normal, with the exception of the innermost unexpanded frond or spear, the exposed portion of which was dried up, brown in colour, and dead. The concealed part, covered by leaf bases, exhibited various stages of rotting.

This part was soft and not firm, slimy, and when removed emitted a most offensive odour, indicating the presence of putrefactive bacteria. No spotting was observed on the leaf bases such as occurs in the Indian disease investigated by Butler[1], although the basal parts of the yellowing leaves were discoloured by large brown irregularly shaped blotches. These did not penetrate any depth into the tissues beneath. The facilities for a careful microscopical examination were insufficient to permit of a detailed study being attempted, but bacteria were found in the neighbourhood of the advancing edge of the rot, and copious mycelium was present in the more distal portions of the central unexpanded frond. Roots and stems, so far as could be ascertained, appeared healthy, and at the time I was of the opinion that infection had commenced some-where on the youngest leaf, and had travelled downwards.

Apart from the examination of individual trees, data were collected in connection with the incidence of the disease which seem worth recording, particularly as they indicate the lines along which the control of the disease should be followed on the east coast of Africa.

Three observations were made which impressed me very much at the time as being connected with the occurrence and severity of the disease. One was the striking fact that whereas the doum palm grew nearly everywhere along the coastal belt, the coconut palm did not, but occurred only in patches; some of these palmless areas which I examined were certainly swampy, but whatever the cause may be, the local Swahili natives asserted that they or their forefathers had tried to grow palms on these vacant spots, but always without success. Now it is largely these areas which are acquired by the European planters, and this brings me to my second observation.

In a number of instances which came under my notice, immature nuts had been planted, a practice which cannot be too strongly condemned. The planting out of young palms raised from immature nuts into swampy ground generally resulted in other diseases besides rot, such as foot-and root-rot, becoming prevalent.

The third and most important point was the effect of sanitation and drainage. On both the plantations mentioned, the manager's house and the huts of the native labourers were, for reasons of health, placed upon the highest and therefore the best drained ground. Palms planted round about the houses and between the huts were much larger, more vigorous and altogether better looking than any others planted at the same time. The former were *always* kept clean, in fact the ground was practically bare of other vegetation; the latter, as recorded above, were growing in rank grass. Furthermore, the former commenced to bear in the fifth year, whereas the latter never showed signs of bearing before the seventh year after planting. Bud-rot could not be found and had never been observed in the more vigorous palms about the houses and in the native encampment, whereas but a short distance away the disease was rife and the palms of Ceylon origin were without doubt the most susceptible.

From these facts, the conclusion may be drawn, that from every point of view, the planting of smaller areas and keeping them constantly clear of weeds is more profitable than laying out very large, sometimes swampy areas which cannot be kept properly cultivated for want of labour, and planting them up with immature nuts.

In conclusion, I would like to direct attention to the similarity between the bud-rot disease of coconuts in East Africa and the Panama disease of bananas, so far as outward symptoms go. In both diseases one of the earliest noticeable signs is the drooping and yellowing of the outer leaves, which in the banana disease Brandes(2) has shown is probably due to the effects of toxic substances secreted by *Fusarium cubense* at the base of the plants.

A similar thing may happen in the East African bud-rot, and a thorough examination for parasitic fungi in the stems and roots of affected palms is much to be desired as far as the east coast of Africa is concerned. At the same time, it is usually unsafe to argue from analogy, particularly in view of Dixon's(3) work on the *descent* of sap, indications of which have also been observed by Brooks(4) and myself(5). The yellowing of the outer leaves in the East African disease *may* be due to the downward movement of toxic substances emanating from an upper and more central seat of infection such as the youngest folded leaf.

References

(1) BUTLER, E. J. "Coconut Bud Rot." *Mem. Dept. Agric. India*, 1909.
(2) BRANDES. "Banana Wilt." *Phytopath.* vol. IX, 1919.
(3) DIXON, H. H. "Transport of Organic Substances in Plants." *Brit. Ass. Report* (Hull), 1922.
(4) BROOKS, F. T. "Silver-Leaf Disease. IV." *Journ. Pom. and Hort. Sci.* vol. III, 1923.
(5) DOWSON, W. J. "The Wilt Disease of Michaelmas Daisies." *Journ. Roy. Hort. Soc.* vol. XLVIII, 1923.

Mr W. Nowell. COCONUT BUD-ROT IN TRINIDAD

About fifteen years ago the U.S. Department of Agriculture issued a bulletin describing at considerable length the researches of Dr J. R. Johnston into the nature and origin of coconut bud-rot as it had appeared in severely epidemic form in Cuba. The disease was attributed to infection of the crown with *Bacillus coli* or an organism closely resembling it. This conclusion has been generally accepted as applicable to losses of the general type described by Johnston taking place in other parts of tropical America and the West Indies.

The position in Trinidad in recent years has been that after removing from the category of bud-rot the cases of the now well defined red-ring disease due to a nematode, there remains a very considerable amount of disease of which the symptoms have a strong resemblance to those described by Johnston. Loss of this kind has a scattered general incidence and in several areas has an intense local incidence amounting to an epidemic. Experience so far goes to show that with most of the scattered cases the usual remedy of cutting out the affected trees as soon as they appear is effective, but, in the special areas mentioned, has no visible effect on the steady march of the disease.

It is more than probable that the difference in incidence is associated with differences in causation, and it is in the scattered type that evidence, as yet indefinite, has been seen of the occurrence of *Phytophthora* bud-rot as described by Ashby in Jamaica. It is more or less characteristic of this type that rotting of the bud occurs before the outer leaves are visibly affected.

The epidemic form has been subjected to careful investigation for the last three or four years, with results almost entirely negative. All attempts at artificial transmission have failed, whether made by means of cultures of bacteria taken from freshly invaded bud material or by the application of small or large amounts of the affected tissues. These have been made on trees under varying conditions, including sections of the same plantation not reached by the disease but certain to be involved later. Transfer of decaying roots has similarly failed.

It has been established by cutting trees not visibly affected but near the margin of advance that the first visible symptom is the browning of the unopened inflorescence, followed by the shedding of young nuts and the yellowing of the outer leaves. In these early stages the central bud is found to be still quite sound and showing no sign of infection, although the tree is already certainly doomed. As the yellowing of the leaves progresses the bud is invaded by bacterial rot and death ensues.

Our researches so far have brought us to the conclusion that in this type of failure, as has been shown conclusively in red-ring disease, the failure of the bud is secondary and the term bud-rot a misnomer.

There is so close a resemblance in the incidence and symptoms of this affection to those of the Panama disease of bananas that of late I have been led to use for it the term coconut wilt to get away from the implications of the use of the word bud-rot. It is exceedingly interesting that as a result of Mr Dowson's investigations in East Africa he has anticipated me by a few minutes in introducing the term to this conference. The conclusions so far reached by Mr Dowson, by Mr Sharples in Malaya, and by my colleague Mr Snell and myself in Trinidad will be seen to have a very considerable and probably significant degree of similarity.

Mr A. Sharples. BROWN BAST DISEASE OF RUBBER TREES

Brown bast is the name given to a disease affecting the cortical tissues of the Para rubber tree. The common manifestation appears on the tapping cut, though it may make its appearance at places where large branches are broken by the wind or at the base of the stem where the bark has been eaten and injured by the gnawing of animals. On the tapping cut, the first undoubted symptom is the cessation of the latex flow from localised portions of the cortical tissues. These "dry" areas, on examination, are found to be discoloured, usually a brownish colour with variations towards a sepia colour, and when a substantial area of the cortex is suffering, the affected tissues are found to be waterlogged. The affected tissues are not killed, but active secondary cambiums are formed in them; as a result of the meristematic activity of these extra cortical cambiums, nodules or "burrs" appear on the tapping surfaces, rendering them uneven and incapable of being tapped.

Occurrence of Brown Bast

This affection, first definitely separated from other bark diseases of *Hevea brasiliensis* about 1917, has been recorded in all rubber-growing countries. During the years 1917–20 the disease became so prevalent as to cause much concern. During the last two years, since restricted tapping has been the rule, the affection has not caused planters much trouble, and as a practical method of control, suitable for present-day practice, has been evolved, less attention has been paid to the disease.

Brown Bast in relation to other Bark Diseases

In 1914, this disease was lumped together with other bark diseases of *Hevea brasiliensis*, and all were supposed to be the result of the attacks of *Phytophthora faberi* (Maubl.). This position resulted from the investigations of Rutgers in Java, following on Petch's work in Ceylon; the latter showed the connection between the fungus demonstrated by Rorer as causing cocoa canker in Trinidad, and that causing a bark disease of *Hevea brasiliensis*.

Under four headings Rutgers included:

(a) Black stripe disease of the tapping cut.
(b) Claret coloured bark canker.
(c) Brown bast in the early stages.
(d) Burrs, the final development of brown bast.

II–2

(*a*) and (*b*) are now known to be caused by different species of the genus *Phytophthora*, while the credit for classing brown bast as a separate disease belongs to Pratt, who was working in Sumatra.

Cause of Brown Bast

The early investigations into the cause of brown bast were directed towards searching for a causal organism, fungal or bacterial. These investigations resulted in no definite finding. Attempts to artificially transmit the disease from tree to tree were next made. Exhaustive efforts met with no success, so Belgrave suggested that the affection should be considered a "physiological disease of non-organic origin." This view obtained substantial support as further work was carried out, but Keuchenius, working in Sumatra, always strongly supported a bacterial cause. His own inoculation results do not lend much support to a bacterial cause, and the extra evidence he adduces is purely negative, being based on certain features which he claims cannot well be accounted for by a physiological theory. The writer is at variance on these special points, and is definitely of the opinion that most of these features can be better accounted for by a physiological theory than by assuming a bacterial origin. After three years' experimental work in the field, the view is advanced that the affection is a purely physiological disturbance resulting from the extraction of excessive quantities of latex.

Tapping Experiments relating to Brown Bast

Rands, in Java, was the first investigator to establish experimentally the connection between brown bast and heavy tapping. His experiments were similar to those carried out in Malaya, and his results practically identical. The heavy tapping experiments he conducted were limited because of the short period of heavy experimental tapping, but were of importance in so far as they demonstrated a method for controlled experiment. The experiments conducted by the writer and Mr J. Lambourne over the last four years have been based on Rands' methods, and much significant evidence has been obtained.

The only sound view to adopt when inoculation experiments designed to establish an organic cause prove a failure is that of a "physiological disease of non-organic origin." Coupled with this was Rands' successful demonstration of the association between heavy tapping and brown bast, which made the foundation for future work a fairly solid one, not only for obtaining evidence regarding the probable influences resulting in the inception of the disease, but for testing the possible effects likely to accrue from the efforts being made to establish permanent, high-yielding trees by propagative methods. During this period of restriction the latter is of far greater importance than the former, for future improvements on plantations appear to lie in a reduction of costs of production, while yields may be successfully maintained or increased by establishing stands of high-yielding trees, with a considerable reduction in the number of trees per acre. Thus if brown bast can be shown to be a reaction following on the extraction of excessive quantities of latex, it may be expected to prove a most important factor when future developments of trees giving high yields are to be considered.

Possible Influences operating on the inception of Brown Bast

Failing to establish an organic cause, the investigation and correlation of factors influencing brown bast was of first importance. These influences were:

    (a) Excessive withdrawal of latex.
    (b) Wounding during the tapping operation.
    (c) Dry and rainy periods.
    (d) Overcrowding.
    (e) Soil conditions.

Up to 1920 there was little definite information upon which to base any conclusions, but it was generally held that the trees yielding most latex were the first to be attacked. As yields appreciably increase during a change from a dry to a rainy period, brown bast might be expected to be more prevalent during a rainy period. Badly developed trees grown under poor soil conditions appear relatively hardy; again the yield factor supplies the probable explanation, as yields from trees grown under unfavourable conditions are comparatively poor.

  The significant investigation was the separation of (a) and (b) influences. This separation is difficult to negotiate in practice, but we have obtained a large amount of indirect evidence to indicate that the wounding influence is comparatively of small importance. In one of our experimental plots, where trials were conducted over an extended period, there was a decided correlation between increase in percentage number of brown bast cases and a sudden increase in yield. In the first two years there appeared to be some correlation between increase in percentage number of brown bast cases and rainfall, but in the third year the increase in percentage number of brown bast cases and yield took place during a dry period.

Special features noted in our experiments were:

(*a*) There appears to be a fairly definite correlation between yield and percentage development of brown bast, without any reference to other influences.

(*b*) A brown bast phase may be divided into a resting and developmental period, *i.e.* a sudden rise period follows on a period when practically no brown bast cases are developed.

(*c*) The sudden-rise periods are so noteworthy that they can be validly classed under the heading of "trigger action" phenomena. A plot of trees under experiment would not show any increase in number of brown bast cases over a period of three to four months, then in a day or two a rise of from 10 to 50 % would be recorded.

(*d*) The passage of brown bast developed on the tapping cut, downwards, is localised in a large percentage of cases when it arrives at a panel of bark of apparently greater age. Further, in a large percentage of cases, if the affection is noted in the very earliest stages, no progress is made if tapping is stopped. With reference to this last observation, it may be stated that this will not be likely to prove of importance in the control of brown bast on the plantations, for the careful and skilled supervision necessary cannot be expected when large areas are under the control of a single European.

Information has been gained respecting Rands' endeavours to develop immune trees by selecting good yielding trees specially resistant to brown bast as judged by heavy tapping, The writer was of the opinion that immediate conditions might influence this matter to a considerable degree, and that with a slight change of conditions in other directions the so-called resistant trees might prove liable to attack. In this connection, a comparatively good yielding tree, which had previously suffered from an attack of brown bast, but had recovered, remained free from the affection over a twelve months' heavy tapping period, while trees giving less latex and not having suffered from brown bast developed the affection. Further, we have shown that plots specially resistant at one period, later behave normally as regards the appearance of the disease. Such observations undermine the possibility of positive results being gained along the lines indicated.

Under controlled experiments, the results obtained lend colour to the view that the various features emphasised as a result of general observations under routine tapping conditions, cannot be trusted to any extent. To test the connection between good yielders and brown bast, towards which most observers have given support, two plots of twenty-five trees, one of good and one of bad yielders as judged by yields under routine tapping, were selected. These trees were then put under brown bast experimental conditions, *i.e.* one cut on a full spiral daily, and yield records taken for a month. At the end of this period, seventeen trees in the good yielding plot were definitely high yielders, and sixteen of the bad yielders were definitely bad. Eight of the good yielding plot, and nine of the bad yielding plot formed a middle class which fell between the above two classes. The final results in the good and bad classes were quite definite; in point of time, over a seven months period, the good yielders showed a much larger number of brown bast cases than the poor yielders, but at the end of twelve months the poor yielding plot showed as many cases of brown bast as the good yielding plot. Susceptibility as applied to good yielders and immunity as applied to bad, are terms which cannot be used happily in the face of such results. The fact that poor yielding trees become affected with brown bast may be considered an argument against an exhaustion hypothesis, but in our experiments it was a common feature that trees about to become affected with brown bast showed either a sudden increase or decrease in yield. Extremely high yielding trees are the first to be attacked, while those that remain unaffected under extreme tapping show a steady yield. The truth appears to be that any tree subject to sudden increase or decrease in yield is liable to an early attack of brown bast, and poor yielders at any particular period may have been good yielders at some time in their previous history. It is extremely probable that the rhythm noticeable in most plant functions is a feature in the tapping of *Hevea brasiliensis* under extreme conditions.

The writer considers that brown bast is a disease of purely "physiological origin," depending mainly on the question of water relations in the plant tissues. Many investigators have suggested an "exhaustion hypothesis," with leanings towards the importance of the extraction of elaborated food materials. Now, if equivalent amounts of brown bast bark and healthy bark are ground up and extracted as thoroughly as possible with petrol, it will be found that there is no significant difference in the amount of extracted rubber; in many cases more rubber can be extracted from brown bast bark than from healthy bark. Thus, as far as caoutchouc content, which is far and away the largest solid constituent of *Hevea* latex, is concerned, there seems no reason to assume that absence of elaborated food materials is of primary importance. The only other large constituent of latex is water, and the result of the disorganisation of water supplies to plant tissues is well known in plant pathology. The latex vessels are rich in osmotically active materials, and doubtless draw rapidly upon the supplies of water transmitted through the neighbouring cortical tissues. In the natural state the water relations between the varied parts of the cortical tissues are balanced, and under light tapping conditions this balance is not seriously affected, as the percentage content of solid and liquid in the extracted latex remains fairly constant. Under a system of heavy tapping the percentage caoutchouc content becomes considerably lower or remains low over long periods, with a resulting increase in percentage liquid content left after coagulation. Thus, under heavy tapping, extra water is extracted because of increased yields of latex, and the position is rendered more critical because the latex extracted contains more water in percentage volume than a similar amount extracted under lighter tapping operations. The greater strain involved in supplying such large extra demands for water must lead to a pathological condition in the transmitting tissues most immediately concerned, and as a result brown bast symptoms appear in the cortical tissues in the neighbourhood of the tapping cut.

The line taken throughout our investigations has been with reference to the disease factor influencing future policy in Malayan plantations. Much publicity recently has been given in the local press to the enthusiastic utterances of supporters of the policy of bud-grafting with a view to increasing the yielding power of individual trees. The claim has been raised of a possible 1500 lbs. per acre from bud-grafted trees as against an average of 500 lbs. per acre for Malayan estates planted in the ordinary way.

This pleasant picture has a reverse side which is not often turned to the light. Up-to-date, the rubber plantations of the Middle East have been fortunate with regard to disease, more especially as no serious leaf disease has yet been found. This indicates that the leaves of *Hevea brasiliensis* in Malaya are well adapted to resist attacks by parasitic organisms.

Our brown bast investigations show that excessive yielding will encourage pathological conditions. It is most certain that a tree of *Hevea brasiliensis* bred from a comparatively high yielder, but expected to be considerably better than the parent from the point of view of yield, will show physiological activities of a different order, and this change in physiological activity might result in important changes of leaf structure. Positive evidence to this effect has already been obtained in Malaya. On one estate bud-grafts were successfully established on three-year-old stocks. As is well recognised, root development is closely correlated with leaf development, and when the stock was topped the only draw on the nutrient supply from the roots was from the few dozen leaves of the bud-graft, which had to accept an excessive supply of nutriment. The result was a development of thin, flaccid leaves on the bud-grafts, which were being rapidly attacked and destroyed by several species of fungi, usually saprophytic.

Some change in physiological activities is a corollary of increased yields in bud-grafts, if the optimistic yields suggested materialise. A slight deficiency in development of leaf cuticle may bring momentous changes, as the parasitic alga, *Cephaleuros mycoidea* (Karsten), which is responsible for serious losses on tea, cloves and pepper, is commonly present as an epiphyte on rubber leaves, and it is only the strongly developed cuticle which prevents the entry of this organism into the tissues. The writer recently investigated the cause of the "black fruit" disease of pepper in Sarawak which has caused a decrease of revenue from $2,733,301 in 1903 to $712,122 in 1920, and found the cause to be this parasitic alga. Further, it is an established fact that rubber growing has been rendered unprofitable in British and Dutch Guiana owing to the attacks of a leaf fungus.

Apart from the brown bast factor, which is definitely connected with excessive yields, it would appear that the risks attendant upon developing high-yielding strains of *Hevea brasiliensis* are sufficiently great as to warrant excessive caution. It may be mentioned that the high-yielding "mother trees," from which the original bud-grafting experiments in Java started, have all developed brown bast since. Big claims have been advanced regarding increased yields as a result of these Javan experiments, but little attention has been directed towards the possibility of transmitting the disease factor, which might effectively put paid to any claims regarding increased yields.

One point remains to be mentioned as a result of Farmer and Horne's publications of the occurrence of degenerate "bast" elements in the phloem of *Hevea brasiliensis* affected by brown bast. As a result, brown bast is quite commonly regarded amongst European plant pathologists as a form of phloem necrosis, similar to that demonstrated by Quanjer in potatoes. The writer has been in close contact with this phase of the brown bast problem since 1920 and can state with confidence that such a conception is entirely at variance with the facts. Lignification of the phloem elements, as demonstrated by the usual reactions for lignin, is a common feature in the renewed bark of healthy trees of *Hevea brasili-*

ensis, which could not be suspected of brown bast. These variations in the phloem elements seem to be quite casual and can have no connection with the disease.

Dr A. S. Horne. FURTHER OBSERVATIONS ON PHLOEM NECROSIS (BROWN BAST DISEASE) IN HEVEA BRASILIENSIS

Since the publication of a preliminary paper on phloem necrosis in *Hevea brasiliensis* by the author(3) the work has received a certain amount of criticism which has emanated chiefly from workers and investigators who possess an intimate knowledge of the conditions under which the rubber tree is cultivated, and the diseases to which it is subject in tropical plantations, and whose opinions are for this reason entitled to consideration. It can be shown, however, that the criticisms are in the main founded upon misapprehensions as to the precise nature and significance of the special anatomical features briefly described in that paper for the first time.

It is believed that there are four counts in all which need to be dealt with by argument:

(a) As to whether the microscopical phenomena observed in the phloem of tissues affected with brown bast have been correctly interpreted.

(b) Do these phenomena bear a definite relation to the brown bast disease?

(c) Is the use of the name phloem necrosis for the disease justifiable?

(d) Are the deductions made by the author relevant?

(a) *Normal phloem.* The phloem of a fairly normally developed *Hevea* tree about six years old contains a number of concentric cylinders of laticiferous vessels separated from one another by tissue zones of varying width. These tissues are intersected by numerous medullary rays: the portions situated between the laticiferous vessels and medullary rays consisting of sieve tubes with their companion cells and phloem parenchyma. In the younger phloem the sieve tubes form the dominant feature, the parenchymatous cells are narrow and less prominent, but in the older phloem the position is reversed, the phloem parenchyma is prominently developed and the sieve tubes much less conspicuous. Scattered throughout the phloem occur lenticular pockets of stone-cells which vary in number and size according to the type of tree examined.

The sieve tubes in transverse sections show rather large outlines, and adjacent sieve tubes are separated by a partition wall usually directed more or less radially, which is in reality a section of a sieve plate. These sieve plates, which are long and almost vertical, exhibit scalariform markings. On this account the sieve tubes exhibit a striking superficial resemblance to certain xylem elements. Upon treating with soluble blue, the areas of the plate between the sculpturing are stained, the coating of callus thus revealed in the youngest tubes being thin and appearing as a thin line on either side of the sieve plate when viewed in transverse section. Sieve tubes of this type extend to the third or fourth row of laticiferous vessels counting from the cambium. They are followed by a narrow belt of tubes which are strikingly conspicuous when stained with soluble blue, owing to the thickness of the callus. The companion cells associated with the sieve tubes are prominently developed. The sieve tubes immediately succeeding these in an outward direction are of a totally different character, the walls bulge inwards as if the tubes are plasmolysed and the sieve plates originally flat are curved, exhibiting in transverse section an outline resembling a more or less compressed letter S. The callus has practically disappeared. The companion cells associated with these tubes are with difficulty distinguishable. The phloem parenchyma is well developed and occupies the space yielded by the collapse of the sieve tubes and their companion cells. The appearance presented by the collapsed sieve tubes suggests that they are no longer functional, but it is difficult to ascertain from material preserved in formalin at what stage they cease to be living. The question of the period of functioning of the sieve tubes and their viability is of considerable importance in interpreting the microscopical appearance presented in tissues affected with brown bast. But these are matters which can be studied only in the tropics, since it is known that edaphic and climatic factors have a profound influence on the anatomical structure of the rubber tree, and as in the case of the development of the laticiferous system, the functioning of sieve tubes will probably vary with the seasonal and other conditions prevalent in different parts of the world where this tree is cultivated.

Brown Bast. (1) Very early stages prior to the occurrence of external or microscopical burr formation.

A microscopical examination of a transverse section of the phloem of tissue affected with brown bast, as evident in authentic material of brown bast received at the Imperial College of Science and Technology, reveals a number of minute, yellowish areas. These often occur generally dis-

tributed throughout the phloem in the tissue regions between the inter-
secting rows of laticiferous vessels and medullary ray elements within a
zone often limited on the cambium side by the third or fourth row of
laticiferous vessels—in a tree of average type—counting from the cambium. The outlines of the smallest yellow patches—those nearest to the
cambium—are suggestive of gum-filled intercellular spaces, but if the
preparations are suitably stained it is found that the apparent cavities
are intersected by sinuous wall outlines similar in appearance but not
in staining reaction to the S-shaped sieve plates present in the collapsed
sieve tubes of the normal phloem. When the phloem is viewed in longitudinal section, more or less vertical yellow streaks corresponding to the
areas present in the transverse section may be observed; these are clearly
due to the presence of discoloured sieve tubes since the typical scalariform markings of the sieve plates can be easily distinguished. The
sinuous outlines observed in transverse section are not restricted to the
region where collapsed sieve tubes are to be found in the normal phloem,
they may be observed nearer the cambium in younger and obviously
plasmolysed tubes: tubes similarly situated in normal phloem would be
considered functional. The narrow belt of sieve tubes with thick callus
pads which forms such a characteristic feature of the normal phloem is
not evident, or only a few sporadic elements of this type are present.
Not only has the callus practically disappeared from the tubes in this
region, but in addition callus is almost entirely absent from the tubes
in the affected region of the phloem. In place of it there occurs a development of a substance which reacts strongly with ruthenium red, a useful
reagent which was discovered by Mangin to possess a strong affinity for
compounds of a pectic nature.

In 1921, Rands(8), working in Java, described a kind of gummosis of
the intercellular spaces of the phloem of *Hevea brasiliensis*. The symptoms described, notably the occurrence of a yellowish substance in the
supposed air spaces and the illustrations given in Rands' paper dealing
with the subject, were strikingly similar to the appearances observed by
the present writer in brown bast tissues. Since at the time of publishing
his paper on phloem necrosis an element of doubt existed as to whether
Rands was dealing with the same or a different problem, the question
was left open and the present writer made it his principal object to show
that similar gummy areas present in the phloem of trees affected with
brown bast examined by him were not appearances due to gum-filled
intercellular spaces but, in reality, sections of yellow necrotic sieve tubes.
At a later date, Rands, during a visit to this country, inspected the
drawings made from the material in the possession of the Imperial College

and agreed that the gummy areas were similar to those which had been under his observation in Java, and expressed the opinion that probably the writer's interpretation of the microscopical details was correct.

The gummy areas nearest to the cambium are composed of one or a pair of necrotic sieve tubes, but passing outwards, *i.e.* in the direction of the bark, they become larger owing to the fact that groups of cells including phloem parenchyma become involved in the tissue degeneration. In the region where the tissue necrosis is more advanced the laticiferous vessels are involved. The latex is yellow. Wound cambiums occur on either side of each row, often completely surrounding a number of elements in the row, and their activity has brought about displacement of the original tissues. The older cells derived from these wound cambiums become sclerotic with the result that plates of stone-cells occur in each zone of cambial activity.

(2) Early stages—microscopical burrs present but prior to the occurrence of external burrs.

Cambial activity has proceeded still further with the formation of new xylem and phloem elements: the latter include laticiferous vessels.

(3) Late stages. These have been dealt with in detail by Saunderson and Sutcliffe(10).

(*b*) The material in which phloem necrosis was first observed was obtained by Prof. J. B. Farmer from British North Borneo. It was collected from trees not definitely known to be affected with brown bast. The material showed no signs of even microscopical burrs. The microscopical features were thought to indicate very early stages of brown bast. In order to make sure Prof. Farmer obtained authentic material from Malaya and elsewhere. In every case phloem necrosis of a similar nature to that found in the original specimens was observed, and the conclusion was reached that the trees from which the Borneo material was collected were affected with incipient brown bast.

The recognition of the identity of the microscopical phenomena described by Rands and the present writer affords additional evidence of the association of the brown bast disease with necrotic phloem. Rands(9) states: "Brown bast is a non-parasitic disease of tapped trees characterised by discoloration and dryness of an area of bark extending below the tapping cut. The discoloration and cessation of latex flow are caused by the deposition of a yellowish gum-like substance in the tissues, impregnating many cell-membranes, filling the intercellular spaces and clogging the latex vessels."

(*c*) The name necrosis has been applied in the past to certain obscure diseases of plants characterised by the occurrence of necrotic tissue elements, and a special name has been employed when the necrosis is confined to particular elements. Thus the expression xylem necrosis has been applied to cases where degeneration of the xylem elements occurs (Küster and others), and phloem necrosis in the case of degenerate phloem (Quanjer). The name phloem necrosis was used to designate the leaf-roll disease of potatoes by Quanjer in 1913(5). That the disease had not been proved to be of an infectious nature at that time may be gathered from the following quotations from a later paper(6):— "This variety (Paul Kruger) under definite conditions of soil and climate must be so liable to the production of pathological bud variations that it is impossible to stop this process of degeneration. In his first paper on phloem necrosis, published in the winter 1912–1913, this standpoint had not yet been abandoned," and "Forced to choose between these two explanations (which for the sake of brevity we shall call the mutation hypothesis and the infection hypothesis) we carried out systematically a series of experiments." The experiments were started in 1913.

Stahel(12), working in Surinam, has recorded phloem necrosis in the Liberian Coffee plant.

The discovery of sieve tube necrosis in *Hevea brasiliensis* places the brown bast disease in the category of obscure diseases in which the occurrence of phloem necrosis is a prominent feature, and the use of the expression does not necessarily imply the agency of a causal organism.

(d) With regard to the conclusions which have been drawn, these do not appear to be at variance, at least in any fundamental detail, with the opinions expressed by investigators of the problem in different tropical countries. Although the view has been taken in some quarters that the disease is due to bacterial organisms, the general consensus of opinion attributes it to physiological causes. In 1921, Prof. Farmer and the writer(2) directed attention to the prevalence of diseased sieve tubes in the young phloem and stated that a necrotic development of this kind, whatever the circumstances bringing it about, must interfere with the normal metabolic processes of the rubber tree. Whilst the possibility of the agency of parasitic organisms was borne in mind, it was clearly stated that such a condition might be due to some ill-defined physiological disturbance of the normal functions within the tree.

During the last few years a great deal of work has been done by Quanjer and others on leaf-curl and allied diseases of the potato. Until 1913, these diseases were attributed to physiological causes. Quanjer, Van der Lek and Oortwyn Botjes have now shown that certain of the diseases of this group, notably leaf-roll, mosaic, crinkle, etc., are of an infectious nature. This conclusion is also reached by Murphy[4]. The investigators working on this problem have never succeeded in finding any causal organism and agree with Quanjer, who expressed the opinion that such a variety of symptoms can only be explained on the hypothesis that there are a number of micro-organisms causing them. According to Quanjer, phloem necrosis (necrosis of the sieve strands) occurs only in the case of leaf-roll, and he finds a definite correlation between leaf-roll and phloem necrosis. This disease was found to be transmissible by aphides by Oortwyn Botjes in 1920[1] and by Schultz and Folsom in 1921[11]. Some work in connection with the transmission of the leaf-roll virus has been done in this country by Whitehead[13]. Taking into account the entirely different habits of the two plants, the parallel with *Hevea brasiliensis* is very striking. In both, phloem necrosis is associated with a specific diseased condition of the plant; in both, sieve-tube necrosis is prevalent, and in neither case is any trace of parasitic organism revealed by microscopical study. Nevertheless, it is not safe to assume from this parallelism that the brown bast disease is caused by a virus. A decision can be reached only as a result of investigations carried out in the tropics by competent specialists working with the definite object of settling the question.

Literature cited

(1) BOTJES, OORTWYN. *De bladrolziekte van de Aardappelplant.* Thesis accepted by the Landbouwhoogeschool, Wageningen, 1920.

(2) FARMER, J. B. and HORNE, A. S. "On Brown Bast and its Immediate Cause." *India Rubber Journal,* p. 25, June 18th, 1921.

(3) HORNE, A. S. "Phloem necrosis (Brown Bast Disease) in *Hevea brasiliensis.*" *Annals of Bot.* xxxv, July 1921.

(4) MURPHY, PAUL A. "Some recent work on leaf-roll and mosaic." *Report of the International Potato Conference,* p. 145, London, 1921.

(5) QUANJER, H. M. "Die Nekrose des Phloems der Kartoffelpflanze, die Ursache der Blattrollkrankheit." *Mededeelingen van de Landbouwhoogeschool,* VI, p. 41, Wageningen, 1913.

(6) QUANJER, VAN DER LEK and OORTWYN BOTJES. "Nature, mode of dissemination and Control of phloem necrosis (leaf-roll) and related Diseases (*i.e.* Sereh)." *Mededeelingen van de Landbouwhoogeschool,* x, p. 1, Wageningen, 1916.

(7) QUANJER, H. M. "New Work in Leaf-curl and allied Diseases in Holland." *Rept. of the International Potato Conference,* p. 127, London, 1921.

(8) RANDS, R. D. "Brown Bast Disease of Plantation Rubber, Its Cause and Prevention." *Mededeelingen van het Instituut voor Plantenziekten,* No. 47, 1921.

(9) —— "Histological Studies on the Brown Bast Disease of Plantation Rubber." *Mededeelingen van het Instituut voor Plantenziekten,* No. 49.

(10) SAUNDERSON, A. R. and SUTCLIFFE, H. "Brown Bast." *The Rubber Growers' Assoc.* May 1921.

(11) SCHULTZ and FOLSOM. "Leaf-roll, net-necrosis and spindling-sprout of the Irish Potato." *Journ. of Agric. Res.* xix, p. 315, Washington, 1921.

(12) STAHEL, G. "De zeefvatenziekte (phloëmnecrose) van de Liberiakoffie in Suriname (koffiewortelziekte)." *Meded. Dept. Landb. Suriname,* 12, 1917.

(13) WHITEHEAD, I. "Transmission of Leaf-roll of Potatoes." *Rep. of the Intern. Conf. of Phytopathology and Economic Entomology,* p. 147, Holland, 1923.

PLANT PATHOLOGY AND MYCOLOGY

THE RELATION OF FOREST PATHOLOGY
TO SILVICULTURE

(CHAIRMAN: PROFESSOR R. S. TROUP)

Dr A. W. Borthwick

In all forms of plant cultivation knowledge acquired by experience has been handed down through the ages from one generation to another. There are abundant records to show that blights and plagues occurred leaving famine and pestilence in their train.

As cultivation became more intensive and the cultivation of pure crops extended over larger and larger areas the balance of nature began to be upset. Certain pests became more specialised and the host plants through being forced, as it were, to lead an artificial life became more liable to attack by disease.

I need not further emphasise the fact that Nations and Empires owe their existence and prosperity, in a large measure, to the good work done by the plant pathologist in making the cultivation of crops possible on the modern intensive scale of cultivation, in tropical, subtropical and temperate regions.

Many years of apathy and neglect have left us with a wonderful and intricate series of problems bearing directly upon the health of our cultivated plants, which as crops yield food, fibre, textile, constructional and other commodities so fundamental to our great industries and so intimately associated with our everyday requirements. The numerous questions which arise in forest pathology are frequently bewildering and baffling unless they are studied and investigated in the right way, and that way is by co-operation between the silviculturist, the mycologist and the entomologist. Disease is not a thing that just happens. There must be some predisposing cause or causes slow or rapid in action which bring about conditions in the host that make attack possible. Normal and abnormal predisposition to disease are the things of fundamental importance. Trees go through various phases of growth from the juvenile stage to the adult. First there is the seedling stage; this is followed by a period of longer or shorter duration, according to species, when the tap and lateral roots are developing and adjusting themselves to the soil, then comes a period marked principally by rapid growth in height and changes from periderm to bark. Towards the end of height growth crown formation begins and usually the flowering and fruiting stage is reached. Increment in girth continues long after increment in height growth has ceased or become negligible and the proportion of heartwood to sapwood increases. Finally, on approaching maturity, the breadth of the annual rings decreases with a corresponding falling off in girth increment. My object in referring to these stages or periods or phases in the life of a tree is to emphasise the fact that each stage of growth is characterised by differences in anatomical structure and in physiological activities, because predisposition to disease is so intimately associated with the anatomical structure and the physiological activities of the host plant. Within this grand cycle of change through which a tree passes, there is a seasonal cycle marked by four stages: 1. The dormant stage. 2. Awakening to vegetative activity. 3. Period of active growth. 4. Cessation and ripening. During this annual cycle there are danger periods when normal predisposition to disease is more pronounced.

It is obvious therefore that there is an intimate relationship between the silvicultural treatment of trees and their predisposition to disease. Silviculturists have learned by long experience and at times by costly failures, that their methods must be copied from nature as far as that is possible. Generally mixed forests are less predisposed to epidemic diseases than pure forests. The clear felling of large areas is more likely to bring about disease in the subsequent crop than clear fellings distributed over smaller areas. Natural regeneration of forests is less liable to produce predisposition to disease than artificial regeneration. However, artificial regeneration is frequently unavoidable and in the afforestation of new areas artificial sowing or planting is the only method available. In such cases the forester must first of all determine the species to plant. He must therefore study what may be called the predisposition to disease of the locality in relationship to the species he selects. For example, stagnation of moisture in soil or atmosphere (mists, fogs, etc.) are very liable to predispose the larch to disease. Frosty localities are equally dangerous to frost tender species. Rainfall and soil moisture also influence predisposition to disease according to amount and season. The presence or absence of other plants which may act as alternative hosts, in the case of species subject to attack by metoxenous fungi must also be noted as potential predisposing agents.

When it comes to the actual planting, care must be exercised in the adoption of the right method, the right season and the right density, as all these things influence predisposition to attack by insects and fungi. The plantation may be pure, it may consist of a permanent mixture, by single stems or groups, it may be a temporary mixture containing certain suitable hardy nurse trees to tide the permanent crop over the earlier and usually more vulnerable periods of growth. The density of planting will naturally influence the period and severity of the earlier thinnings and here great skill and judgment are required, because thinning not only alters the light conditions and therefore the assimilating and transpiring surface, above ground, but it has a marked effect on the water content, temperature, and available food material in the soil. Mistakes in an operation so far-reaching in its effects may easily bring about serious predisposition to disease.

In addition to the grand cycle and seasonal phases of tendency towards disease through which all trees and plantations pass, there is undoubtedly a difference among individual plants which varies from extreme suscepti-bility to disease to greater or less resistance and finally almost complete immunity. It is possible to recognise races or geographical types of Scots pine which show a greater resistance to the leaf cast fungus than others. The same is true of the larch in relation to the larch canker, and the spruce in relation to the spruce leaf-rust. This fact is of great importance in connection with the origin of the seed we use in our forest nurseries. It is evident, then, that the silviculturist can do a lot in the way of eradicating predisposition to disease in his growing stock, but in order to achieve this purpose, he must be constantly in the closest touch with the mycologist and the entomologist. A diseased condition in a planta-tion may be brought about on a given soil with a low water content, by over-crowding in planting or by neglect of timely and sufficient thinning. The lowered water content of the trees which results from this, though not necessarily fatal to the trees, may predispose them to insect attack and the insect attack in turn may bring about abnormal predisposition to fungus disease.

In such and similar cases the mycologist is expected to prescribe some simple remedy indicated by the symptoms shown by material in the last stages of what I may be allowed to call a compound disease. Such material may and generally does show a large number of secondary symptoms so bewildering in their complexity that in many cases it is impossible to say which is primary and which is secondary. Still in many cases the mycologist has succeeded in disentangling the ravelled threads, but even then all he can do is to recommend perhaps some form of spray or the removal of all diseased and therefore infectious material from the plantation, and its destruction by burning. This is obviously only a tem-porary makeshift. The thing to be done is to ascertain the predisposing conditions, be these normal or abnormal, and by removing those conditions which bring about predisposition, the disease will cure itself; if the silviculturist follows the advice of the pathologist these primary predisposing causes will disappear. How is this to be done?

I submit for the sake of discussion that the forest pathologist does not as a rule get a fair chance. The scope and nature of his work is not properly understood and appreciated. The principal aim in forest pathology should be the study of all those conditions, temporarily or permanently inherent in the plant or in its organic and physical environment which bring about predisposition to disease. The silviculturist by consultation with the pathologist would then be in a position to know what conditions he should endeavour to avoid creating. The study of the development of a fungus in the tissues of a host plant and the effect which the growth and spread of the mycelium has upon the parts locally affected and the general health of the tree is not the final aim of the forest pathologist. It is necessary by means of artificial infection to discover the mode of attack. Artificial inoculation is not always successful at the first attempt and it is essential to avoid being led into error by such pitfalls. We must know the condition in which the spores must be as regards ripeness and age, and also the conditions necessary as regards moisture, temperature, aeration and light for their successful germination. Successful inoculation in the laboratory must be confirmed by inoculation in the forest. A subject exposed to inoculation in the laboratory may also be exposed to a myriad of predisposing influences. It is the same in the case of insect attack. We may observe an insect at work, we may study its mode of feeding and reproduction and the damage done to the host plant or plants, but this is not sufficient; we must find out whether or not some preexisting disposition due to silvicultural treatment has made the attack possible.

I have already indicated that apart from a fixed hereditary tendency to disease shown by certain races or types, there is no fixity in other predisposing influences. Each outbreak of disease must be studied on its own merits. Soil, climate, flora and fauna vary so much from place to place, that the outbreak of one and the same disease in different localities may be due to different disposing causes. Hence extensive and comparative experimental work and study in the forest is necessary and this work must be done with scrupulous care; it must be confirmed and tested not only in the forest but also in the laboratory by exact scientific methods and standards of comparison. No detail, however trivial it may at first sight appear, should be overlooked and neglected. Intensive work in the laboratory and frequent visits to the forest by the trained observer is necessary in order that disease and the conditions which bring it about may be avoided. We cannot hope to eradicate disease entirely, but we ought to be in a position to prevent new outbreaks and to lessen the duration and severity of existing attacks.

It is of infinitely greater importance that the silviculturist with the aid of the pathologist should be able to keep his woods and plantations in a healthy state, than that the pathologist should only be consulted when a disease has already half ruined a valuable crop of trees because, even though a check may be administered or a cure effected, the total potential increment of such a rescued crop has inevitably been reduced for all time.

A crop of trees may pass successfully through the first phase of growth, that is, practically from the time of planting until the branches of neighbouring trees begin to meet and a complete shading of the soil and its surface flora is formed. It is now that the keenest struggle and competition takes place among the individual trees and from now until the end of height growth we find a continuous segregation of the trees into classes among which we can almost at all times recognise a class of stems, which contains hopelessly suppressed, dead, dying or diseased trees. The utmost vigilance and skill of the forester is required to bring his crop successfully through this critical phase in its development. Diseases of the cortex and roots which may have been contracted during the first phase of growth now begin to show up with greater prominence. A cortical disease may result from wounds caused by wind or snow pressure tearing off branches, hailstones, lightning, sun scorching, insect attack, damage by birds or squirrels, and finally by carelessness in the cutting and removal of thinnings.

It is of importance that the progress of such cortical and root diseases should be kept under strict observation and control, because a fungus may live in the cortex for many years without showing any very apparent damage, but a time comes when cambial activity slows down and the power of the tree to occlude diseased patches is lessened, and therefore more pronounced symptoms of damage become apparent, and often at this stage the time for effecting a remedy or rather measures of prevention has long since gone by. In forestry it is not practically possible to apply individual remedies; in the nursery this may be possible, but it is not so in the forest. We may do something to help trees already attacked by using every method to stimulate their vegetative energy and rate of growth, but this is usually more difficult and costly than the timely adoption of measures to banish as far as is humanly possible all causes disposing to disease.

In a country like ours, where many of our most valuable commercial species have been introduced from other countries, it is doubly essential that fundamental problems of the nature I have indicated should be the subjects of close study by the silviculturist, the mycologist and the entomologist. To begin with, we know very little or next to nothing regarding their silvicultural treatment. Still less do we know how they will respond in different localities and environment to treatment based on past experience of our indigenous or at least longer established species. New diseases may be introduced with an alien species, or the alien species may be specially sensitive to and suffer badly from an indigenous disease to which our native species are more or less immune. Already, by apathy and neglect, we have lost from our future forests two valuable species, namely, the Silver fir (*Abies pectinata*) and the Weymouth pine (*Pinus strobus*), the first through neglect to control the *Chermes nusslini* and the second (including all other five-needled pines of the *Strobus* group) by the epidemic spread of the stem rust fungus *Cronartium ribicolum*. Both these species have valuable silvicultural characteristics which would have made them most desirable constituents of our woods and plantations, either singly or in mixture with other species, and they would not only have thereby improved the growth conditions for other trees, but their timber value would have been measurable in hundreds of thousands of pounds per year. The loss of these two species, or our inability now to grow them as forest crops, is equivalent to the loss of a valuable national asset. Such a warning should make us doubly careful to preserve and maintain the constitution and robust growth of these other more recent and equally valuable members of woodlands and forests. The same precautions are necessary in other parts of our world-wide Empire. Where the replacement of the virgin forests by artificial means occurs, and especially when accompanied by the introduction of exotic species, too much care and attention cannot be given to forest protection against disease.

At the outset I endeavoured to indicate how complicated and difficult and even impossible it was to arrest or cure a disease once it has got a firm hold on a plantation. Primary and secondary symptoms are apt to become so involved, that it is not always possible to know what cure to apply. We usually know so little as to when and how the primary attack occurred. It is only after continuous extensive and comparative study of the life habits of the disease-causing organism, and those special conditions of the host which render attack possible, that we can adopt really sound and practical preventive measures. It has always seemed to me that the mycologist and the entomologist were pretty much in the same position as a man would be in who was shown a single picture or section from a long cinema film and asked to explain the whole story, part of which it represented. He might make it into a drama, a tragedy or a comedy. The man who was shown a greater number of specimens from different parts of the film would be in a position to come nearer the truth, but the man who saw the whole of that film animated and in motion from start to finish would have no difficulty in interpreting the meaning of each of the component pictures. It is the same in silviculture and pathology. It is only by the constant observation and record of nature's laws at work, and by close co-operation between the silviculturist and the pathologist, that we shall ultimately be able to write the complete story of the inter-relation between silviculture and pathology.

Mr W. E. Hiley

It is the function of the forest pathologist to study all those agencies which militate against the good health of the forester's crops. It is the function of the silviculturist to grow these crops in good health with due regard to economy and the maintenance of soil fertility. Thus in very large measure it is true to say that pathology is the science of silviculture and silviculture is the application of pathological knowledge.

Historically, however, this is not so. In the past silviculture has been based on tradition which has grown empirically. The reason for particular forms of practice has not mattered; so long as they worked, that was sufficient—I am speaking of a past generation—and it was only when catastrophic losses occurred, such as those caused by insect or fungus epidemics, that calls were made upon biologists for investigation. The pathologist was thus the handmaiden of the silviculturist, called upon to give explanations—perhaps even advice—when required, but not expected to make suggestions at other times.

A change has come over this relationship in the last few decades. The difficulties of silviculture are so great; the danger of applying in a new locality a form of practice which has proved successful elsewhere is so well known and the number of species which have to be dealt with is becoming so large, that silviculturists are more and more demanding of biologists that they shall study the reasons for success and failure of different forms of silvicultural treatment in order that their possible extensions and normal limitations may be known without the disastrous expense of trials and failures.

I have defined Forest Pathology as the "study of all those agencies which militate against the good health of the forester's crops." These agencies are partly parasitic and partly non-parasitic and, in my opinion, the latter have received less than their due share of attention in the past. I intend, therefore, to confine myself to non-parasitic agencies in my contribution to the discussion this morning, and I propose to consider three examples which illustrate the manner in which the study of pathological physiology may assist the silviculturist.

(1) The percentage of viable seed that produce plants varies with the species and method of cultivation. With conifers, nursery practice may produce between 10 and 20 per cent., but in virgin forest not more than one seed in a million can grow to become a tree. The losses are due to vermin, insects, fungi, but in still larger measure to drought, frost, overheating and lack of light. One of the most important of these agencies, and one which until recently was unsuspected, is over-heating.

It was first shown by Münch[1] that the temperature of the surface millimetre of soil under direct insolation might rise to over 60° C. By heating up seedlings in an incubator he found that about 55° C. was the limiting temperature that they could withstand. When seedlings were growing in a soil with overheated surface constrictions appeared in the hypocotyls and the seedlings fell over. The hypocotyl was killed at the soil surface and was soon attacked by soil fungi so that if the seedling was submitted to a mycologist some fungus was generally credited with its death. Münch attributed a large proportion of seedling losses to this cause, and in England very many casualties are now associated with it. In America Toumey and Neethling[2] have found that the same cause is responsible for losses up to 70 per cent. in seedlings of larch and balsam fir in exposed seedbeds.

This type of damage can be reproduced in the laboratory by placing a coil of fine copper tubing round the hypocotyl of a seedling (without touching it) and recording the temperature in the middle of the coil with a thermo-couple. By this means it has been found that exposure to a temperature of 42° C. for 30 minutes is frequently sufficient to cause a constriction in the hypocotyls of spruce seedlings. This is too low a

[1] *Naturw. Ztschr. Forst- u. Landw.* xi, 1913, p. 557; xii, 1914, p. 169; xiii, 1915, p. 249. See also Hartley, *Journ. Agr. Res.* xiv, 1918, p. 595; xv, 1918, p. 521.

[2] Yale University, *School of Forestry Bull.* No. 9.

temperature to cause immediate death, and that the constriction occurs
before death is shown by the fact that when subsequently placed in
water the seedling may recover and the constriction disappear. It is
probable that a temperature of 42° C. alters the permeability of the
plasmatic membranes allowing the water in the cells to escape into the
intercellular spaces and to be absorbed by the still actively osmotic cells
above and below. On cooling, the plasmatic membranes may again
become operative and the collapsed cells again become turgid.

The importance of maintaining shelter over seedlings is now widely
realised by silviculturists. Not only are artificial shelters freely used on
seedbeds in nurseries but natural shade is recognised as being desirable
in obtaining natural regeneration of spruce and other species (cf. Wagner's
Blendersaumschlag). Incidentally, this fact may enable us to question
the soundness of a tendency which is noticeable in Eastern Canada.
Foresters in Eastern Canada find that the systems of extraction employed
in Sweden are on the whole applicable in Quebec, and many of them hope
to adopt also the Swedish methods of management and silviculture. The
cheapest method of regeneration, which frequently proves successful in
Sweden, is to leave scattered mother trees, singly or in clumps, from which
the surrounding almost bare areas may be naturally seeded. In Sweden
they appear to have little trouble from high soil-surface temperatures
and can adopt this method which in Southern Germany would generally
fail. We must remember, however, that the temperature of the surface
soil under direct insolation is determined principally by the angle of
incidence of the sun's rays. In Sweden, at a latitude of 60°, this can never
be more than 38°, whereas in Quebec at a latitude of 45° to 50° it is very
much higher and I have observed in Quebec that on bare ground coniferous seedlings only succeed in establishing themselves on the north side
of boulders and logs. This suggests that in Quebec methods of natural
regeneration will have to follow the practice of Southern Germany rather
than Sweden.

(2) One of the most important decisions that the silviculturist has to
make is the choice of species for planting on specific sites. The importance
of this will steadily increase as forestry more and more outgrows the
conception of conserving natural forests and approaches in intensiveness
the cultivation of agricultural crops. It may appear to some that the
adaptability of species to new sites is outside the scope of pathology, but
I venture to think that the study of the edaphic and climatic requirements of trees is a legitimate branch of pathology and that when we
come to investigate the health of whole woods and plantations as opposed

to particular trees a knowledge of environmental conditions is essential. Also many parasitic diseases are encouraged by unsuitable growing conditions so that the parasitologist is also called upon to study these conditions in order to estimate the importance of factors which predispose trees to parasitic attack. In studying the edaphic and climatic requirements of trees the experimental method is extremely slow and expensive. Quicker and equally reliable results may be expected from the statistical analysis of observations on existing woods. Schütze[1] has applied this method to the relation between the quality class of pine woods and the phosphoric acid content of the soil, and came to the conclusion that the latter might act as a limiting factor in tree growth when it fell below 0·5 per cent.

The investigation of the relation between tree growth and climatic conditions is greatly facilitated by the fact that most trees in our climate preserve a record of their annual increment in the thickness of their annual rings and, in the case of many conifers, in the height elongation between the annual whorls of branches. The annual growths, whether in thickness or height, can be compared with the weather records for the corresponding years and a statistical analysis of the influence of particular climatic factors can be made by calculating all relevant correlation coefficients.

In this manner Schwarz's[2] figures show that the ring thickness of Scots pine at Eberswalde in Prussia is principally determined by the air temperature from January to March. A cold winter is followed by poor growth and *vice versa*. A similar relation has been found with balsam fir in Quebec, but in England, with much milder winters, this factor appears to be of no importance.

A negative correlation has been found between May-June temperatures and growth (annual rings or height) of *Pinus ponderosa* and *Quercus* sp. in America, Scots pine in Prussia and Corsican pine in England. This means that a cool May and June give rise to a good growth, a relationship which appears to be independent of rainfall. In the arid climate of Arizona, however, rainfall appears to be a dominating factor. So far this study of limiting factors has only given us a few disjointed data. If applied over a much larger field it might, however, elucidate many of the problems of the relative success and failure of species on new sites.

(3) Another set of important silvicultural problems is connected with the thinning of woods. The object of scientific thinning is to allow such

[1] Quoted in Henry, *Les Sols Forestiers*, 1908, p. 178.
[2] For references see Hiley and Cunliffe, *Ann. Appl. Biol.* x, 1923, p. 442.

a relation between the crown space and height of the trees that the largest possible increment of good quality timber is produced. There is reason to believe that when plantations are allowed to become too dense the health of the trees is adversely affected. This becomes very noticeable in the case of larch, which is especially liable to canker in over-dense woods.

Observations on the larch have shown that, in trees whose crown space is small in relation to their height, so that their food supply is restricted, annual rings may be missing from the lower part of the main stem. In such trees the cambial activity may become confined to the upper part of the tree (and, perhaps, the roots), though the height growth is not at first diminished. Lundh[1], in a recent paper, has shown that pruning away the lower branches of the crowns of Scots pine slightly increased height increment and annual ring thickness above halfway, but greatly reduced the thickness of annual rings in the lower half. This appears to be analogous with the observations on the larch.

Though the evidence on this subject is not yet conclusive, there is reason to believe that the water supply of a conifer is conducted principally in the youngest annual rings in the main stem and that, as the tracheids become filled with air, they become useless for such conduction. If this is so, the failure to make new wood down to the bottom of the stem of a tree must greatly reduce the water supply of its crown and *ipso facto* reduce its vigour. Observations on the manner in which the relation between crown space and height in individual trees affects the distribution of cambial activity should thus enable us to elucidate one of the factors influencing the optimum density of plantations.

Dr J. W. Munro

In entering this discussion on Silviculture in relation to Forest Pathology, I should like to say that I do so not because I think Forest Entomology is merely a branch or offshoot of Plant Pathology, but because my experience has impressed on me the need for close co-operation between the entomologist and the plant pathologist. This is especially true in forestry where local conditions play so important a part in affecting the health and growth of the crop.

My aim in this brief contribution is to indicate certain problems the solution of which depends on the close co-operation of the pathologist and the entomologist, without which the entomologist and the plant pathologist are both helpless. The tendency of the entomologist is to lay

[1] *Medd. fr. Statens Skogsförsöksanstalt*, Häfte 21, 1924, p. 49.

stress on insects as agents causative of disease, of the pathologist to lay stress on such agents as frost, drought or fungi, while, in many cases, the real cause is a combination of these factors brought about by bad silviculture.

A good example of this combination of a number of factors in producing disease and losses has been seen for some years past in our oak woodlands. The oaks have been defoliated by caterpillars, the chief being that of the Green Tortrix Moth *T. viridana*. Defoliation is complete in June and, in July, a new foliage appears which is rapidly attacked by the oak mildew *Microsphaera*. Oak, like most broad-leaved trees, has considerable recuperative power and reserve and may withstand one or two years' defoliation by the Tortrix, but the loss of its second foliage through mildew over two or three years exhausts the tree and creates conditions favourable to further fungus infection and decay. In this case the sequence of pathological factors is fairly clear and one may reasonably cite the Tortrix moth as the prime injurious agent, always assuming that the planting of extensive areas of pure oak is sound silviculture.

Another instance where insects and fungi combine occurs in our larch plantations. In recent years one of the insects most frequently submitted as killing larch has been the longicorn beetle *Tetropium gabrieli*. This beetle lays its eggs under the bark, between which and the sapwood its larvae tunnel, finally pupating within the sapwood. The tunnel systems under the bark are extensive, the tree stem is often completely ringed, but *Tetropium* is not the first cause of trouble. In every case I have investigated, or which has been carefully reported on, it has been found that *Tetropium* attacks only those trees which are already reduced in vigour and in about 70 per cent. of such cases the fungus *Armillaria mellea* has been found. So frequent is the association of the beetle and the fungus that I am inclined to regard the beetle as a friend of the forester warning him of the presence in his woods of the honey fungus.

The importance of insect attacks as indicators of pathological forest conditions has scarcely received the attention it deserves, although Robert Hartig laid stress on this aspect of forest pathology. Nowhere in forest entomology, as he pointed out, is there greater need for the co-operation of the silviculturist, the pathologist and the entomologist than in the bark-beetle problem, and it is interesting to find that the advances made in our knowledge of the bark-beetles and their relations to the forest all point to the truth of Hartig's dictum that in studying forest insects it is essential also to study forest conditions. Probably no group

of forest insects has received so much attention from the entomologist
as have the bark-beetles. Lindeman, Nüsslin, Fuchs, Hopkins and
Swaine have placed our knowledge of the taxonomy, the morphology,
and the biology of the bark-beetles on a high plane but from the forestry
standpoint the entomologist is pulled up in his progress for lack of the
assistance of the pathologist. As Escherich has pointed out no group of
forest insects is more dependent on forest conditions than the bark-
beetles, and the whole question of bark-beetle control is dependent on
the proper assessment of these conditions and their re-adjustment to
normal with the aid of pathology and silviculture. Even in our limited
British bark-beetle fauna we find a range in bark-beetle economy from
those species which, like *Ernoporus fagi* and *Dryocaetes autographus*, live
only in fungus-infested trees, to those which, like *Myelophilus*, are capable of attacking trees in which the only abnormality predisposing to
attack is some slight deficiency in the soil, temperature, or light conditions under which the trees are growing. An illustration of the importance of these factors, soil, light, and temperature, in predisposing to
Myelophilus attack was well seen recently in the New Forest. In Knight-
wood Inclosure there, Scots pine and Corsican pine are growing side by
side in groups. The Scots pine was badly punished by the bark-beetle
Myelophilus; the Corsican pine was almost unscathed. Now, as Guillebaud
and his colleagues have shown, the rate of growth of Corsican pine in the
south of England is 50 per cent. greater than that of Scots pine (hence
the Corsican's immunity), but in Scotland no difference in rate of growth
is apparent and there the Corsican suffers equally with the Scots pine.

So far I have dealt chiefly with insects and pathology. I should like
briefly to refer to an aspect of silviculture in relation to entomology
which has, so far as I am aware, been somewhat neglected. I refer to
the influence which the forest environment produced by different systems
of silviculture may have upon insect outbreaks. It is a generally accepted
axiom that extensive planting of pure woods (*i.e.* woods consisting of
one species of tree) favours insect outbreaks, and this is generally attributed to the fact that the pure woods afford a continuous uninterrupted
range of host-plants to the insect or disease, while in mixed woods the
range of host-plants is interrupted or discontinuous. A further consideration is that pure woods tend to exhaust the soil more rapidly than
mixed woods. These factors are not, however, the only ones affecting the
prevalence of insects or diseases and apart from them the type or system
of silviculture may affect the prevalence of insects or diseases in three
main ways.

(1) It may act on the crop itself, being favourable or unfavourable to its vigorous growth, and may thus, when unfavourable, produce in the crop a predisposition to insect or fungus attacks. A good example of this has recently been cited by Schollmayer Lichtenberg in the Austrian *Mittelgebirge*, where the undue prolongation of the rotation in first quality spruce stands resulted in an extensive bark-beetle outbreak.

(2) The silviculture practised may act directly on insect pests themselves, favouring or hindering their increase. Thus the system of clearfelling favours the large pine weevil, *Hylobius*, and the bark-beetles, but is unfavourable to the lesser pine weevils, *Pissodes*.

(3) Finally, the system adopted may act in a more subtle way, not necessarily on the crop nor directly on the insect pests themselves, but on that important but little understood group of insects predaceous or parasitic on our forest insect pests. The inter-relation between our insect pests and their parasites and the local forest conditions is too complicated to be dealt with here, even if it were fully understood, but I should like at least to call attention to the question and to suggest that this subject of the relation of the forest conditions to the occurrence of useful, predatory or parasitic insects may yet prove of great importance. An indication that the type of forest conditions existing in any locality may affect the presence or absence of important parasites of forest insect pests is perhaps to be found in the Balsam bud-worm problem in eastern North America. Tothill has found that the most important enemy of the budworm is a hymenopterous parasite of the genus *Phytodietus*, and further that while in the mixed woods of British Columbia and the better types of spruce and balsam forest in the Eastern Provinces this parasite is present and effective, in the cut-over lands which have degenerated into poor forest of pure balsam this parasite is absent and the bud-worm is extremely injurious. The bud-worm problem is by no means completely understood, but I should like to suggest that here is a field for study where the entomologist, the ecologist, and the silviculturist may hopefully work together. There is no doubt that the change in the forest flora brought about by various systems of silviculture directly affects the forest fauna and indirectly the forest insect problems.

In conclusion I should like, as an entomologist, to emphasise the need for co-operation between plant pathology, entomology and silviculture. This need is felt not only by myself but by my fellow entomologists in Canada and the United States, and the tendency in Europe too, as shown by the work of Escherich, Boas, and Tragardh, who are pioneers in forest entomology, is to interpret insect phenomena in terms of forest conditions, normal or pathological.

The co-operation of the entomologist with the pathologist and silviculturist in no way involves the submergence of any of these branches of science. On the contrary, my experience is that such co-operation increases the value and individuality of each and makes one realise that sound forestry practice can be attained only by utilising the knowledge gained in these distinct branches. Further, I think that by co-operation the pathologist and the entomologist can better insist on the need for fundamental research on broad ecological lines, which is the greatest need in forest entomology and pathology to-day.

PLANT PATHOLOGY AND MYCOLOGY

FUNGAL ATTACKS ON TIMBER

(CHAIRMAN: SIR DAVID PRAIN, C.M.G., C.I.E., F.R.S.)

Prof. Percy Groom, F.R.S.

(*Abstract*)

Economic loss caused by wood-attacking fungi, as well as the range of conditions under which the latter grow, are reflected in a list of sites of the attacked woods: (standing trees and stumps—sapwood and heart-wood: felled timber—submerged wood, piles, shipping; paving, railway sleepers, posts, fences, outhouses; buildings, unheated and heated; warm moist mines, conservatories; cold stores, intermittently and permanently cold). This wide range of habitats facilitates physiological and structural research. Contrast between fungi respectively parasitic on parenchyma and attacking dry heartwood; types of fungi and habitats transitional between these extremes. Fungal floras of parasites on trees and of forms inducing dry-rot in felled timber.

Research required may be treated under nine headings.

(1) *Identification* (including life histories)

Need for recognition by means other than sporophores.

Morphology and histology of the fungus. (a) Macroscopic: form and colour of the cultures from spores and emerging from wood, and of the more mature mycelium on and in wood. (b) Microscopic: structure of mycelia and hyphae, and calibre of the latter, outside and inside wood, and the degree of constancy of these.

Physiology of the fungus. Cardinal temperatures of growth, and immunity of certain woods as aids to identification.

Structure, macroscopic and microscopic, of attacked wood as affording diagnostic characters.

(2) *Relations with Water*

Amount of water in wood directly and indirectly (by influencing the amount of oxygen available) determining the species present, the form of the mycelium, and the production of reproductive cells and organs. Production of water of respiration. Quantitative investigations required on these matters, and on rate and degree of desiccation as a lethal agent.

(3) *Relations with Oxygen*

Fungi attacking wood are aerobic, yet investigations are required as to the amount of oxygen required by different species and available in different cases. Part played by anaerobic bacteria.

(4) *Relations with Temperature*

Cardinal temperatures for growth and so forth, as indicators of identity; possible changes of these points by "direct adaptation." Lethal doses of heat (temperature × time) supplied at extreme temperatures.

(5) *Relations with Light*

Influence, direct and indirect, on the production of sporophores and spores, on the form and structure of the mycelium, and on germination of spores.

(6) *Relations with Nutritive Medium*

Investigation of the food-material upon which these fungi are actually feeding when found on and near wood, and upon which they are capable of feeding, and finally upon the enzymes that they contain.

(7) *Infection*

Differences of the conditions for infection and growth respectively. Certain species incapable of infecting sound wood by means of spores; the investigation of substances (acid bodies) and their concentrations capable of promoting or inhibiting germination. Species capable of infecting sound wood and rendering the wood susceptible to spores of the types previously mentioned: causes of these characters. Unknown factors preventing germination. Adventitious aids (*e.g.* seasoning cracks) to infection by spores.

Co-operation of fungi with each other, and possibly with bacteria, in attacking wood.

(8) *Immunity*

True and false immunity. Range of different species of timbers attacked by a single species of fungus. Differences in the durability of the sapwood and heartwood of the same and different kinds of timbers. Research on the causes of immunity.

(9) *Preservation of Wood*

Sanitation: investigation of the sources of infection.

Fungicides: differences between those suitable for use on the living tree and felled timber respectively. Need for quantitative investigations on lethal powers, penetration, durability, and so forth.

Dr A. W. Borthwick. PARASITIC DRY ROTS

Loss from timber rot occurs in the forest as well as in buildings

The study of rot in timber may be divided into two parts:

(a) Butt and stem rot in the forest.

(b) Rot in timber used for structural purposes.

How far initial attack in the forest on standing trees or felled trees prior to their removal may continue, or how far such attacks may open up or expose the timber to subsequent attack by other fungi, is a question of fundamental importance and a subject which requires further investigation. It cannot be too strongly emphasised that diseased timber can and does infect healthy timber. Hence every effort should be made to ensure that only healthy timber leaves the forest and that all places where timber is accumulated or stored should be kept clean as regards infectious material, and lastly careful attention should be given to keeping structural timber free from rot and decay once it is in position.

Proper care in the growing, harvesting and after treatment of timber will help to prolong its durability. The annual loss in standing and structural timber caused by various forms of preventable rot amounts to a high percentage of our available supplies and demands a vigorous campaign of study and research in the interests of national economy.

It is surely better to expend capital and to employ labour in connection with all operations involved in the proper care and preservation of timber, than to neglect these precautions and ultimately be forced to expend money and labour on the renewal of timber structures and erections the duration of which could have been prolonged if the proverbial stitch in time had been made.

It is not so much the actual amount of woody substance consumed by rot that counts; it is the position and distribution of infected patches. The value of a large bole or tree-trunk may be ruined by the presence of local decay caused by dead and broken branches or other local wounds. The size of timbers which can be cut from a large trunk so affected may therefore be no greater than those to be obtained from a younger or smaller tree. The same principle holds good in the case of baulks, beams, joists, spars, poles and props of all kinds. Local decay in a beam or joist where it rests on the brick, stone or iron, of the framework of a building, renders it useless, though the wood of every other part of the beam is sound; similarly, a telegraph pole, flagstaff, gatepost or a fence stob, if rotted at the ground-level, is rendered not only useless but dangerous.

In addition to actual physical deterioration, there are several forms of fungi which render timbers unsuitable for special uses by discoloration. In the case of sycamore, which is a valuable timber, and which is used to a large extent for the manufacture of dairy utensils, household utensils, and various articles for ornamental purposes where the white satin gloss of the wood is essential, mistakes in handling may, and often do, render the timber useless. It should be made widely known that in the case of this species late felling in spring is inevitably followed by a blue discoloration of the wood. Beech is another timber subject to discoloration, even after having been sawn up into planks, if not stacked suitably and with care. The timber of these and other species should be carefully "pinned" so that a free circulation of air is permitted. Recently a case came to light where the larger part of the product of a winter's output of a portable sawmill was lost, through the sawn material having been piled plank on plank without any attempt being made to secure a free circulation of air.

Prevention of timber rot is not only possible but it is a practical proposition; by practical I mean that it is worth while, from the economic standpoint.

While further research is desirable and necessary, we do not make anything like proper use of the knowledge we at present possess. First, good silviculture is important so that nothing but sound healthy timber leaves the forest.

Second: Care in handling, storing and seasoning is necessary.

Third: Artificial impregnation by antiseptics, coating with protective paint or other medium.

Fourth: Proper ventilation in buildings.

Fifth: The avoidance of contamination and infection through unsound timber being allowed to come in contact with sound timber. How often do we see this last precaution neglected!

I feel convinced that a great stride could be made towards prevention of rot in timber if it were more widely known that all forms of rot in timber are caused by living organisms and that infection may readily be conveyed to sound timber. Most people know the infectious nature of foot-and-mouth disease, cholera, etc., and speak of epidemics of these and similar disease-causing organisms in the animal kingdom, but few people appreciate the fact that the phenomenon of rot in timber is of a similar nature. We have become accustomed to a chronic epidemic of timber rot, and seem to take it for granted that this is in the order of things, and must be accepted like changes in the weather, without knowing that means are available to check the epidemic.

Research has shown how prevention can be applied, but further intensive investigations are necessary in order to make prevention more effective. The time when primary infection occurs, the inter-relation between timber-destroying fungi and their reaction to different forms of treatment deserve close attention.

Butt and heart rot in the forest and plantations of all ages occurs more frequently and is to be found over a much larger area than is generally supposed to be the case. We have reason to suppose from evidence already available that the soil, and especially the subsoil, has a lot to do with the phenomenon. The length of the rotation under which crops of different species are grown, or rather can be grown, on certain soils is a question that has to be very carefully considered in this connection. A study of blown trees is very instructive. When the tap-root, or deeper sinking secondary roots, cannot penetrate sufficiently deeply into the subsoil, one generally finds butt rot in the stem when cross cut from the root. Certain forms of stem and heart rot may occur high up on the bole of the tree where they have undoubtedly originated. However, one is brought to the conclusion that heart rot in the standing tree is generally brought about by the extension of butt rot, which in turn has originated from unfavourable root conditions in the soil.

SYSTEMATIC BOTANY AND ECOLOGY

THE BEST MEANS OF PROMOTING A COMPLETE BOTANICAL SURVEY OF THE DIFFERENT PARTS OF THE EMPIRE

(CHAIRMAN: PROFESSOR F. O. BOWER, F.R.S.)

Dr Arthur W. Hill, F.R.S.

It seems very fitting that in this year 1924, when the attention of the home botanists and other inhabitants of these islands is drawn so insistently to think about the Empire, that we botanists at this Imperial Conference should turn our thoughts to the Empire's vegetation not only from the purely floristic but also from the economic point of view.

In the past, as is shown by the well-known work of many British collectors, too numerous to mention, and by the invaluable labours of the writers of the floras of many parts of the Empire, British botanists have taken the lead in making known the richness of the Empire's botanical resources.

Can we say that to-day our younger botanists are carrying on the great traditions handed down to us by Banks, the Hookers, King, Kirk, Harvey, Bentham, Trimen, Forbes, Hemsley, Bayley Balfour, Thiselton-Dyer, Prain and H. H. W. Pearson in the domain of the systematic study of the floras of the British Empire?

It may be urged that the botanists at home—at Kew, the British Museum and Edinburgh, for instance—are carrying on faithfully the great works of their eminent predecessors, but what can we say generally of the provision that is being made for a proper study of the plants in the field in our overseas possessions, in order that we may learn as fully as possible not only about the range and variation of species with details of their relations to soil, altitude, and other ecological particulars, but also about their economic possibilities?

Speaking generally I venture to say that if, for instance, we wish to glean some general idea of the vegetation of Australia we have to turn to the writings and accounts of the botanical travels of American, German, or Czecho-Slovakian botanists, who have had opportunities denied to, or neglected by, the younger botanists of the Empire. The same remarks apply to our West Indian Islands—some of our oldest possessions—for a fuller knowledge of whose floras we now have to turn to our colleagues in the United States, also to the Falklands and Antarctic Islands where we have to consult the recent work of a Swedish botanist, while other examples could be cited with reference to East and West Africa and other domains.

As regards India, Ceylon, the Federated Malay States, Hongkong, and the Union of South Africa, I think we may claim, with no small degree of pride, that British and overseas botanists have upheld the best traditions of British botanists, but I will leave the discussions of the botanical affairs of these regions in the competent hands of overseas representatives who are with us and who have intimate knowledge of the countries and of the problems that await solution.

What shall I say of those rich fields, Trinidad, British Guiana, British Honduras, Kenya and Uganda and other regions, of some of which we know but the merest outline of their floras?

Why is it that we are not able to send our younger botanists to explore and study in our own possessions when foreigners can go in and reap the fruits which should be for our own gathering? Younger botanists of to-day, I venture to suggest, may be compared, not inaptly, I think, in some respects, with a celebrated character portrayed by Charles Dickens in *David Copperfield*. Need I say that I refer to that versatile but impecunious gentleman Wilkins Micawber! Mr Micawber offers more than one interesting feature of similarity with the botanists of to-day. He realised, like the ecologist, that careful survey of a problem on the spot was essential, and without unduly using him as a text for my remarks, I would only refer to the Medway coal trade, on which, as you will remember, Mr Micawber thought it might be desirable to embark. He surveyed the situation: in the words of Dickens—"We came," repeated Mrs Micawber, "and saw the Medway. My opinion of the coal trade on that river, is, that it may require talent, but that it certainly requires capital. Talent, Mr Micawber has; capital Mr Micawber has not."

There are, I think, three reasons why we have to regard the present state of affairs with concern and regret.

Talent, like Mr Micawber, our botanists undoubtedly possess, though they may not always have had their botanical talent directed towards the domain of the systematic study of plants, which many of us—regarding systematic botany in its widest sense—consider, and I think quite rightly, of paramount importance.

The second and very serious reason is the difficulty from which

Mr Micawber habitually suffered, and that is the lack of adequate funds for prosecuting our botanical survey work in the field in our overseas possessions.

While, thirdly, it must not be forgotten how vast are the territories over which Great Britain now exercises control.

For many foreign botanists our colonies offer a ready field for their activities and of course we gladly welcome their assistance, but I feel that we should all be glad to see some means afforded to our younger men in the matter of finance which would enable us to send out botanists to our colonies to study their floras and assist in rendering as complete as possible the botanical survey of the Empire.

This is, I think, the really crucial matter, for though we know fairly well what is needed and though we could produce men capable of undertaking the work without much difficulty, it seems almost idle to formulate schemes, however good and however necessary, unless at the same time we can put forward practical proposals for bringing them to effect.

What exactly does "a complete botanical survey" mean?

It certainly indicates a programme of work extending over many years, since a *complete* survey includes not only a full knowledge of the phanerogamic flora, but also of the higher cryptogams, the mosses, lichens, algae and fungi.

Then there is the economic aspect, especially on the physiological side, where so much remains to be discovered. Those plants, for instance, which yield such products as rubber, oil, resins, fibres, etc., we find exhibit physiological variations or differences within what we are accustomed to consider pure "species" on taxonomic lines, but how little do we yet know about the cause and real meaning of these physiological variants or races?

Yet from the purely economic standpoint very great issues and interests are involved, and I submit that it is our duty as botanists to enlighten the world of commerce, as far as may lie in our power, with regard to plants in their relation to man and their relation to conditions of soil and climate.

If we consider, in passing, the work of British botanists at home and all they have taught us in recent years with reference to species, subspecies and varieties, as well as the growth of ecological work, I think, that if ever it should be found possible to carry out work on similar lines throughout the British Empire, "the World itself could scarce contain the books that would be written!" But there is little fear, alas, of such an "intensive study" of the flora of any part of our overseas possessions, except possibly in New Zealand and some parts of Australia and South Africa, for many years to come.

When we look over the sheets of specimens which have been collected by botanists and travellers in various parts of the Empire it is sad to find that though they may be excellent specimens, they too often lack those particulars as to altitude, soil, climatic conditions and other general details which are so necessary for forming a correct idea of the flora of a region and gaining an insight into ecological matters of importance.

Our collections are mainly enriched in three ways:

(1) By trained botanists working at a particular flora, who give full details of all that is essential—such collections unfortunately are relatively few.

(2) By "botanical collectors," especially in India and China, who as a rule collect good material, but since they are rarely trained botanists, generally omit the information most needed to render the specimens of real value for survey purposes.

(3) By travellers and private individuals—their collections, though perhaps not very comprehensive, are often better than those of a paid staff, as they have been made as "a labour of love" with the keen eye of the naturalist and with full attention to important details.

Were it not for the collections made by such private individuals, often doctors, political officers, foresters, lady travellers and others interested in natural history in East and West Africa, India and other countries, our national collections and our knowledge of the floras would scarcely be half as complete as it is to-day.

What, then, is our chief need to enable us to achieve the objects we have in view?

First and foremost we should seek to send out as collectors expert botanists, who are not only good systematists, but who have also a keen interest in ecology.

This is the ideal to be aimed at, and it is an ideal which before the war was attained by Germany with regard to her colonies in East Africa (now the Tanganyika territory), and in the Cameroons. With our traditions in botanical matters and our great responsibilities with regard to the vegetation of the earth it ought to be possible for this country to set to work on similarly comprehensive lines.

Such a policy as I have indicated requires funds, and the question is where are the funds to come from?

The West African colonies have generously come forward to finance the work of the preparation of a flora of West Africa which is now being carried out at Kew. India also maintains an Assistant at Kew for dealing with the magnificent Indian flora in our herbarium and for the assistance of Indian botanists in the identification of plants submitted to Kew for identification from India.

South Africa until recently supplied the funds for an Assistant for helping the South African botanists in the determination of critical South African specimens, and I have hopes that it may not be long before this most necessary "liaison" officer will be reinstated at Kew by the Government of the Union of South Africa.

Similar assistance from the governments concerned will have to be sought if we are to take up, as is our desire, the preparation of the floras of Trinidad, British Guiana and British Honduras.

It is well, I think, to indicate what has already been achieved at Kew with regard to the preliminary work of a botanical survey.

The colonial and Indian floras and such ancillary works as *The Dictionary of Economic Products of India* and *The Useful Plants of Nigeria*, afford good evidence of this Imperial function of the Royal Botanic Gardens.

The proposal that a series of Colonial floras should be compiled at Kew was made by Sir William Hooker as long ago as 1863, and it is of interest to compare his proposals which were made public in the *Annals of Botany* for 1902 (p. lxxxiii) with the results that have so far been accomplished.

Colonial and Indian Floras

Sir W. Hooker's proposals	Volumes	Present position	Volumes
1. Australian Colonies including Tasmania	8	*Flora Australiensis*	7
2. South African Colonies	10	*Flora Capensis*	10
3. British North America, Pacific to Atlantic	2	(Not prepared)	
4. West Indian Colonies	2	*Flora of the British West Indian Islands.* (A new work is now needed.)	1
5. New Zealand	1	*Handbook of the New Zealand Flora*	1
6. Ceylon	3	*Handbook of the Flora of Ceylon*	5 and 1 quarto volume of plates
7. Hongkong	1	*Flora Hongkongensis*	1
8. Mauritius and the Seychelles	1	*Flora of Mauritius and the Seychelles.*	1
9. British Guiana	2	(Not yet commenced)	
10. Honduras	1	(Not yet commenced)	
11. West African Colonies	2	*Flora of Tropical Africa*	10
12. British India	10	*Flora of British India*	7

In addition to these the following floras have already appeared, or are in course of preparation, and have been prepared wholly or partly at Kew:

1. *Materials for the Flora of the Malay Peninsula* 6 volumes
2. *Flora of the Malay Peninsula* ...   . 5 „
   (in preparation, 4 vols issued)
3. *Flora of the Upper Gangetic Plain* ...   ... 3 „
   (1 vol. to complete)
4. *Flora of the Presidency of Madras* ...   ... 6 parts
   (under continuation)
5. *Flora of the Presidency of Bombay* ...   ... 2 volumes
6. *Flora of Bihar and Orissa*   ...   . 4 parts
7. *Flora Simlensis* ...   . 1 volume
8. *Flora of Aden*   ...   . 1 „
9. *Index Florae Sinensis* ...   . 3 volumes
10. *The Commercial Products of India* ...   ... 1 volume
11. *The Useful Plants of Nigeria* ...   . 4 parts
12. *Sketch of the Forestry of West Africa* (Moloney) 1 volume
13. *The Flora of West Africa* (in preparation)
14. Floras of Bermuda, St Helena, Ascension, Tristan da Cunha were published in the *Scientific results of the "Challenger" Expedition*
15. *Flora Vitiensis*   ...   . 1 volume
16. *Flora Antarctica* ...   . 2 volumes
17. *Flora of New Zealand*   ...   . 2 „
18. *Flora Tasmaniae* ...   . 2 „

Financial assistance for the preparation of the floras at home, however, is only a part of what is needed if the botanical survey of the regions is to be properly carried out.

We need, as I have said, the trained botanist in the field who will supply the information and the collections for the workers at home.

We ought to be able to revert to the old and admirable practice initiated by Sir Joseph Banks of sending out "Kew Collectors" to the colonies, but instead of sending out the "gardeners" as in old days —valuable as that then was—we now need to send out the botanist, and I think that the funds for this purpose—as I ventured to say in my paper to the Agricultural Conference at Jamaica in February last—should be found by the Home Government.

In addition to the work which could be performed by these expert botanists, much useful help could be given, as it has already been given, by the agricultural and forestry officers in the various overseas possessions, but here again there is a great need for providing these officers with a botanical training as well as with materials in order that their collections and notes may be of the fullest value.

Let us suppose for a moment that my somewhat Utopian suggestions have become matters of actual fact. What then is the position and how far can we say we are prepared to complete our botanical survey?

We are, I think, confronted with several important questions which are the concern of the herbaria.

Critical examination and identification, geographical classification and the incorporation of all the available material represent very large demands on the resources of the home botanical institutions.

Until, however, such work can be fully carried out the labours of the trained men in the field are of very little value to the community at large.

Again we are confronted with the financial difficulty. At Kew, had it not been for the labours of several private workers—for instance, Brandis, C. B. Clarke, Drummond and Gamble—much of our Indian material would probably not be in the well-arranged condition in which it is now, for with our staff fully occupied as it is, I regret to say that many families monographed by De Candolle, Engler and others have not yet been properly written up. Though Kew has many deficiencies in this respect it is some consolation, though no adequate justification, to know that our other great herbaria, through lack of funds, are in no better condition in this respect.

I have been speaking mainly up to the present from the point of view of the home botanist, but the overseas botanist must also be considered, as, after all, it is mainly in his interest and in the interest of his country that our efforts should be directed. Not only in India, Ceylon, Singapore, Hongkong, South Africa, etc., where magnificent collections of herbarium specimens exist for the guidance and for the studies of the botanists in those countries, but in the smaller colonies herbaria are equally necessary since their welfare and progress depend so greatly on an accurate knowledge of their vegetable products.

In order to render their collections as valuable as possible I venture to submit that

(1) They should be able to send over all their critical material to be accurately determined at home so that the duplicate specimens in their own herbaria can be accurately written up[1].

(2) It should be made possible to send out from the home institutions reference sets as far as possible of type material or of specimens carefully compared with the types to serve as the nucleus of these local herbaria.

In this connection I may allude to the desirability of instituting a collection of really good photographs of type plants which could be exchanged between the various botanical centres of the Empire, the negatives naturally being retained in the department to which the specimens belonged. Similarly, it would, I am sure, be of great advantage to the colonial herbaria if photographs of classical type specimens could be provided from Kew or from the British Museum, or from herbaria on the

[1] The three great Indian herbaria—Calcutta, Coimbatore and Dehra Dun—and also the herbaria at Cape Town and Pretoria, S. Africa, habitually send their critical material for verification or correction to Kew or to the British Museum.

Continent. Such a scheme would involve some outlay, but the cost would be amply repaid by the considerable saving that would result on costly journeys to compare specimens with the types.

It might also be considered whether in cases where two or more type specimens of classical specimens occur—as, for instance, of Burke's or Zeyher's South African plants—it would not be advisable to send one to the colonial herbarium where such types would be of value.

Kew, the British Museum and Edinburgh possess a certain number of duplicate type specimens collected by the earlier botanists and travellers, and if any such duplicate types could be presented to the herbaria of the colonies or dominions from whence these specimens came, the gift would be highly appreciated by the overseas botanists and would be of very great value to them in their endeavours to help forward the botanical survey of their particular portion of the Empire.

(3) There should be instituted an arrangement whereby workers could be temporarily exchanged. This would be of great benefit both to the home and to the colonial institutions since the home worker would gain enormously from being able to study taxonomic and ecological problems in the field, while the colonial worker would benefit greatly by being able to study in a large herbarium and library, and would thus learn how best he or she could assist in the preparation of the written "Floras" which, I think, must in the majority of cases be undertaken at home.

With regard to the exchange of workers. I may refer here to the official stay of our mycologist in the West Indies a few years ago, which was not only of value to the mycologists in those islands, but of very great value to our officer, as she was able to gain an extensive knowledge of tropical plant diseases caused by fungi in the field, and was enabled to appreciate the problems with which mycologists working in the tropics are confronted.

The visit will also bear further fruit when the hand-list of West Indian fungi, which is in preparation, comes to be published.

I may also refer to the advantages on both sides of the recent visits paid to Kew by the mycologists from Zanzibar and from Pretoria, as well as of the value of the recent visit of Mr Pillans, of the Bolus Herbarium, Cape Town, to study our South African collections.

Our American botanical colleagues are, as might be expected, fully alive to the advantages to be gained by pursuing their studies in visits to botanic gardens and herbaria outside their own country. They are always welcome visitors at Kew, where, no doubt, they gain much that is to their own advantage, while at the same time they leave us all the richer from the information and the inspiration they are able to impart.

With regard to the publication of any results of botanical survey work, it would be of great help to workers both at home and overseas if descriptions of all new species could be published in one or two well-recognised scientific journals, instead of being scattered, as is too often the case, in a good many periodicals, often of merely local interest, some of which are difficult of access. It is to be hoped that our overseas colleagues will do their best in this respect so that the results of their labours can always be easily accessible.

It would also be of value if a series of short preliminary papers, showing the present position of our knowledge of the botanical resources of each dominion and colony could be prepared, in which it could be indicated not only how much knowledge we possess about the flora of a region but also how far we are lacking in our knowledge of the vegetable products of any particular portion.

There is an excellent series of colonial handbooks giving a great deal of useful general information about each of our colonies, and I feel it would be of considerable use if a series of botanical handbooks could also be prepared. Such handbooks would be of service not only to the professional botanists but also to the administrators and political officials in the colonies, many of whom, I am glad to say, are only too willing to help us when they know something of the problems we are trying to solve.

In conclusion, since this is a meeting happily attended by so many of our colleagues from overseas, it may be considered fitting if I say a word with regard to the two great National Herbaria of Kew and the British Museum.

I am not going to enter into or invite discussions upon the history or upon the work of these two Institutions, but all I should like to say now is—and I do so with the approval of Dr Rendle—that Kew and the British Museum are working together in the friendliest co-operation.

We both realise that there is far more taxonomic work to be done than we are capable, even jointly, of tackling, but we are both doing our utmost to undertake the work which falls to our share.

At the British Museum at present they are engaged on the completion of the "Flora of Jamaica" and quite rightly so, since they possess the Sloane collections from that island, while we at Kew are engaged on the preparation of the flora of West Africa.

As we give all the help we can to the Museum in connection with their Jamaican flora, so they are always ready to give every facility to Dr Dalziel and Mr Hutchinson in their West African work, and with this policy of co-operation which so happily exists, I think you may rest assured that we at home will do our best to help forward, as far as may lie in our power, the botanical survey of the Empire the completion of which we are so anxious to accomplish.

Lt.-Col. A. T. Gage

As the immediate problem for discussion—within strict time limits—is not either the extent of the field or the area of it so far delved, but how to get more diggers and more and better spades and picks, I refrain from trespassing on your patience by attempting a sketch of the very varied and extensive Indian section of the field or an account of the diggers of the past and their work. It would in any case be presumptuous and superfluous on my part to make such attempt, for both sketch and history are available by master botanists in the shape of Hooker's "Sketch of the Flora of British India," written in 1904 for the *Imperial Gazetteer of India*, and King's "Sketch of the History of Indian Botany" delivered as his address to the Botanical Section of the British Association at the Dover meeting in 1899.

Although on the assumption that Science excels its devotees—who have nevertheless created it—we may for our immediate purpose disregard the life and work of individuals, a survey however brief of the general means by which botanical work in India (or anywhere else for that matter) has been carried out in the past, is likely to yield suggestions for the future. It is therefore significant to recall that the study of Indian vegetation on scientific lines began by the banding together in India towards the last quarter of the eighteenth century of a small body of men, who called themselves "The United Brotherhood." The aim of the Brotherhood was the promotion of botanical study and its keynote mutual help.

Although of necessity in those days many of the members of the Brotherhood were in the service of the Honourable East India Company, the Brotherhood, as such, neither rejoiced in nor suffered from the patronage of Government. ·Since those early days, although the Honourable East India Company and its successors have, according to their various and varying lights and purses, aided the investigation of the vegetation and the vegetable resources of the territories entrusted to them by the establishment and upkeep of botanical gardens and herbaria and by expenditure on publications, a very large proportion of the additions to such limited knowledge as we possess of Indian vegetation has been contributed by men independent of Governments but of the faith of the far-off and long agone Brotherhood, even if they had never heard of it.

We recognise with gratitude the valuable and almost indispensable assistance rendered by Governments in the past and in the present, but

in view of the fact that such aid, in comparison with the vast extent of
the field, is little more than a means of meagre subsistence to the science,
we need not adulterate our gratitude with any unduly exhilarating sense
of favours to come. Until Government itself becomes a branch of
Science—a prospect that is scarcely within our present horizon—it
would seem well to place any great accession of help from Governments
amongst those things that we may continue to hope for but not expect.

This may appear a somewhat jaundiced view, but that as far as India
is concerned it is not entirely due to hepatic congestion, may be supported by the fact that twice within the last twenty years, schemes have
been submitted to the Government of India by its Board of Scientific
Advice for the establishment of an official Botanical Survey Department,
with staff, income and equipment on a scale reasonably adequate to the
work in front of it. Both schemes were very carefully thought out by
special committees of the Board of Scientific Advice, and the first scheme
had the advantage of the advice of the three last Directors of the Royal
Botanic Gardens, Kew. The first scheme was submitted in 1907, the
second in 1919. On both occasions the department of Government
immediately concerned expressed sympathy with the proposals, but it
is in accordance with the sorry scheme of things, over which poets have
lamented and philosophers pondered, that the department which was
possessed of sympathy had no control of the cash, and the latter was not
available. Of such are the tears of things.

It has been mentioned as significant that scientific botanical investigation in India was started by an unofficial union of enthusiasts. It is
scarcely less significant that about the time that the second attempt to
persuade the Government of India to enlarge its so-called botanical
survey to dimensions less microscopic than its past or present proved
fruitless, faith in the development of the Science was invigorated by a
sort of re-incarnation of the eighteenth century Brotherhood, under the
more prosaic designation of "The Indian Botanical Society." The
coming into existence of this Society received no small stimulus from the
shortly preceding appearance of the *Journal of Indian Botany*, founded
and financially supported by a distinguished officer of the Indian
Imperial Forest Department. The Society, with the assistance of grants
from most of the Indian Universities, has recently taken over the conduct
of this *Journal* as its own organ. In recent years there has been a great
increase of interest amongst Indians in the vegetation of their own and
other countries, whilst wealthy Indians have endowed research institutes
and the Universities travelling scholarships.

This necessarily compressed, hasty and very incomplete account of the general means employed during the more than 150 years since Indian Botany in a scientific sense began purposely concentrates attention on the spirit and fact of union amongst botanists that distinguish both the beginning and end of the period, rather than on the extraneous aid that the Science has received during the period, not because such aid is to be despised or rejected, very far from it, but because union amongst botanists, if sufficiently comprehensive, is most likely to bring about an increase in such aid. Time permits only of allusion to the beneficent results to the Science that have flowed from the constant correspondence and association over the period between botanists in India and those in other parts of the Empire and the rest of the world. I am much more ignorant of other parts of the Empire than I am even of India, but probably it is not a misassumption that knowledge of the vegetation of other parts of the Empire has increased largely through means similar to those employed in the case of India. If association has been found, so far as it has gone, so helpful in the past, it would seem to indicate that at least one way of promoting not merely a complete botanical survey of the British Empire but the advance of all botanical work throughout it, would be the extension and intensification of the means employed hitherto. I therefore suggest for consideration by the Imperial Botanical Conference the possibility of the formation of an Imperial Botanical Union or Association.

Unless and until the Conference considers this suggestion worth discussing, it would seem needless to enter upon details, but a few general remarks are submitted as a contribution to a basis for possible discussion. The limits of the Union or Association should, I think, be wide enough to include not merely students of all branches of Botany, but also those who, while not claiming to be botanists, are interested in any way in the green mantle of the Earth, for the more comprehensive the Association would be, the more likely would it be in a position to afford help, to exercise influence and enlist extraneous aid. Unlike other Unions we hear of, it would be far from seeking either the restricting of work or of the number of workers or of their hours of work. It should not in any way conflict with existing societies by publishing results or reviews of work, such as are published in the Journals of existing societies. On the contrary, the suggested Union should, if founded upon a sufficiently comprehensive basis, be of assistance to existing societies by broadening the interest in the Science and cultivating ground for the raising of more strictly scientific recruits. The main aim of the Union should be the co-ordinating and complementing of botanical investigation in all lines and the spreading and cultivation of interest therein. It might publish at intervals Bulletins giving information of what work is in progress or in contemplation, where and by whom, where help is required, where and how it could be given, or the general conditions of botanical investigation or instruction in different parts of the Empire. It might serve to bring about more co-operation than exists at present between workers in different lines so that a systematist, a morphologist and a histologist might join forces in investigating the same group of plants. It has to be recognised that money is indispensable to the development of the Science and that the money is not in the pockets of botanists. One way of attempting to obtain a supply of the indispensable would be to advertise, in the best sense of the term, the supreme importance of plants to mankind and the vast vegetative resources of the British Empire and their incalculable economic value. From this point of view an Association such as is suggested could help to spread the knowledge necessary to appreciate the very great importance of botanical investigation in all lines, and given this knowledge and appreciation multitudinous opportunities would present themselves to those in possession of the means for the endowment of botanical instruction and research throughout the British Empire.

Dr I. B. Pole Evans, C.M.G.

I have been asked to give a short account of my views on the best means of promoting a botanical survey as regards South Africa.

The Union of South Africa is in the fortunate position of having a botanical survey which is now six years old. She claims, therefore, that she is already taking part in a botanical survey of the Empire.

The Botanical Survey of South Africa was promoted by far-sighted South African statesmen and supported by a small band of zealous botanists whose only objective was to labour in a vast and unexplored field. It also received unstinted support from scientific officers in closely related spheres of work, viz. Veterinary Research and Forestry. To the late General Botha, South Africa's first Premier—a keen agriculturist—our Survey really owes its existence. Since his time it has enjoyed the personal interest and support of the late Premier, General Smuts, himself a no mean student of our science. The loyal band of botanists to whom the Survey owes much includes such well-known names as Louise Bolus, J. W. Bews, Rudolf Marloth and Selmar Schonland. Goodwill and co-operation permeated "the Survey" from its inception and the gratifying results so far obtained are due to this.

The actual organisation of the Survey consists of a Director assisted by a small committee. The Director is the Chief Government Botanist for the Union, while the committee consists of Government officials and independent botanists. The latter are mostly professors of botany in the different South African Universities. Committee meetings are usually held once a year at different centres in the Union, and at these the various matters affecting the Survey are discussed and settled. At other times business is done by circulating correspondence through the Director.

For the actual carrying out of the work of the Survey the Union has been divided into five administrative areas each of which is in charge of a resident botanist. This officer acts in a purely honorary capacity, but receives a daily allowance from the Government when he is engaged on field work or other business connected with the Survey. His transport and travelling facilities are also defrayed by the Government. This officer is also empowered to grant travelling facilities to workers assisting him in his own area.

The administrative areas at present adopted are:

1. The Western and South-Western portions of the Cape, including the Cape Peninsula, under Dr Marloth at Cape Town.

2. The South-Eastern portion of the Cape, under Professor Schonland at Grahamstown.

3. The Eastern area, including Pondoland, Griqualand East, Natal and Zululand, under Professor Bews at Pietermaritzburg.

4. The Central area, including the Orange Free State and Basutoland, under Professor Potts at Bloemfontein.

5. The Northern area, including British Bechuanaland, Griqualand West, the Transvaal and Swaziland, under the Director of the Survey with headquarters at Pretoria.

Each of these areas has its own regional herbarium for which the botanist-in-charge is responsible, and where specimens of all collections made within that area under the auspices of the Survey are stored and preserved. From here duplicates of all material collected are sent to the National Herbarium at Pretoria. There the collections are worked through and then any new and doubtful material is sent to Kew for critical examination. At Kew the Survey had its own botanical assistant to attend to these collections and to keep it up-to-date generally so far as South African material and literature were concerned. This officer unfortunately resigned in 1921, and up to the present the Government of

South Africa has found itself unable to fill the vacancy. Provision has, however, been made in the current (1924) year's estimates for the post, and it is hoped that it will shortly be filled.

Such is briefly the organisation of the Survey at present. With regard to the future the Committee feels that much remains to be done if the country is to obtain the best results from the Survey. It is felt that the present system of voluntary workers must be supplemented by whole-time officers, and with this end in view it is proposed to station young trained South African botanists under the experienced botanists now in charge of each administrative area, so that in time they may be well qualified to take charge of the work and devote the whole of their energies to it. This principle has been agreed to by the Union Government, and a young South African botanist has already been stationed in the South-Western area at Cape Town under Dr Marloth, while provision has been made in the current (1924) year's estimates for a similar assistant for the South-Eastern area at Grahamstown under Professor Schonland. In due course it is proposed to deal similarly with the remaining administrative areas. Only when this is brought about will it be possible to undertake a systematic mapping of the vegetation and many other important matters connected therewith.

With regard to the Survey's assistant at Kew it is felt that the post should be occupied by a rotation of South African trained botanists who have already served as assistants under the Survey. Each officer would be required to spend two years at Kew, or any other of the great herbaria of the world as occasion might demand. This would ensure that the Survey's assistant at Kew or elsewhere would have a field knowledge of the native flora and not be solely a herbarium worker. At the same time it would enable South Africans to take up some research work and complete it at Kew or elsewhere when their turn came round to act as the Survey's assistant overseas.

In this connection it is felt very strongly that Kew ought to have on its staff, as it has done in the past, a permanent assistant for South Africa, for the growing activities of the Union Survey will make this even more necessary than before. The work accomplished by such men as Mr N. E. Brown, Dr O. Stapf and Mr J. Hutchinson speaks for itself, and it is a matter of regret amongst South African botanists that they have never had an opportunity of seeing or meeting such distinguished colleagues in the country to whose botanical knowledge they have contributed so much.

The promotion of a complete Botanical Survey of the different parts of the Empire would be the biggest step made by botanists in modern times, for a properly organised and adequately supported Survey must lead to scientific and economic results of far-reaching importance. To accomplish this, it will be essential to enlist the interest, sympathy, and support of British and Dominion statesmen. This can best be brought about by indicating the scientific and economic value of such a Survey. The broad scientific problems might be outlined, and the main economic results likely to accrue might be stated. The Director of Kew might well draw up a memorandum showing what Kew with her limited resources has actually accomplished up to the present for the different parts of the Empire, while the botanists in the Dominions and Colonies might in the same way draw up memoranda showing what scientific and economic results of value they have obtained from botanical exploration both within their own spheres of activity and beyond. Such data ought to be quite sufficient to convince the most prejudiced that a profitable field for investigation still awaits enquiry, and that its exploration is intimately connected with the prosperity of the Empire.

The chief scientific aim of the Survey would be to extend our knowledge of the vegetation and floras of the different parts of the Empire, so that as much light as possible might be thrown on the origin of the different floras, their relationships with other parts of the world, the conditions which have influenced them in the past, and their agricultural, industrial and economic possibilities. This would entail a systematic mapping of the vegetation units, a detailed study of the floras, and the publication of local floras, memoirs and handbooks. For this the establishment of central herbaria with botanical museums in each Dominion would be a *sine qua non*.

On the economic side one of the principal aims of the Survey would be the free interchange of plants likely to be of commercial and agricultural importance; such as would lead to cheaper food supplies, land development, and the creation of new industries.

No one who is acquainted with the work of the office of Seed and Plant Introduction of the Department of Agriculture at Washington and knows what botanical exploration has done for the United States can fail to realise what could be accomplished in a similar manner in the different parts of the Empire if such work were undertaken by a well-organised Botanical Survey of the Empire. From what I know of South Africa, its potentialities and its possibilities, I am convinced that on this one aspect alone an Empire Survey would more than justify its existence. It remains therefore in this short paper to indicate briefly how such a

Survey could best be organised and administered from a South African point of view.

The headquarters of such a Survey would naturally be at Kew, where the foundations for such a structure have already been well and truly laid. Here the services of a Director of the Imperial Botanical Survey would be required to guide and co-ordinate the work in the different parts of the Empire. In addition, so far as South Africa is concerned, I think a permanent botanist—styled the Assistant for South Africa— would be required, who would be responsible for all material submitted from the South African Survey, and under whose personal guidance and direction the various Survey assistants from South Africa from time to time would work.

The organisation in South Africa would remain much on the same lines as at present, but much greater attention would require to be paid to plant introduction stations than has been done in the past. In a country like South Africa, whose area is so vast, whose physical features are so varied, and whose climate and rainfall are so diversified that distinct vegetation regions result, no one station or Botanical Garden could possibly fulfil all the essentials required for the acclimatisation of plants from all other parts of the Union or from overseas. The Union of South Africa is fortunate in possessing botanical collections of living plants under such different climatic conditions as Kirstenbosch, Grahamstown, Pietermaritzburg, Durban and Pretoria, where it is possible to grow specimens of the many remarkable types of the rich South African vegetation. In the National Botanic Gardens, Kirstenbosch, where the Bolus Herbarium is now housed, South Africa has a garden of remarkable interest and great promise, though unfortunately many typical hardy indigenous South African plants are unable to adapt themselves to the climatic conditions prevailing there. Many of the plants which are unsuited to the conditions at Kirstenbosch, however, flourish in the drier climate of Pretoria, where a fine botanical collection is being cultivated in the grounds of the laboratory. Thus it is possible in one or other of the South African Gardens to have the different aspects of the flora of the country thoroughly well represented.

At least one plant introduction station in each vegetation region of the Union would be desirable. For this purpose use might well be made of the existing Botanic Gardens and Agricultural Schools already established in South Africa. If such work were properly conducted and came directly under the supervision of botanists in charge of the Survey it would be a sure means of bringing untold wealth into the country. Not only would these stations be used for the introduction and study of plants from overseas but the indigenous economic plants of the country would also receive close attention there.

To attain the desired results in this direction Kew would naturally be the clearing-house for the Dominions. Promising economic material would be forwarded by the Director of the South African Survey to the Director of the Imperial Survey with all the necessary information regarding it, and this in turn would be apportioned out by the Director at Kew to the different parts of the Empire where it was considered most likely to succeed and to be useful. In the same manner South Africa would receive plants worthy of trial from other parts of the Empire through the Director at Kew. The exchange of herbarium material, museum material, and botanical literature could be carried out in the same way.

Such are the broad outlines on which a complete botanical survey of the different parts of the Empire could be promoted. The details, such as means whereby further use might be made of the botanical staffs of the various Universities and of private individuals who were willing to participate in the work, would naturally have to be left to the discretion of those in charge of each Dominion Survey in consultation with the Director of the Imperial Survey.

In conclusion, I think that it would be highly desirable to hold a meeting or conference of the chief officers of the Empire Survey at least once every three years and it would be greatly to the advantage of each Dominion concerned if such conferences were held in the different parts of the Empire in rotation.

Prof. S. Schonland. SOME DIFFICULTIES OF THE BOTANICAL SURVEY OF THE UNION OF SOUTH AFRICA

(Abstract)

The following points were briefly dealt with: vastness of the territory, large parts botanically quite unexplored, difficulties and expenses in travelling, absence of reliable topographical maps, only incomplete data available as regards Meteorology, Geology and Soil-chemistry, very few workers, diversity of the flora even in comparatively limited areas in the coast-belt, uncertain seasons and consequently plant-successions difficult to study (especially in the S.E. area), rapid changes in the flora owing to the influence of man, etc., taxonomic knowledge of South African flora very incomplete.

SYSTEMATIC BOTANY AND ECOLOGY

CORRELATION OF TAXONOMIC WORK IN THE DOMINIONS AND COLONIES WITH WORK AT HOME

(CHAIRMAN: PROFESSOR A. C. SEWARD, F.R.S.)

Dr J. Burtt Davy

Introductory

This Conference appears to be agreed that an organized survey of the botanical resources of the British Empire is urgently needed. In such a Survey the Imperial Government has good reason to take a prominent part, as it depends so much on the overseas portions of the Empire for the supply of raw materials for manufacture, and of foodstuffs. I would go further, and say that unless the botanical institutions at home undertake a large share of the work it will be impossible to conduct a mutually beneficial survey. By a mutually beneficial survey is meant one which will bring to the notice of this great centre of consumption the vast resources of available material ready to hand in the Dominions and Colonies overseas.

It is important that we should recognise the wholesome sense of self-reliance developing in the overseas Dominions, which extends to matters scientific as well as to those political and industrial. But there is danger of the pendulum of public and parliamentary opinion in Great Britain swinging too far and throwing the whole burden of responsibility for the investigation of their natural resources upon the Dominions and Colonies, before these are strong enough in population and financial resources to carry it, and before the Mother Country can afford to cease fostering their trade.

The scope of such a botanical survey is wide, as is indicated by the agenda of this Conference. Of several equally important items, taxonomy is one. A previous generation of botanists has been accused of pursuing taxonomy as the sole aim of botanical science, instead of as a means to the true end. I do not think our great taxonomists really made any such mistake—they were too deeply imbued with true learning and wisdom. But they found that without names to handle their plants, and without adequate descriptions by which to distinguish between allied species—the plant of the desert and its close relative of the steppe or the rain-forest—no proper morphologic, phylogenetic, ecologic, or economic study of plants could be carried out. And they must have found

—as we do to-day—that to distinguish between species, with these points of view in mind, often involved a complete revision of a genus or other group of plants. As the number of known species multiplied, it became necessary to prepare local floras, without which it proved practically impossible to name the species except by constant reference to the large home herbaria, involving delay and expense. So the botanists of half a century ago, and rather more, with true scientific spirit, decided to concentrate on laying the foundation for future botanical work by defining and classifying the known species of plants. It is possible that in their endeavour to be thorough in this great undertaking, their disciples lost sight of the wood in the dense forest of trees, for towards the latter part of the last century a reaction against taxonomic botany set in, and swept through the schools and universities in such a wave that —with very few exceptions—it has almost ceased to form part of the curriculum. But the tide has again turned, and I have heard able ecologists complain of being hampered in their work by the dearth of really qualified systematists available for the critical naming of their material. This is very significant, and I hope that from this time on we shall pursue a more sane policy, combining taxonomy with phylogeny, ecology and the other branches of our subject.

If I appear to have digressed somewhat from the subject of the Correlation of Taxonomic Work in the Dominions and Colonies, with work at home, which I have been asked to introduce, it is because an explanation of the *raison d'être* for taxonomic work seemed necessary.

The subject is so wide that it is impossible to cover the ground in the time to which the introducer is necessarily limited with such a full agenda as we have before us, therefore Dr Rendle and Prof. Craib have kindly agreed to relieve me of certain specific aspects of the problem.

The part left for me to deal with, is the share of this work which must necessarily fall on the home institutions. The discussion of this aspect of the question must necessarily involve suggestions the adoption of which—if found practicable—would, it is believed, greatly facilitate co-operation by overseas workers and reduce the number of obstacles to efficient correlation of work on both sides of the waters. These suggestions are based on extended experience in botanical institutions in several countries, both British and foreign. It is hoped that they will prove helpful in the evolution and development of herbaria both overseas and at home.

The development of a survey of the botanical resources of the Empire is retarded by our lack of accurate knowledge of the floras and vegetation of much of the area comprised. The preparation of up-to-date series of Handbooks to the Flora and to the Vegetation of the several Dominions and Colonies is an essential feature of the Survey and an urgent necessity.

It is desirable that some degree of uniformity of plan should be followed in the preparation of such handbooks, and it would be to the interest of both the overseas and the home communities if the organisation and publication of the series were carried out at the Centre of the Empire. This is one of the directions in which the Home Institutions could be of great service.

The series of publications suggested includes:

1. A preliminary series of short papers showing the present position of our knowledge of the botanical resources of each Dominion or Colony. Might we not treat the Report of the Proceedings of this Conference as the first instalment of this series? These introductory papers should include, also, preliminary annotated lists of the plants known to occur in the particular areas under consideration, where such do not already exist.

2. A series of Handbooks to the local floras of the several political areas, or of those groups of political areas which are allied phyto-geographically. This subject is discussed more fully farther on (see p. 217).

3. A series of Handbooks to the Vegetation of the Empire on the lines of Dr Cockayne's *Vegetation of New Zealand*. If we had had the vision to organize such a series some years ago, it is reasonable to suppose that Dr Cockayne would not have found it necessary to send his manuscript to Berlin, instead of to Great Britain, for publication.

4. A series of Monographs of plant Families for the larger continental areas. These might be issued as fast as they could be prepared, irrespective of sequence in the series, but numbered according to a uniform sequence of Families as adopted for the Survey. We have examples of this method of treatment in Dr Britton's *North American Flora* and Prof. Engler's *Das Pflanzenreich*.

5. A serial publication in which descriptions of all new species from within the Empire could be published, or reprinted, promptly. It should also include revisions of genera and other taxonomic notes. Its commercial value would be increased by the addition of newly acquired information on economic plants. Such a publication would be of great value to overseas botanists, as it would materially reduce the time, labour and expense of getting at the original descriptions of new species. Every botanical department within the Empire should be able to subscribe for it, and no botanical institution outside the Empire could afford to be without it.

Every botanist within the British Commonwealth of Nations who describes a new species elsewhere, should recognise it as a duty and a privilege to send a reprint of his paper to the editor of this proposed central publication. I would suggest for the consideration of the Continuation Committee of this Conference, in conjunction with the responsible authorities, that the *Kew Bulletin* or its *Additional Series* might be enlarged to meet these requirements.

The Survey Publications would not be complete without two other series, which, as they are outside the scope of this paper, I will merely mention in passing. These are:

6. A series of Handbooks to the Economic Plants of the Empire. A beginning has already been made by the publication of *The Useful Plants of Nigeria* and the *Commercial Products of India*. The scope of each Handbook should include "economic plants" in the widest sense of the term (*i.e.* useful plants, noxious weeds, and poisonous plants), together with a summary of the crops grown or capable of being grown either (*a*) for domestic use, or (*b*) for export. A volume classifying this information by products, and giving—where known—the latest recorded figures of production and acreage, would provide invaluable information as to sources of supply of raw material and foodstuffs, available in times of emergency or otherwise.

7. A serial publication dealing with ecological botany. The *Journal of Ecology* already supplies this need in admirable form as far as it goes. With the co-operation of all the botanists of the Empire, in a co-ordinated Survey, the number of available contributors and subscribers would be greatly increased and the general interest of the publication would be widened, so that there should be opportunity for its enlargement.

The Preparation of Handbooks to the Floras

It is with the Handbooks to the Floras of the several States, Provinces, Colonies, Protectorates and Territories of the Empire, that this paper is particularly concerned, and they are fundamental to any botanical survey. Without the means of naming our plants accurately, sound useful ecological work is impossible; catalogues of the economic plant products prepared under such conditions, must be unreliable and often misleading, and any account of the geographical distribution of species would be worthless owing to the uncertainty of the specific identity of the plants discussed.

I should like to take the opportunity that is afforded me by this public gathering of botanists, to express my deep sense of the tremendous debt of gratitude which the Colonies and Dominions owe to Kew—and may I couple with it the Botany Department of the British Museum—for the invaluable assistance they have rendered in the investigation of the botanical resources of outlying portions of the Empire.

We owe to Kew in particular a great debt of gratitude for the series of Floras elaborated by, or at the instigation and with the continued support of, her several Directors and staff. These include: *Flora of British India* (7 vols.); *Flora Simlensis* (1 vol.); *Flora of the Presidency of Bombay* (5 vols. and 1 quarto vol. of plates); *Flora Hongkongensis* (1 vol.); *Index Florae Sinensis* (3 vols.); *Flora of Mauritius and the Seychelles* (1 vol.); *Niger Flora* (1 vol.); *Flora of Tropical Africa* (11 vols.); *Flora Capensis* (10 vols.); *Flora Australiensis* (7 vols.); *Flora Antarctica* (2 vols.); *Flora of New Zealand* (2 vols.); *Handbook to the Flora of New Zealand* (2 vols.); *Flora of Tasmania* (2 vols.); *Flora Vitiensis* (1 vol.); *Materials for the Flora of the Malay Peninsula* (6 vols.); *Flora of the British West Indies* (1 vol.); *Flora Boreali-Americana* (2 vols.); and the Floras of Bermuda, St Helena, Ascension and Tristan da Cunha published in the *Challenger Expedition Report*.

Some of these Floras are out of date, but they remain the foundations on which are being (or will be) constructed the Local Floras of the smaller constituent areas, which are now required. Some of these are already completed; others are in course of preparation; and we hope that others —long desired—will soon follow if this Conference gives the necessary urge. Those in course of publication and preparation include: *Presidency of Madras* (5 parts); *Upper Gangetic Plain* (3 vols.); *Bihar and Orissa* (5 parts); *Malay Peninsula* (5 vols.); *Jamaica* (at the British Museum, Natural History Department); *Upper Guinea*; and the *Transvaal with Swaziland*.

But for many of the Colonies, Protectorates and Mandated Territories we have as yet no Floras. Among other regions for which Descriptive Handbooks are needed may be mentioned: Kenya with Uganda; Tanganyika Territory; Nyasaland with Northern Rhodesia; Southern Rhodesia and the Bechuanaland Protectorate; the S.W. Protectorate; the Sudan and Somaliland; Samoa; some of the Australian States; British Columbia and other parts of Canada; British Honduras; British Guiana; Trinidad and Tobago; and the Leeward and Windward Islands. Until these Floras are completed, the other branches of the Survey of these areas must remain at a standstill. For this piece of work we need the money for preparation and publication, and well-trained systematists to carry it out.

In those cases (*e.g.* Fiji, Mauritius and the Seychelles, and the British

West Indies) where the existing Floras, after having served a good purpose as foundations, are now hopelessly out of date (owing to the acquisition of more complete material), new and simplified Handbooks are required. These should be compressed within reasonably small compass, to eliminate unnecessary bulk and reduce cost. To this end the principle of Descriptive Keys to genera and species, and the omission of all unnecessary duplication of descriptive matter, for both genera and species, might be followed to advantage; space thus saved could be devoted to economic and ecologic notes, local names and pertinent biologic notes, and illustrations of economic or otherwise interesting species.

It is desirable to cite the original *"type specimen,"* the range of distribution, and the collector's numbers of selected specimens, for each species. It adds to the value of the work if the date of publication of the accepted name and of each synonym cited is given; but as the reference to place of publication can be obtained from the *Index Kewensis*, it need not be cited in a Handbook.

From some of the Colonies very little material is available, and extensive collections are yet required before their floras can be written. The Home Institutions could help in the botanical exploration of such areas by calling attention to those regions from which more material is urgently required, and by encouraging expeditions thereto. For the Crown Colonies, Protectorates and Mandated Territories, we usually have to rely on the Home Institutions (with such financial and other assistance as the Colonies concerned can render) in the preparation of Floras.

The self-governing Dominions should be encouraged and assisted—as required—to prepare and issue their own Local Floras, in line with, and to form part of the general series suggested above.

The preparation and publication of these Floras should be encouraged as a sound public investment. It is one of the means of stimulating immigration, land settlement and the development of agriculture, forestry and commerce, thus helping to place the particular region in question on a sound financial basis. To wait until a young Colony is in a position to support such an expenditure unaided, is to put the cart before the horse, and to delay its progress.

The Importance of the Large Home Herbaria in Floristic Work

At present, and for many years to come, the preparation of Floras for the Dominions and Colonies can be done satisfactorily only by constant reference to the Home collections, owing to the concentration of type material in these herbaria. The earliest collections of plants from overseas were made by travellers, or by collectors sent out for this purpose from Europe, and were taken home by them; there were no public herbaria in the young Colonies in which the collections could be housed. These specimens formed the nuclei of the larger European herbaria and naturally attracted to them the collections of other early travellers and residents in the Colonies, until such time as the colonial populations grew large enough to take a practical interest in the formation of local herbaria.

The necessary result is that the "*type specimens*," on which we rely for the interpretation of the—so often inadequate—published descriptions of species, are largely housed in Europe. In the case of the British Empire, these types are to be found chiefly at Kew and the British Museum; these two herbaria contain the collections acquired by the two Hookers, and by Bentham, Banks, Aiton, and others, on which foundation the series of Colonial Floras, already referred to, has been constructed. It is on this account that colonial botanists, *e.g.* Mr Gamble, Mr Ridley, Mr Haines, Mr Fawcett, Dr Dalziel and the present writer, engaged in preparing local Floras, find ourselves obliged to do the work in England, instead of in the particular country concerned, and without access to the living specimens. Not only is this because experience proves that it is practically impossible to write a Colonial Flora without consulting the type specimens, but also because of the great wealth of collateral material and literature now accumulated in these great National collections, and the invaluable libraries attached thereto.

How the Home Herbaria can Assist

It must be obvious, if we think about it at all, that the purpose of a paper on the Correlation of Taxonomic Work at Home and in the Dominions is to bring out suggestions for its more efficient development; and, further, that the individual who deals with the subject has a very invidious task to perform. In self-defence I must disclaim any hand in the choice of my subject; had I realised as fully as I do now all that it meant, I do not think I should have had the temerity to accede to the request. I can only conclude that the Organising Committee of this Conference selected me for the task because no one else was willing to undertake it.

We botanists from overseas have received so much, and such invaluable assistance from Kew and the British Museum, and such kindness and courtesy from the individual members of their respective staffs, that we feel for these great parent institutions a deep sense of filial affection and gratitude. May I, therefore, express the hope that the members of this Conference, the majority of whom are connected with home institutions, will accept my remarks in the spirit in which they are offered, that is to say, as being suggestive and in no sense critical.

The Functions of the National Herbaria. The large National Herbaria may be said to have three primary functions:

1. The maintenance and development of working collections in the best possible condition for reference and study by the botanical public.

2. The maintenance and development of complete reference libraries of botanical works, and other allied publications required for consultation by working botanists.

3. The naming of collections from Colonies not possessing public herbaria, and of *critical* material from any part of the Empire.

In the next few pages certain directions in which the several home herbaria might render very great service to overseas botanists are suggested.

Rapid Incorporation of Accessions and of Accumulated Material. Collections should be incorporated as promptly as possible after receipt. If there is much unclassified material accumulated where it is not accessible to the public (as was the case after the late war and the succeeding period of reconstruction) every effort should be made to incorporate it as soon as possible. If the accumulation is great, it may become necessary to stop other work and to incur extra assistance in order to clear up the arrears, and to render the collections accessible to workers. Even though named only generically, such material will be of great use to writers of Revisions, Monographs or Floras, if placed in the genus covers; left unsorted in the stores it is useless to anyone. Where there are such arrears due to the war, there is a legitimate claim on Government for special temporary expenditure for the purpose of catching up.

Collation of the Herbarium with Monographs, Revisions and Floras. Trained systematists capable of carrying out floristic and monographic work are never numerous. Theirs is a profession with limited possibilities for employment, carrying small salaries, and requiring long training. Few, therefore, are attracted to it, and the work waiting to be done is always far in excess of the available workers. The time of the latter is correspondingly valuable, and should not be wasted over sorting and writing up material to make it available for floristic or monographic work. Such preparation, though requiring care and accuracy, is to some extent mechanical, not calling for that mature judgment and knowledge of species required for monographic and floristic work; it could be performed quite well by permanent second-class technical assistants with a little suitable training and experience; only such exceptional conditions as those imposed by a great war should make it necessary for a public herbarium to expect this type of work of visiting botanists.

The collation of the collections with published Monographs, Revisions and Floras is one of the most important duties of the staff of a herbarium. Where this has not already been done the collation with existing standard works should be brought up-to-date as soon as possible, and it should be part of the routine work of the staff to collate new publications immediately they appear, so that the herbarium sheets will reflect the latest authentic information. In some of our herbaria this has been done to a certain extent, but owing to the war there is, in some herbaria, a serious accumulation of arrears of this kind of work, which calls for attention before other things are undertaken.

Collation with Monographs and Floras denotes the marking off in the library copy of the particular book in question, of those specimens cited therein which are represented in the herbarium, and the addition to the herbarium sheet of the name adopted in that work, with a reference to volume and page.

In the Kew herbarium many of the specimens cited by De Candolle in the *Prodromus*, Harvey in the *Flora Capensis*, Hooker in the *Flora of British India*, Bentham in the *Flora Australiensis*, and by Engler, Diels, Gilg, Knuth and other monographers, have been written up in this way, and it has added very much to the value of the collections, proving of the greatest assistance in determining what particular plants the authors had in mind; it also benefits the herbarium by acting as a check to the type material which it contains. Unfortunately, this work has not always been recognised as an essential part of the duties of the staff, but has been left largely to the conscience of the individual assistant, with the result that it has not been completed.

Where an assistant collates a Monograph or Revision with his particular herbarium, he is adding to the value of the collections, and is saving the time of every subsequent writer of a Flora in whose area members of that particular group are found; otherwise, each of these workers must collate for himself the specimens with the Monograph, at great loss of time. It ought not to be necessary for men sent home by Dominion and Colonial Governments at considerable expense, and for private workers sacrificing time and money at great inconvenience to themselves, to spend their time in this way. These men are here for the purpose of studying types and preparing local floras in the interest of science and the botanical public, and the whole of the limited time at their disposal should be available for that work. By maintaining their collections, in this way, at a high standard of efficiency beneficial to themselves and to local workers, the Home Institutions can render invaluable assistance to visiting botanists, and materially further the botanical survey of the Empire.

Indication of Type Specimens. The practical value of an herbarium, to visitors at least, is greatly increased by the definite indication on the sheet that a particular specimen is the type or co-type. With all new species such indication should be made at the time they are described, and with all the old types as soon as this work can be undertaken. At the British Museum this is done by means of a small neat gum-label with the words TYPE SPECIMEN in *black* type; at Kew and at Paris a similar label in *red* type is used. But in these herbaria only a small percentage of the types are at present so indicated. At the same time that Monographs and Floras are collated with the herbarium, it would be an easy matter for the assistant to indicate the types and co-types.

Where more than one specimen is cited under the original description, it is desirable that the author should select and indicate on the sheet at time of publication which particular specimen he recognises as the type of the species; this prevents possible confusion or doubt in the case of subsequent segregation.

Cross-references on Herbarium Sheets. Trouble and loss of time are caused by trying to find species which have been sunk or transferred to other genera. This difficulty would be avoided if the person who made the change, or the one who collated the herbarium, were to leave the original species cover in its old place, noting on the outside the species or genus to which the contents have been transferred. This is done at Kew, with very great advantage, and might well be adopted as the regular policy of all herbaria.

Geographical Information on Sheets. Certain valuable collections incorporated in our public herbaria were not provided with locality labels by the collectors, the geographical and habitat notes being contained in separate catalogues. Some of these collections—*e.g.* those of Drège and Burchell—contain valuable *types*, and much time is lost in referring from the sheets to the catalogues to obtain the locality-records for working out geographical distribution. Where this is the case, an appalling amount of time is wasted—over and over again—by herbarium staff and by visitors in hunting up data; in several instances the geography of the species has been omitted from the literature on that account. It would be of great assistance if a technical assistant could be put on to the work of systematically writing up these particular collections once for all. This, perhaps, comes next in importance to the collation of the herbarium with standard works.

Field Notes and Economic Information filed with the Specimens. The increased interest in the botanical resources of the Empire, and the need for rapid collation of information, call for some improvement in the system now in general vogue in herbaria. We need to keep in mind the necessity for promptness and efficiency in supplying information of economic or horticultural interest, and to avoid becoming obsessed with the idea that the collections, however important, are an end in themselves. The aim should be to develop the herbarium cases into a card catalogue of information—as complete as possible—concerning the plants, instead of a mere collection of specimens. This would result in much saving of time to visiting botanists and also to the staff.

Useful information as to colour, habit, habitat and native lore are furnished, sometimes, by correspondents in remote parts of a new or little-known country. The collection of such information should be encouraged.

In the case of herbaria under control of Government departments much useful information of economic interest is buried in the correspondence files, where it soon becomes impossible to trace it owing to the time involved in searching the files, or in compiling a cross-index; when this correspondence is eventually destroyed the information is irretrievably lost. By mounting, on a herbarium sheet, the original letter or an extract made by the correspondence clerk, it can be placed at once with the specimens of the species concerned, and is then accessible for reference at a moment's notice. This method is more effective and time-saving than cross-indexing the correspondence, and it is encouraging to find that at Kew it has been followed to a certain extent; the desirability of adopting it as the established policy of every herbarium is worth consideration. Notes and illustrations of value taken from periodicals such as the *Kew Bulletin, Journal of Botany, Icones Plantarum, Botanical Magazine, Gardeners' Chronicle, Journal of the Royal Society of Arts, Consular Reports*, etc., could be treated in the same way, and to advantage.

My friend and former colleague, Mr N. E. Brown, A.L.S., realised years ago the value of MSS. notes on economic, horticultural and botanical points furnished by correspondents, and where and when he could make time to do so, he extracted and attached them to the herbarium sheets. The Kew collection is rich in such notes, and many workers bear testimony to their great value. But for Mr Brown's forethought in this matter, much valuable information furnished by African collectors, *e.g.* the late Mrs Barber, Colonel Bowker, Sir John Kirk and others, would be irretrievably lost, for native tribes change rapidly with the advance of civilisation and soon forget about the native foodstuffs, fibres, dyes, etc., used by their forbears, so that much information obtainable from them by travellers may be lost long before any thorough ethnological survey has been made.

When economic notes on species are filed separately (*e.g.* in a separate room or in a separate building), so much time is lost in connecting them with the specimens that in practice it is often omitted under pressure of work.

Filing of Illustrations and Original Descriptions with the Specimens. Where separate illustrations of plants can be obtained of a size not too large for the purpose it has been found advantageous to mount and file them in the herbarium covers (in a separate species cover, where this is considered desirable), rather than in a separate collection. An admirable practice formerly in vogue at Kew, which might with advantage be carried out in all herbaria, is that of preserving on a herbarium sheet the original description of new species, especially those published in periodicals.

All of this is work that could be done by a second-class Technical Assistant, and if carried out systematically, as part of the regular herbarium routine, immediately on the publication of new species, the time involved would be but a fraction of that at present spent by staff and visitors in passing to and fro between the herbarium and library to compare specimens with figures or original descriptions. This latter constitutes a very important item with short-time visitors from overseas.

A method suggested to me by Prof. Craib is that of filing the mounted illustrations of any one family on shelves reserved for that purpose in the last cabinet of the particular family. With certain woody plants, the drawings may become damaged if filed in the same species-cover. The advantage of this alternative method is that thereby the illustrations are protected from injury by irregularly shaped specimens such as those of conifers and palms, while still being placed in proximity to the specimens to which they relate.

Adoption of a Geographical Subdivision in the Collections. The geographical subdivision of the herbarium is much more useful for floristic work than the monographic. Visiting botanists are subjected to adverse criticism for not making greater use of certain collections. As one of the offenders I may mention that a well-known American botanist remarked to me that, owing to the lack of geographical subdivision in a certain herbarium, it took him twice as long to accomplish the same amount of work there as in those herbaria where there was a thorough geographical subdivision. Personally I should like to see the geographical arrangement carried still further than is done in any of our herbaria, but I doubt the practicability of so doing at present.

Collection and Incorporation of specimens of Cultural Plants. I wish to emphasise the need for herbaria, both at home and overseas, to include in their collections representative specimens of the cultural plants of the various countries concerned, particularly those of agricultural and forestal interest, but not omitting the various morphological forms of leading horticultural species. These are of value in facilitating the rapid naming of plants sent in for determination. But even more important is their historic value, as a record of the morphological changes which have taken place in our crop plants, through selection, etc., over a period of years. For lack of such a record, our knowledge of the changes which have taken place in historic times is peculiarly meagre. The work of making such a collection (which should be incorporated with the general collection) ought to be undertaken methodically, in accordance with a well-laid plan. The plants desired should be listed, and marked off the desideratum list when incorporated, and the work should be continued till all the desiderata have been acquired.

I am indebted to Professor Wright Smith for the suggestion that this is a favourable opportunity to bring forward the value of this aspect of herbarium work. It is one which has always interested me, and when attached to the staff of the United States Department of Agriculture at Washington I strongly advocated the plan of representing in the herbarium the cultural plants of the country, but did not remain at Washington long enough to see the policy carried out.

Representation of all Empire Species in the larger British Herbaria. A concerted effort should be made to secure representation in one or other of the three large British herbaria of specimens or representations of every species known to occur within the Empire. This would involve securing duplicate types, or photographs or drawings of types, or material critically matched with the types, of all Empire species of which the types are deposited in foreign or colonial herbaria. To be at all satisfactory this work must be done systematically and thoroughly, not sporadically.

In discussing this matter with Professor Wright Smith he has made the very excellent suggestion that each of the overseas herbaria might undertake to obtain for itself from the European herbaria, representations of the types of species in which it is particularly interested. He further suggests that in undertaking to do this, the overseas herbaria should—in the interests of botanical science—see that a representation of every such type is deposited in one or other of the three large British herbaria.

In making these suggestions, he points out that they form part of a broader problem, for consideration by an International Conference, which would provide that the publication of a new species, to be valid, must include the deposit in one or other of ten leading herbaria (*i.e.* one in each of the following cities or countries: Kew or the British Museum, Paris, Berlin, Vienna, Petrograd, Leyden (or elsewhere in Holland), Brussels, Rome, Washington and Japan) of a representation (*e.g.* co-type or photograph or authenticated drawing of the type) of such new species.

I venture to suggest that we go a step further and urge the next International Botanical Congress to provide that valid "publication" of a species must *include* representation in all of the ten herbaria aforesaid, by authentic specimens or other representations, of such new species. It is beyond the province of this paper to discuss the broader suggestion at this time. But it would be a valuable piece of constructive work for this Conference to recommend to the botanists and institutions represented here that authentic material (or authenticated illustrations) of all new species described by them, shall hereafter be deposited in one or other of the great British herbaria. In the interests of science, it should be a matter of pride and honour with us to see that all new species described by us are represented in the herbaria of Kew or the British Museum because of their unrivalled advantages as central clearing-houses for information. The day should have passed when any institution worthy of recognition as scientific would take pride in possessing a unique collection of types not represented elsewhere. Not in that way is science fostered, but rather by a wide distribution of authenticated material, which reduces the danger of loss by fire and otherwise. How much confusion and nomenclatural trouble has been caused through the destruction or loss of types!

Types in Continental Herbaria not represented in Britain. Overseas botanists working here on Colonial Floras, particularly those of any part of Africa, are handicapped by the number of species of their particular areas which have been described by continental botanists during the last fifteen to twenty years, and which are not represented in Britain. In the case of many genera it is impossible to revise them for the local

Floras without seeing this material. But to divide one's time between the British and Continental herbaria involves too great a loss of both time and money, and it is necessary, in most cases, to compare the specimens side by side. Furthermore, if the prestige of our British herbaria is to be maintained it is absolutely necessary to fill these gaps. This can be done by securing co-types or portions of type specimens, by matching our unnamed material in the Continental herbaria, or by obtaining drawings of the types.

The Home Institutions could assist the Dominions, and also add greatly to the value of their own collections, by securing a complete set of illustrations of types preserved in Continental herbaria, of those species, at least, which concern any part of the British Empire. Copies of these could be supplied to the Dominions concerned, at their expense. This would be no light task and would require comprehensive and sound organisation. But if the work were well organised, divided between several institutions (as is done by the National Observatories which co-operate in making the star maps of the heavens) and spread over a series of years, it would not be as serious an undertaking as might appear at first sight. It would be of such very great value that it would well repay the outlay. As in the case of duplicate types, a system of exchange might be inaugurated, and the sale of sets of photographs, etc., would help to cover the cost. As in the case of the distribution of types and co-types, the work should be done methodically and systematically.

In like manner overseas botanists should supply to the Home herbaria photographs or drawings of types deposited in the overseas herbaria.

I would suggest for the consideration of this Conference the desirability of strengthening the hands of the Home Institutions, by pressing for more adequate support to enable them to send members of their staffs from time to time to the Continental herbaria, to match unnamed material, to select and arrange to secure duplicate types and drawings or photographs of types where specimens cannot be obtained or matched. To be really efficient we must be constantly alert to the possibilities of improvement. For this reason, also, it is of practical importance that not only the administrative heads, but also the scientific staffs of our large herbaria, should be given the opportunity to visit the large Continental herbaria from time to time.

Scientific institutions in which the work is of a highly specialised nature, and which are isolated from other scientific institutions, and from men working at other branches of science, tend under the very best of conditions to develop one-sidedness and eccentricity in their staffs. It is easy to overlook the fact that there may be more than one way of doing a thing, and that the requirements of one particular branch of science change with increase of knowledge in other branches. "What has been must be, because it always has been," is an attitude as fatal to growth as it is contrary to the whole trend of evolution. The natural development of our particular branch of science then becomes checked and dwarfed or malformed from lack of that stimulus to expansion which naturally follows from contact with like institutions.

Such interchange of visits cannot be undertaken without money, and I do not suggest a large annual outlay; but this should be made part of the annual programme of the Department concerned. One member of the staff of each of the larger institutions might be sent abroad every year. If a well-thought-out scheme were organised, and the work done systematically, a good deal of ground could be covered in a few years.

Protection of Types. The value of duly authenticated material in the preparation of floristic and monographic works, suggests that steps should be taken for the greater protection of type specimens against loss by fire and from other causes. The experience of many herbaria is that the care bestowed on types is largely a personal matter, not adequately provided for in regulations but varying inversely with the personal interest taken by the particular curator. This is most noticeable in young countries, but the older herbaria are not free from danger. The preservation of the valuable collection of types belonging to the California Academy of Sciences was secured by the fact that they were stored in separate cases in which they could be removed bodily in the event of fire. It was due to this fact that Miss Eastwood, the indefatigable curator, was able—though at no little personal risk—to remove them to a place of safety, at the time of the great San Francisco fire. The lesson should not be forgotten, and I, personally, favour the segregation of actual types, dividing the herbarium into a working collection and a purely reference collection, the latter to be preserved in easily removable cases; this is now done, I believe, in several herbaria in the United States.

Distribution of Duplicate Types. The concentration of both type and co-type in the same herbarium is dangerous on account of possible loss of both by fire. The concentration of types in the European herbaria has been a hindrance to the development of taxonomic botany overseas. It is also true that the preparation of local Floras and Monographs, which for a long time to come can be carried out satisfactorily only in the home herbaria, is hindered by the absence of types or co-types of species described and published overseas. By the exchange of co-types between the Home and Dominions' herbaria, floristic and monographic work would be greatly assisted. There is a good deal of such material in the Home herbaria which could be readily spared.

I would urge upon the Home authorities the importance of the distribution of duplicate types and critically compared material to the Dominions concerned. The work of selection should be done by a systematist of experience; the expense of making the selection ought in fairness to be borne by the Dominion concerned. It is important that if this work is undertaken—as I earnestly hope it will be—it should form part of a well-planned scheme of exchange of co-types, and should be carried out systematically; to attempt a sporadic distribution would, I fear, prove unsatisfactory. Lists of desiderata, and of the types and co-types offered, should be furnished to each of the herbaria concerned, and the co-types actually exchanged should be checked off on these lists, which would in time form a valuable permanent record of the location of type specimens relating to the particular floras of the Empire.

Drawings and photographs of types, though less valuable than type specimens or co-types, as they often fail to show the important or characteristic points, are nevertheless of great value. And where co-types are not available, drawings or photographs are a valuable aid in determining the particular plant which the author had in mind when he drew up his description.

Visitors' Notes. The critical herbarium notes of visiting botanists, particularly of monographers and authors of Floras, may be of great value to subsequent workers, as indicating the particular plant the former were writing about. To ensure that such workers will authenticate their notes, it is suggested that each visitor working for any length of time in our larger herbaria, be furnished with a rubber stamp or small printed "determinavit" slip, bearing his name, with the definite request that he attach one to each sheet on which he comments.

In the older collections there are many sheets in the handwriting of the Hookers, De Candolles, Bentham, Asa Gray, Harvey, Sonder, C. B. Clarke, and numerous monographers and other workers of importance, but bearing no signature. The handwriting of many of these men is still known to a few—but very few—living botanists, and it would be invaluable if some means were adopted by each institution of identifying the authorship of such notes before it is too late.

Towards this end, and for the preservation and identification of the handwriting of present and future botanists, it would be well if each herbarium were to adopt an Autograph Book, in which each visiting botanist would leave a specimen of his handwriting; these could be arranged alphabetically, in an indexed and bound volume, and the book should be additional to the ordinary visitors book which is signed daily.

Alphabetical Guides to Families in the Herbarium. Many modern botanists have been brought up on the Engler and Prantl System, and are not sufficiently conversant with the Bentham and Hooker sequence to follow it readily. It matters little what system is followed in the herbarium arrangement, but it is important that visitors who are using it should be able to find their way about with the least possible loss of time, and without taking up the time of the staff to find things for them. It would be of great assistance if alphabetical indices to the Benthamian and Englerian Families, with references to the particular alcoves and galleries in which they are housed, were placed in readily accessible positions and at frequent intervals throughout every herbarium.

Library Classification. In some cases great saving of time of visitors and increase of efficiency of staff could be effected by a rearrangement of libraries. The books should be classified by subjects, and sets of those books constantly used for reference should be kept as near to the specimens as possible. Botanical libraries should be properly card-catalogued and cross-indexed. In the classification the geographical arrangement of books and pamphlets would be of great assistance to both visitors and staff. The lack of cross-references to titles of periodicals often causes much loss of time in finding the one sought. All pamphlets should be catalogued both by authors and by subjects. The author-catalogue of *all* books and pamphlets in the library (including travels, books of reference, etc.) should be continuous, not subdivided. But extensive subdivision in the subject-catalogue is to be commended.

Necessity for Increased Assistance

For a whole generation at least the cry has been for more men qualified to carry on floristic work. To-day the supply of available material seems to have grown less, rather than greater. To secure the right type of man some incentive to promotion and opportunity to make a name as a scientist are required.

Scientific Assistants. In public herbaria the first duty of the staff, as employees of the public, is the maintenance and development of the collections for use by the botanical public. This necessary routine involves a considerable amount of drudgery. It should not be forgotten that the very best training in systematic botany is obtained in the routine work of "laying out," "sorting in" and "matching" plants in a large herbarium. It is training which no assistant—however high his other qualifications or scientific knowledge—should miss. Experience shows that only after novices have had some two or three years' work of this character do they become efficient herbarium assistants.

But the efficiency of the staff cannot be maintained at a high level if it sees no outlet for its energies beyond the daily routine of "laying out" and "sorting in"; ambition must be stimulated at the same time it is kept within the bounds. If the Senior Assistants could be encouraged to take part in the preparation of Handbooks to Colonial Floras, and of Revisions and Monographs, and the revision of Floras now out of date, their interest in taxonomic work would be stimulated and they could, thereby, also further the work of the Botanical Survey of the Empire.

Research Assistantships. The establishment of a certain number of permanent Research Assistantships, to which would be eligible—as vacancies occurred—those Junior Assistants who have proved themselves capable while serving an apprenticeship in the routine work of the herbarium, would act as an incentive and be conducive to greater efficiency and provide a certain number of qualified men for floristic work. But the drudgery of the ordinary herbarium routine might be minimised by a regulation that all Assistants be allowed to devote their official time one day a week to some piece of floristic or monographic work approved by the head of the Department.

But the dearth of well-trained systematists capable of preparing floristic works, and the number of these latter for which we are waiting in order to carry out ecological surveys intelligently, suggests the urgent need of releasing the time of trained assistants from the more mechanical routine work, in order to prepare the requisite Handbooks. There is no prospect of much if any increase in the number of private individuals coming forward to volunteer for this service; in fact the number is likely to be less than greater, for our modern systems of education do not seem fitted to produce naturalists.

Technical Assistants. Our larger herbaria, both at home and abroad, suffer loss by not making use of "Second Class" or "Technical" assistants. It is not economical, not a "business proposition," to put trained University men on the more mechanical routine work of the herbarium, while collections lie year after year in the stores waiting to be named. Each trained Assistant should have one or more Technical Assistants for whose oversight and direction he is responsible. I know that difficulties attend such appointments, especially in the Government Service, but they can be overcome. Such Technical Assistants should be removable if they do not earn their "salt"; but, on the other hand, efficiency should be encouraged by security of office and increase of emoluments to those who prove worthy, with a possibility, also, of qualifying to cross the "bar" which separates their grade from the one above, as vacancies in the latter occur.

I have been impressed by the fact that there are in this country persons with a taste for natural history and a knowledge of plants, who have not the opportunity for taking a University degree. Among such there are those who would be thankful for even the routine work of museums and herbaria (such as mounting, laying out, sorting in, matching common species, etc.) as long as it afforded the opportunity to work among natural history material. Such people, when once trained, may become invaluable to an institution (I have in mind the late Mr Shrubbs of the Botany School, Cambridge, known to several of you), and I would offer as a suggestion that the establishment of *permanent* Second-grade Technical Assistantships, to which this type of person (not necessarily possessed of a University degree) would be eligible, would relieve the pressure of work on the First-grade Assistants, and tend to increase the efficiency of the staff.

Financial Assistance. The suggestions outlined above, if carried out in their entirety, would probably require the augmentation of the staffs of our public herbaria.

I do not know how far our home institutions are seriously cramped for lack of funds. But if this is the case, this Conference ought to strengthen the hands of the responsible authorities in their efforts to secure more financial support. Their importance demands the full support of the Treasury.

Co-operation of Universities and Unattached Systematists

As the taxonomic work involved in any programme for a Botanical Survey of the Empire is necessarily great, as the work is so much in arrear, and as trained workers are so few, the necessity for co-ordination of the work and the co-operation of all available workers at home is imperative. Why should not this Conference form a strong Continuation Committee of Taxonomists, to organise and carry on this part of the Survey? If such a Committee be formed, I would suggest that it endeavour to organise the work in such a way that each public herbarium and each University having a herbarium and with one or more systematists on its staff, should be invited to undertake a definite piece or pieces of work

—floristic or monographic, or perhaps in the nature of field investigation. In any such arrangement the overlapping of work should be carefully guarded against, for workers are too few to justify duplication. I cannot but suppose, also, that there are several unattached systematists in this country—the names of several come to mind as I write—who would doubtless welcome the opportunity to take part in the work of an Imperial Botanical Survey if the way were made clear for them. We need to *co-opt the services of every available trained worker willing to assist*. It is conceivable that some of the Universities would require grants of public funds to enable them to undertake work of this character, if only to provide assistance for mounting, poisoning, and sorting away material. But if the survey were well-planned it ought to be possible to obtain small grants for this purpose.

In conclusion may I summarise my plea, that as the late war found the British Empire woefully behind in its knowledge of its own resources, as this knowledge cannot be obtained at a moment's notice, and we do not know when we may urgently require it, we ought to organise our forces in a systematic effort to acquire it.

Let us take stock of our resources, see what we can do by co-ordinating existing agencies, plan our campaign to cover a series of years, and then come forward with a definite, clear-cut scheme, backed by all the forces at our command, and boldly ask for what funds are required. Now is the time to come forward with a comprehensive programme. Applications for money for small details of scientific work, which bear no relation to any definite scheme, are turned down, when a well-planned practical programme, offering good value for the proposed expenditure, appeals to the man of affairs. We cannot do big things without money; it is equally certain that we cannot get the money without a precise scheme to lay before those on whom we depend for the funds. Nothing succeeds like success. If we plan our work well and demonstrate our ability to "deliver the goods" we shall command the confidence of the Governments concerned, and also their continued and increased support.

Dr A. B. Rendle, F.R.S.

I have been asked to indicate how the overseas dominions can most effectively co-operate with the Home Institutions in taxonomic work.

This may be considered under three heads: (1) Collection, (2) Conservation, (3) Determination.

1. *Collection of specimens.* In South Africa a definite botanical ex-

ploration of the country is being organised, but this is not the case in all the overseas Dominions. In our work on the flora of Jamaica we find that a number of species are known only from single specimens collected more than 100 years ago by Dr Wright, when resident in the island; and Mr Ridley recently, when on a short visit to the island, brought home, in a small collection made near Kingston, at least one new species (a plant quite conspicuous). It is evident that the botanical exploration of one of our oldest overseas Dominions is by no means complete; in fact areas of great interest have scarcely been touched.

Obviously the first essential is the development of an interest in its own flora by the overseas Dominion, and the organisation of the interested units by some central authority, government official or otherwise. The ideal arrangement is the marking out of definite areas to be assigned to individual collectors who are resident throughout the year so that a complete representation of the flora is obtained. But more effective work might be done if systematic efforts were made to get into touch with persons interested in the flora in various parts of a Dominion, and offering tactful suggestions as to the best methods of work. The highly trained botanist may be a poor collector, but a little botanical knowledge will greatly increase the value of the work of an amateur who may have a natural aptitude for collecting. Some of the best specimens we have received at the Museum have been from this type of collector, as some of the worst have been from the trained botanist. Given some central authority whose duties include the taxonomic investigation of the flora of his area, it should not be difficult to co-ordinate much help which is at present largely running to waste for lack of knowledge.

In an area which has been inadequately explored everything should be collected which is in the proper condition for collecting and the different stages of the various plants should be included. Careful notes should be made as to locality, altitude, habit of the plant, time of flowering, colour of the flower and fruit, and any other item of interest. The collector should aim to present as complete a picture as possible of the plant for the benefit of the worker. If he can add a photograph of the plant growing so much the better. Instructions should be given as to methods of preservation of the specimens; lack of care may spoil a large number. It should be emphasised that ten good specimens carefully collected and dried are more useful than a hundred casually collected or poorly preserved.

To summarise, collection should be systematically carried out in definite areas, and the specimens, with the notes, should give as complete a representation of the plant as possible in as good a condition as possible.

If the specimens, when dried, are to be dispatched by post or otherwise instructions should be given. Much damage is done to specimens in transit by inadequate packing. Bundles should be very tightly tied, and protected by oiled paper or canvas.

2. *Conservation, etc.* For future reference each specimen should bear a number; specimens taken from the same plant should of course bear the same number, but unless there is absolutely no doubt as to identity specimens from different plants should bear different numbers. The number should be written on a tag and tied on to the plant before it is placed in the press. Notes should be entered in a small pocket-book against the number on the plant.

The herbarium label will be prepared later and will bear the number of the plant and a transcript of the notes.

Specimens from overseas sometimes arrive in a dilapidated state owing to insect or fungus attack. The value of good work in collecting may be discounted by inadequate precautions at headquarters.

3. *Determination.* It should be possible to name the majority of the specimens collected at the one or more herbaria in the overseas Dominion. And it goes without saying that each Dominion should aspire to a complete representation of its flora. The attitude should be that of the British Herbarium to the county floras. But as early collections in overseas Dominions were in most cases sent to Europe, need for comparison with the types in the home herbaria will continually arise. Whenever possible duplicate specimens should be sent home, one to be retained, the other to be returned. If desired there would probably be no difficulty in obtaining a photograph of a type specimen. When new species are described by overseas botanists co-types should be sent for preservation in the home herbaria, and similarly when species are recorded hitherto unknown for the flora. The representation of the overseas flora in the home herbaria should give a general idea of the distribution of the species; the detailed distribution must be left to the local herbaria.

It is also important that a comparison be made between the species of other, perhaps even widely separated, areas; for instance, the flora of British North America with that of the Old World temperate area, and this can be done adequately only in a great general herbarium.

Help in this should be obtainable at home, but it would add greatly to the efficiency of the overseas worker if he were encouraged to visit frequently the home herbaria, to consult types, compare critical material and obtain a broader outlook on the components of his own flora.

Prof. W. G. Craib

The part of the subject with which I have been asked to deal is the inter-relationship of the individual in the Dominion or Colony and the Home Institutes.

Experience on the staff of the Kew Herbarium supplied me with details of taxonomic work as understood by systematists on this side; a brief tenure of the post of Curator of the Herbarium and Library at Calcutta helped me to realise the many difficulties to be surmounted by systematists away from London, and at the present time I am once more practically in the position of the Dominion and Colonial botanist with a very incomplete and often unsatisfactorily determined herbarium and with a limited library at my disposal. In addition I have had placed before me from time to time the difficulties of many Indian systematists. On such personal experience I would base a few remarks from both points of view.

I think we are all agreed that the field for work is a vast one and that in spite of the enormous amount of collecting that has already been carried out there are not too many workers at the present day. At the outset, therefore, we naturally ask from what sources we can draw our recruits. In the first place we have those who are to proceed to an overseas herbarium or to an overseas botanical-survey post. Their selection for such posts should be a sufficient guarantee of their being adequately trained and equipped. There is, on the other hand, the private individual who botanises as a hobby. In the past the science has been deeply indebted to such and every encouragement should be given by the home institutions. In addition to these we have the large number of University graduates who go overseas every year to posts in, *e.g.*, the Forestry and Agricultural services. Special training in systematic work is given to many of these, but others proceed to their allotted posts without any knowledge of systematic work beyond what can be provided in the Universities in an already overcrowded curriculum. Would it not be possible to provide a course in systematic work to such of those men as desired it before proceeding overseas? A suitable course might be one of say six or nine months in one of our larger herbaria spent in the routine work of sorting out and laying in specimens and finally in systematic work either general or restricted to some particular genus. I feel certain that a large percentage of the men would accept eagerly the chance of such a training during a probationership.

Such a training, besides being of possible real economic value afterwards to the individual, would set the men on the lines of a hobby which has been adopted very successfully by so many of their predecessors. In addition, it would supply that personal link with the home staffs which is so frequently lacking. Further, it would bring home to the individual, by his own experience, the truth of the points so fully elaborated by Dr Rendle.

In many of the overseas establishments there are now local herbaria more or less fully representative of the flora of the district and with the specimens more or less satisfactorily named. With the help of such an herbarium some part at least of the collection can be named, but there will be in practically all cases a residue which must be sent to a home institution for naming. The less fortunate officer who finds no local herbarium would be forced to build one by sending home all of his first collection for naming. I feel that the stage now reached is a critical one and probably worthy of fuller examination as to the possibilities. The enthusiastic collector has sent home specimens which he has failed to name, specimens which have very often been collected under difficulties and dried at some considerable expenditure of time. He awaits eagerly the reply from home. The institutes at home should do everything possible to encourage the collector and at this stage no greater encouragement could be given than the prompt naming of his collection. As soon as possible after receipt of the specimens they should be examined critically and named as far as possible. From most regions there will be a residue of doubtful material. The reasons for leaving such specimens unnamed should be communicated to the collector and where necessary additional material should be requested. By prompt and careful naming of collections received the home herbaria will be effectively maintaining the interest of the collector, and it must always be remembered that the home herbaria are themselves the ultimate gainers by a fostering of the ready collector.

The picture, however, is incomplete without at least a glimpse at the other side. However impatient the collector may be to have his plants critically examined he must remember that the staffs at home are very limited and that they have to deal with similar collections from all parts of the world, many of them being official and therefore entitled to precedence. Although somewhat outside the scope of the present paper, I feel that I must, in passing, refer to the present staffing of our national herbaria. Even allowing that these herbaria through mutual working arrangements minimise overlapping and consequent duplication, the staffs are inadequate to cope with the work and the increase of staffs, where any has occurred, has not been commensurate with the considerable increase in the volume of work requiring attention. In this connection one wonders why scarcely any systematic work is now done in our Universities; formerly, the Universities produced a fair amount of systematic research work, but now that is almost entirely centralised in the national herbaria.

In conclusion, I should like to refer to one book which is, probably more than any other, an essential reference book to the taxonomist whether as a solitary worker or as a member of the staff of an institute —the *Index Kewensis*. The compiling and seeing through the press of the *Index* is no light undertaking and I am certain that it is not generally recognised outside Kew what the work entailed really is.

The *Index*—and this is not generally appreciated—is a private venture on the part of the authorities at Kew, and is additional to the work which has to be carried out by that institution. Consequently, it might appear ungenerous to offer any criticism in connection with it. Its essential value is, however, in its being kept consistently up-to-date by the punctual appearance of the quinquennial Supplements, and this, owing to the general disturbance during the last decade, it has not been possible to maintain. The accumulation of arrears has to be worked off in addition to the examination of an ever-increasing mass of current literature, and it would appear essential that some additional assistance be rendered to Kew to enable this monumental and unique work to regain its former efficiency. No doubt the best way would be to enable the Kew authorities to overtake the work in hand by the addition of temporary staff, and this needs help. Is it too much to ask that those institutions, both at home and overseas, which have so greatly benefited by this publication, now come forward and offer what assistance they are able?

SYSTEMATIC BOTANY AND ECOLOGY

SURVEY AND STUDY OF VEGETATION, AND TRAINING IN ECOLOGICAL FIELD WORK

(CHAIRMEN: PROFESSOR F. W. OLIVER, F.R.S., AND MR A. G. TANSLEY, F.R.S.)

Mr A. G. Tansley, F.R.S. INTRODUCTORY REMARKS

The occasion of the first Imperial Botanical Conference seemed the ideal opportunity for discussion among botanists from different parts of the Empire on the very important subject of what could and should be done to advance and consolidate in a systematic way our knowledge of the plants and vegetation which form one of the greatest imperial assets. I say plants *and* vegetation advisedly, because a mere knowledge of the species which exist in the different Dominions, Colonies, Dependencies and Protectorates is obviously insufficient. In addition to and alongside of the description and cataloguing of species it is necessary to study the aggregates of plants which we call units of vegetation or plant communities, their composition, origin and behaviour, both in a "natural" state and when subjected to the influence of man and his animals. I think everyone who has had practical experience, in the field, of studying and of dealing practically with vegetation will agree that this is so. It is not necessary to have a complete knowledge of the species composing a vegetation unit in order to begin the study of the unit as a whole. If it were so, no vegetational study would be possible in those regions, and they are the vast majority, in which the species are incompletely known. It goes without saying that the better the species are known the more successful vegetational study will be, but the two can and should proceed side by side, and should be as little divorced as possible, for they are closely interdependent. For that reason it is most desirable that field botanists should be capable of dealing with both aspects of botanical survey, and that we should aim at a training which includes taxonomic botany and also the study of vegetation; though we have to recognise that, of individuals, some are more interested in the one, some in the other, aspect of the subject. The description and recording of species has naturally hitherto occupied most of the time and energy of workers, but though a vast amount still remains to be done in this field, it is necessary now that we should take the other aspect—the study of vegetation as such—consciously and seriously into the account, and bend our energies to promoting it.

The fundamental and far-reaching importance of the practical scientific study of vegetation as such is now beginning to be well recognised here and there, more especially in the United States and New Zealand. We ought to try, as an imperial duty, to spread this recognition throughout the Dominions, Colonies and Dependencies of the whole Empire.

I will not take up the time of the Conference at this stage by going into any details, but will merely mention the divisions into which the problem seems naturally to fall:

(1) The ascertainment, record and publication in summarised form of what has been already done. This is, I think, an essential preliminary task, which could be undertaken at once. Dr Chipp will, I believe, put forward proposals for the summarising, correlation and publication of all the serious work which has already appeared in print. Such a summary, with full references to original sources, would be an indispensable basis for present and future work.

(2) The desirability and possibility of drawing up a scheme or schemes, sufficiently elastic but sufficiently definite, to serve as a guide to workers as to what should be aimed at. If such schemes were to be at all detailed they would have to be worked out for each Dominion and for each Crown Colony or group of Colonies by the botanists on the spot, the Directors of the local Botanic Gardens, and so on, because they alone have the necessary local knowledge.

(3) The very important question of the preliminary training of workers before they enter on their work. Many, perhaps most of the field workers —foresters, officers concerned with grasslands, etc.—who work in the Dominions are trained there, but the officers who go out to take up botanical work in the Crown Colonies and Protectorates mostly get their preliminary training in Great Britain, and it is very important that the home universities should have at their disposal some authoritative expression of opinion as to the lines on which such training should be conducted. In the opinion of many people well qualified to judge, current botanical training is at present seriously defective from this point of view.

All these topics will be dealt with from different angles by those who will speak after me, and we shall conclude to-morrow afternoon with a general discussion, and I hope with resolutions[1], which will give some practical effect to the conclusions of the Conference. It has been suggested that we in the home country ought to have done our best to thrash out the question beforehand and to present definite recommendations to the Conference. But it seemed to me premature to take

[1] See pages 384, 385.

this course, and much better to give the members of the Conference the widest opportunity, within the limits of the available time, to hear various views and to discuss the subject unfettered. In my opinion, however, we ought not to separate without formulating conclusions and taking the necessary steps to give them practical effect.

Dr T. F. Chipp. SUMMARY OF WORK DONE IN THE CROWN COLONIES, INDIA, ETC.

The presentation of this review has proved a real difficulty. There are very many Crown Colonies and there are many studies embraced in a vegetation survey. The difficulty has been to avoid a mere citation of bibliography or an enumeration in a dictionary form of the Colonies themselves. By steering a middle course I have endeavoured to put this *résumé* before you in as interesting a manner as possible. May I emphasise, please, that the object of this paper is to summarise the work that has already been done in the Crown Colonies. It does not contain any expression of personal opinion or observations which are reserved for a later paper.

The Crown Colonies are a series of political entities of the Empire stretching right round the world within the tropics. In no case does one contain within itself any complete vegetational zone, but each stretches across or is cut out of a main zone of tropical vegetation and may comprise every type of vegetation to be found within the tropics—the heavy tropical evergreen forest of Malaya, West Africa and the West Indies, open parkland and grass savannahs of Ceylon, Guiana and East Africa, the so-called "alpine" vegetation of the Central African mountains and the deserts of the Sudan.

It is, therefore, convenient when treating of the Crown Colonies to consider as a unit for discussion all which contain or are included in the same zone or zones of vegetation. In this manner we may group the Colonies as follows: An Eastern group comprising the Malay States, Straits Settlements and Borneo; Ceylon; a North-East African group comprising the Anglo-Sudan and Somaliland; an East African group of Uganda, Kenia and Tanganyika; a Central African group of Nyasa and Northern Rhodesia; a West African group of Nigeria, the Gold Coast, Sierra Leone and the Gambia; and a West Indian group comprising the British West Indian islands, British Honduras and British Guiana.

It is quite realised that very many Crown Colonies and territories have not been mentioned above, but the time limitations of this paper preclude consideration of all but the larger.

As it has been necessary to limit the areas under discussion to the major groups of Colonies so the sources of information consulted have only been traced for approximately the last ten years.

These sources may be classed under five headings. Government publications form the bulk of the literature and provide information in such records as Colonial Reports, Official Handbooks to Colonies, military reconnaissances, periodicals and bulletins of scientific and technical departments, special bulletins and publications by individual research officers. Another source is the records of special scientific missions or research promoted or undertaken by Universities, Societies and individuals. Some of these have been organised and carried out locally, others have been sent out from the home country with a naturalist and sometimes a botanist attached. Then there are a few compilations effected at home by authors who may or may not have a field knowledge of the country they describe, but who have access to libraries. The diaries and narratives of travellers such as hunters, missionaries, political officers or traders who record their impressions of the countries they traverse provide valuable information, especially when they are accompanied by photographic illustrations.

Lastly, there is the more modern record of the cinematograph film such as Angus Buchanan's "Across the Great Sahara." These portray the vegetation more vividly than pen pictures and provide a permanent record of the state of a country at definite periods.

In glancing over these sources of information one cannot fail to be struck by the large number of non-British who have worked in the Crown Colonies, as, for instance, in the Eastern group, German travellers as Schimper and Karsten, and Italian—Beccari—or American workers tracing the southward extension of the Philippine flora. In the North-East, East and Central African groups we find the work of Germans and also Swedes. In West Africa we are largely indebted to the French, though here it is their work in the area surrounding the British Colonies which has been so helpful. In the British West Indies and America we have the work of Danes and Swedes, and the most recent and complete ecological work by Americans.

The field of work is large and the workers few, so that I think we should rather welcome this outside aid gratuitously rendered by other botanists.

When we come to a consideration of the organisations and institutions which have been more directly responsible for the prosecution of the work we find that local Governments have helped in a certain measure. Their handicap is that they find it extremely difficult to provide funds for what appears to be purely scientific work. It is only when they can realise that such work has a practical application and, though it may not produce any immediate revenue to the country, will be of ultimate advantage, that they can be induced to finance the workers and pay the cost of publications. Foreign Governments have assisted, as is instanced in the missions sent by the French to West and Central Africa and Guiana, and those organised from the Berlin Botanic Gardens to East and Central Africa, and from New York to the West Indies and Africa. Amongst Societies, the work of the Royal Asiatic Society and the Royal Geographical Society deserves special mention, and also the assistance rendered to expeditions by the British Museum. In recent years the Royal Botanic Gardens, Kew, has confined its activities to working through local individuals in the countries concerned rather than to organising special expeditions, and this has been the source of much valuable information.

We now come to an examination of the results that have been produced. These it will be convenient to group under two main headings, those that provide a general summary of the vegetation of the respective countries, and secondly such special studies as deal in detail with a part of the subject or country only.

GENERAL SUMMARIES

Reconnaissance or Linear Surveys

West Africa. All the British West African Colonies have been reported on generally, but linear surveys are chiefly found in reports of military reconnaissances, in the exploration of forest officers, and in occasional descriptions by visitors and officials. Some of these records provide useful data with which to compare the state of the country at different periods and for noting the disappearance of the forest. Complementary studies to these have been carried out in the surrounding country, which is of the same vegetational entity, by French workers, notably Dr A. Chevalier, who has traversed the country in many directions. There are also German and British accounts of Togoland, now partly mandated to Britain. On the whole, our information is sufficient to indicate the character of the chief zones of vegetation and in many cases the more particular nature of circumscribed areas.

East Africa. The East African group of Colonies has more records of what may be termed linear surveys than any other group of tropical Colonies. This is chiefly due to the attraction of this part of the world to the traveller and sportsman and also to the energetic studies pursued by the Germans. As may be expected, however, the records either centre around the high mountains or treat of the better known trans-continental tracks from the east coast to the edge of the great forest. There are many interesting parts which have not been traversed or are unrecorded. On the whole, therefore, we have not such a good general idea of the distribution of the vegetation as we have of the west coast, but, on the other hand, certain areas are better known.

Central Africa. Our knowledge of Central Africa is scanty and is confined to the records of Livingstone, Sir John Kirk, Sir Harry Johnston, and Major E. J. Lugard, and some German accounts. Much information is required before a summary can be presented that will form a basis on which future workers can build.

North-East Africa. Of the North-East African group we have very scanty linear survey records; part of the country is unknown and part desert.

Eastern. To Mr Ridley and travellers such as Wallace and Beccari we are indebted for such linear surveys as exist. Mr Ridley's account of the change from the Malay to the Siamese flora in a journey in the North-West Malay States is an interesting example. The records of the Forestry Department contain a mass of information to which future workers will require to refer.

British West Indies and America. The British West Indies and American Colonies have a few excellent examples of the valuable record of linear surveys as Sir Daniel Morris on British Honduras; but more especially Richard Schomburgk's travels in British Guiana in 1840–4 provide such excellent material that a map of the country showing the principal belts of vegetation could be drawn up from this account.

Ceylon does not show any recent records of linear surveys.

AREA SUMMARIES

Physiognomic

It is in the matter of physiognomic area summaries that we have most information. Travellers and recorders seem impressed by the marked changes from forest to grassland, swamp country to parkland, areas clothed with vegetation to desert, and the varying types of vegetation found in the greater altitudes. With the information available we are very well able to construct the main physiognomic zones of the different groups and Colonies, and in most cases maps have been given showing the general boundaries between them.

There are many general summaries of the Colonies such as those by Schlick, and Zon and Sparhawk, which, although dealing with forestry, give sufficient indications of the rest of the country, or Shantz and Marbut's recent work on Africa. Colonial Reports and Reports of Congresses and Conferences, such as the British Empire Forestry triennial Conferences are of assistance in this direction. Of illustrations those shown in the *Vegetationsbilder* are certainly the best. The *Journal* of the Royal Geographical Society provides many excellent photographs in its various papers of travel, though there are on the whole few references in the text that are helpful.

The *West African* group has been summarised chiefly by British forest officers and the French. *East Africa* has, in addition to the summaries of forest officers, several accounts of the vegetation of the mountains by travellers. *Central Africa* has been described by Sir Harry Johnston, and the Bangweolo area by R. E. Fries. *North-East Africa* is described but poorly, Somaliland perhaps best by Drake-Brockman, and the Bahr El Ghazal by Dr Christy.

In the *East*, early travellers such as Wallace and Beccari, and later Miss Gibbs, have summarised the vegetation of British North Borneo. The Malay botanists have several contributions, and the Federated Malay States Forestry Department in 1920 published a map of British Malaya, 8 m. = 1 in., graphically representing in colours the known forest and agricultural areas of the country.

The British *West Indian* and *American Colonies* are known in general chiefly through reports of forest officers which have been summarised in Zon and Sparhawk's recent work on the Forest Resources of the World.

Summaries of *Ceylon's* vegetation have been compiled by Trimen and, later, Willis.

Floristic

In some cases there are valuable floristic summaries, but of these there are only three groups that have received such consideration as require special mention.

West Africa has several such records to its credit, chiefly by the French, but also by British forestry officers. In the *Eastern* group local botanists as well as visitors have materially assisted. In *British Guiana* we have Richard Schomburgk's records, and compilers in this country have worked out the results of expeditions to the Falkland Islands.

These are only the chief instances, for floristic surveys may be considered to have been carried out directly or indirectly wherever collections have been made in which the collectors have recorded sufficient field notes to enable the vegetational type to be recognised.

Special Studies

Under this section I have grouped those treatises that deal with some special aspect of ecology. Many of them are short papers only and appear scattered in various periodicals. In these cases reviews such as the *Review of Applied Mycology*, of *Applied Entomology* and *Botanical Abstracts* will greatly assist the worker in his search.

The Habitat Factor. There is a mass of detail in connection with some aspects of this subject, chiefly the meteorological. In the majority of cases the data extend for a few years only as stations are opened and closed in various localities. Even in those established for some years one has to be very cautious in working on the records which are often kept by semi-educated natives. I have personally come across instances of broken instruments, where the last record had been repeated for months, a dash which may signify no reading taken or nil, a relative humidity of 120, temperatures in low altitudes in the tropics at below freezing-point F., and the total number of degrees of heat obtained by adding the maximum monthly averages during a year solemnly published in Government Gazettes. All these point to the fact that such records must be treated with caution.

In *West Africa* we have a masterly treatise on climatology by M. Hubert, who unfortunately has only published the first volume of his *Mission scientifique au Sudan*. It contains much original work and is the best appreciation of this subject. British forest officers have dealt with other habitat factors as well as the French, and the Government geologists, chiefly Mr Falconer in Nigeria and Dr Kitson in the Gold Coast, have contributed to our knowledge of the physiography and of the soil.

In *East Africa* Professor J. W. Gregory and others have made several presentations of the physiography which are most interesting. The climate is mentioned in Professor Troup's work on the forests and in Government publications.

Our knowledge of *Central Africa* is chiefly summarised in Engler's work, and in *North-East Africa* Dr Christy records his observations in this connection on the Sudd formation, and Mr Hamshaw Thomas and Dr Buxton some very interesting work on desert conditions. Sir John Russell's recent visit to the Sudan provides further information on the reaction of habitat factors.

In the *Eastern* group the Government records and reports of technical officers provide much information, but we have nothing to compare with the records published by the Dutch from their stations in Java.

For *Ceylon* we are chiefly dependent on Trimen's work, a special piece of work by Lewis in 1920, and Stockdale's recent observations on soil erosion and forest destruction.

Floristic Surveys

Under this heading we approach the taxonomic study and are enabled to get much assistance from purely taxonomic work. There are some twenty records of definite pieces of floristic study that have been carried out, apart from mere enumeration of collections.

East Africa has most records to its credit, and of these the majority are provided by Engler and T. C. E. Fries. A pretty instance is that of the Flora of Aldabra, a group of islands in the Indian Ocean, the records of which were prepared at Kew by Dr Hemsley and others. Very few, if any, of our larger territories have been reviewed in such a thorough manner.

In the *North-East African* group we again find Engler's work dealing with Somaliland, and in connection with the Upper Nile one or two studies of the Sudd formation by British authors.

In the *Eastern* group the botanists of Malay have several contributions. Mr E. D. Merrill, whose recent transfer from the Philippines deprives the East of one of its greatest botanical authorities, has worked on the Borneo flora through its connections with that of the Philippines and published a valuable critical appreciation of its floristic elements and their connections. Dr Stapf's work on Mt Kinabalu provides another instance of a detailed study of this region.

There are really very few contributions when one considers that much help might be expected from systematic workers. These latter, however, almost always confine themselves to an examination of species, and rarely give or are in a position to give any idea of ecologic grouping. He who reports on the vegetation generally, on the other hand, has not the knowledge of the constituent species, and so this work falls between both parties and is sadly behindhand.

Biotic

Although there are only a few treatises dealing specifically with the biotic factor, nearly every work in some way or other alludes to the great changes wrought by man in his shifting-cultivation methods and pasturage of domestic animals. Indeed, it is chiefly this factor with which we must couple the concomitant grass or bush fires which are rapidly changing the vegetation in many parts of the world. In the *Eastern* group, where a humid island climate prevails, the effect is more gradual and localised, although we can find, as in the state of Negri Sembilan in

Malay, that the natural forest covering of the country has given place to a uniform covering of the Para rubber tree. In *British Guiana*, where the population is comparatively small, this factor does not assume such importance, but in the densely populated areas of *West Africa* dominated by the neighbouring great desert its effect is so paramount that it is no exaggeration to say that the nature of the soil covering will be completely altered in the course of a few generations. Even where it is replaced by a secondary growth during a period of respite the secondary growth is of an incomparably poorer constitution.

In *North-East Africa*, again, the arboreal vegetation of Egypt is considered to have been removed within historic times and chiefly since the introduction of the camel. Sir John Russell's recent report on his visit to the Sudan bears additional testimony.

The most interesting records in this direction are perhaps Busse's account of the annual grass fires in tropical Africa, Stigand's *Lost Forests of Africa*, and Buxton's *Animal Life in Deserts*.

Physiologic

There are four publications of special observations on the physiological relation of plants to each other and their habitats that may be mentioned. Crowfoot's studies of desert flowers in Egypt; Hamshaw Thomas' observations on plants of the Libyan Desert; an American botanist, C. H. Farr, on the ferns of the Rain Forest of Jamaica, and Mr Burkill's examination of a piece of secondary jungle in Singapore. The workers in the field, and such work can only be done in the field, are remarkably few and, in fact, we can hardly consider the study of this subject yet begun.

Physiognomic

Similarly, special studies showing the physiognomic relationship of plants are remarkably few. Engler's work on the mountain flora of tropical Africa must be mentioned, and also that of Hope, A. F. Broun and Dr Christy on the Sudd vegetation of the Sudan.

Crops

In coming to the examination of literature on cultivated plants in field crops or plantations or plants growing more or less naturally in the forest, but tended on account of their economic properties, we find that the Agricultural Departments of the various Colonies provide a mass of information. This is only to be expected, for ecological work that has a direct application to a plant of economic value, such as rice, rubber or tea, will immediately command attention and funds for the prosecution of its study.

COMPLETE ECOLOGICAL STUDIES

There is one other heading under which it is convenient to group our information, and that is works containing a complete ecological study of a piece of country. It is interesting and a little mortifying to find that for the countries with which we are concerned there appears to be only one work that can rightly be classed under this heading, and that is by an American, Forrest Shreve. The work in question is entitled *A Montane Rain Forest*, and was published by the Carnegie Institute in 1914. This is a complete ecological study illustrated by excellent photographs of the vegetation of Jamaica.

The foregoing *résumé* is based on information gained over several months' examination of records during which at least two hundred publications have been consulted and abstracted.

There is a mass of information relative to our subject, especially that of the preliminary survey of linear reconnaissances and area summaries. The time now appears ripe to consolidate this information into a series of permanent records so that present and future field workers can start with a clear idea of the situation.

Systematists have carried out floristic surveys and appreciated their results in the enumeration of plants, analyses of the flora with its connections and in recording habitat factors.

Government archives in all countries contain records extending over long periods which would amply repay investigation, and physiognomic survey summaries of all countries have been published already, but of the Federated Malay States only do we possess a modern differentiated map.

Of special studies apart from crop investigations there are perhaps only a dozen for the whole Crown Colonies, of which Forrest Shreve's is the most complete piece of work.

For organising centres of Colonial Groups we have the botanical institutions at Singapore for the Eastern, Peradeniya for Ceylon, Amani for East Africa, the Imperial College of Tropical Agriculture at Trinidad for the West Indies and America, whilst the proximity of West Africa to the home country enables its work conveniently to be based here. These are potential centres, but they require to be made "live" and their work correlated to obtain the results we require.

The problem to be faced is a financial one. This has been overcome by the Americans, who are able to send such expeditions as Shantz and Marbut's to Africa and publish such works as Shreve's on Jamaica.

INDIA AND BURMA

In considering the work recorded for India and Burma we find we have quite different conditions from those that have obtained in the Crown Colonies. In fact, for our purposes India may be considered at least a generation ahead, and in comparison we may liken the state of work in the Crown Colonies to that which took place in India during the later part of the nineteenth century.

India has been fortunate in having been worked by many botanists of first-class reputation who have served chiefly in the Medical or Forestry Services. Later, the institution of Universities has brought professional teachers into the field who have been scientifically educated in the home country and who possess the inclination and leisure to continue their studies in the field.

Of these two classes we find that the former have dealt chiefly with general studies or investigations based on a knowledge of systematic botany. The latter, both themselves and students locally educated by them, have carried out investigations in special ecological studies.

There are four papers to which I would like to refer by way of example which do not find their counterpart in the Crown Colonies. The first is Hooker's classical essay in the *Flora Indica*. This may be considered as a starting-point for ecological work and provides a summary of the situation to that date which students need to appreciate before beginning.

Two other papers are of interest; that of H. B. Grigg, a civil servant who published a Manual of the Nilgiries in 1880, and Part II of the Rev. R. F. Mason's compendious work on Burma, which contains a botanical *résumé* by W. Theobald and was published in 1883. Works of this nature are getting particularly valuable as a sufficient period of time has elapsed for considerable changes to be effected in the natural covering of the ground, and such works provide us with records of the virgin state of the vegetation which are now getting impossible to secure.

Another work to which I would refer is Mr R. S. Hole's *Manual of Botany for Indian Forestry Students*. This is a distinct work bearing in many points on the elemental problems of Indian ecology and is of great assistance to student beginners.

Again, we have a distinct piece of ecological work in Mr Hole's Presidential Address to the Fifth Indian Scientific Conference, in which is shown the bearing of Plant Ecology on problems of economic interest. This able summary is an instance in which the scientific worker can demonstrate his value to the administrative services who are financially responsible for providing funds for such work.

One of the outstanding features of the Indian work is the great assistance, particularly in the case of publications, rendered by the various natural history societies. Of these the *Journal of the Bombay Natural History Society* and the *Journal and Proceedings of the Asiatic Society of Bengal* provide many important records of ecological summaries of different parts of the country. The *Records of the Botanical Survey of India*, which originally recorded many of these summaries, has at a later date begun to show papers involving special ecological studies such as Saxton and Sedgwick's "Plants of Northern Gujarat." Modern ecological papers generally, however, appear in the *Journal of the Indian Botanical Society*, which contains a most interesting series of special studies.

The Forestry and Agricultural Departments in their ordinary and special bulletins provide many papers of all kinds, but naturally include a proportion of studies of special crops.

Separate publications as apart from Journals are numerous, and range from pamphlets to such monumental works as Troup's *Silviculture of Indian Trees*, the introduction to which provides a concise survey summary of the vegetation in general.

The ecological studies so far carried out, however, are but a fraction of what remains to be done when the whole country is taken into consideration. We find that most of the work, especially the general reconnaissances and summaries, concern the northern part of the country and more especially the mountain country. Of special studies there are just one or two treatises at most. India, then, would appear to be provided with more and better equipped workers, and has more investigations recorded than have the Crown Colonies at the present day.

Mr A. G. Tansley, F.R.S. SUMMARY OF VEGETATIONAL WORK AND PROBLEMS IN THE DOMINIONS[1]

Canada. The outstanding feature of the vegetation of Canada is the fact that by far the greater portion of its surface is, or was originally, covered by a coniferous forest consisting of comparatively few species (*Picea mariana* and *canadensis*, *Abies balsamea* and *Pinus banksiana*

---

[1] Nothing more can be attempted here than to turn the attention of the audience to the vegetation of each Dominion in turn.

with *Betula papyrifera* are the most widespread), and on the whole wonderfully uniform from the Atlantic coast to the Mackenzie River and even to the Pacific coastal mountains. North of this lies the Arctic tundra, and to the south, especially in the southern portions of Ontario, there begins to be prominent the deciduous forest vegetation which indicates the transition to the typical Deciduous Forest of the eastern United States. In the Rocky and Selkirk Mountains new species of Conifers appear, and on the Pacific slope in British Columbia, and stretching northwards into Alaska, two other types of coniferous forest, characterised respectively by Douglas Fir and Sitka Spruce, are developed. This continuous area of coniferous forest is, however, interrupted by the northward extension of the Prairie-Plains region of the United States, which crosses the international boundary between about 96° and 114° W. longitude and extends northwards to about N. latitude 53°. This climatic grassland region is separated from the coniferous forest by a belt of aspen forest which, when cleared, gives some of the best agricultural land.

Advance in our knowledge of Canadian vegetation seems to have been carried on more vigorously in the past than at present. In the early days of exploration and settlement the plants were described in Europe from specimens sent home by collectors. Sir W. J. Hooker's *Flora Boreali-Americana* (1833) formed an indispensable foundation for systematic work. In 1860 only the St Lawrence basin was at all well known, but in that year the foundation of the Botanical Society of Canada and concurrent exploring expeditions gave a new starting-point for the development of purely Canadian work. The nature and distribution of the forests were excellently summarised by Prof. John Macoun in 1894. Much of the more recent specifically vegetational and ecological work on vegetation common to the United States and Canada has been carried out by United States botanists, such as Ganong, Transeau, Cooper, and Nichols, while the Harvard (Gray Herbarium) school of systematists—notably Fernald—has contributed much to the more accurate knowledge of East Canadian plants. One of the great Canadian industries based on vegetation was and is, of course, lumbering, but during the last 30 years or so the prairie region has become of world importance for its grain production. It is confidently to be expected that the application of modern methods of vegetational study will enable both forest and grassland to be dealt with in a more scientific and economical manner, and it is much to be desired, as a first step, that we should have, as soon as possible, a summary of what is known to date of the nature, distribution, and succession of the vegetation.

South Africa[1]. Here we have the greatest contrast to Canada in climatic conditions and vegetation. From the northern Transvaal to Cape Agulhas we range from 23° S. latitude (within the southern tropic) to 34° S., practically still within the subtropical belt, while the Bechuanaland Protectorate and Damaraland extend far into the Tropics. A comparatively small portion of this vast area is forest land—parts of the northern tropical region (much of which is, however, savanna), the so-called "bush" of the eastern Transvaal and Natal, practically confined to the region of summer rainfall, and the hard-leaved (sclerophyll) vegetation, largely shrubby, of the south-western region of winter rains. From Natal the annual rainfall decreases westward to the Kalahari desert and the Atlantic coast, which present severe desert conditions. Between the desert and the forests of Natal and of the eastern Transvaal we have the great Veld region of climatic grassland.

After the great pioneers who founded the knowledge of South African botany, especially Thunberg and Harvey, our knowledge of the flora and vegetation has been developed by various workers such as Macowan, Bolus and Medley Wood, and later Marloth, Sim and Schönland. Bews has been the first to apply modern points of view to South African vegetation, especially in regard to succession, of which he has shown the economic importance, particularly in a "new" country. During the last few years a Union Botanical Survey has been organised, under the direction of Dr Pole Evans. Its aims are very wide, excellently drafted, and include the taxonomic and ecological knowledge of the vegetation and its practical application to specific economic problems. The Survey has already published several memoirs by various authors. Owing to the wide extent of its natural grasslands the primary industry immediately based on the vegetation is the raising of stock. The Union Survey is co-operating with the Department of Veterinary Research in the attempt to deal with such questions as stock diseases due to poisonous or unsuitable plants in the Veld pasture, and with the Forestry Department in questions affecting forests. Afforestation will probably be vigorously pursued in the future, and may help to solve some of the very difficult social problems confronting South Africans. An important feature of the Survey is the voluntary help of professional botanists at various University centres, who receive in return free passes over the State railways and grants towards other travelling expenses. Whole-time Survey officers, employed partly in herbaria and partly in the field, are temporarily attached to these centres.

[1] Data based on material supplied by Prof. J. W. Bews.

Australia[1]. The continent of Australia has a range of climatic conditions somewhat similar to that of South Africa, but much more of it —about two-fifths—lies within the Tropics, and this is the region least known botanically. Again, the vegetation is largely determined by the seasonal incidence of rainfall, the region of summer rains, as in South Africa, being the east and south-east and of winter rains parts of the south and south-west. Again, the eastern (and northern) coastal regions, especially the slopes of the coastal mountains, bear the heaviest forest, while the south and south-west contain sclerophyll vegetation of the Mediterranean type. Unlike South Africa, however, much of this vegetation is more or less open forest, dominated by *Eucalyptus*, this distinctive tree-type being extremely widespread and represented by numerous species. The general tendency is that of progressive xerophytism as one passes inland, and from east to west. Much of the interior and parts of the west coast are desert and semi-desert, and great areas, whose climate is somewhat less severe, are occupied by savanna or by scrub. The interior of the continent is devoid of high relief, so that local variations of climate, due to topographic features, are insignificant. The tropical and subtropical forests of the eastern belt are confined to the eastern slopes of the coastal mountains, and do not cross the divide, even at the highest points. Towards its southern limit, in New South Wales, the subtropical forest is sharply limited by edaphic factors, occurring only on deep basaltic and alluvial soils. Within a few yards one may pass from dense forest on basalt to open Eucalypt forest on the poorer clays and sandstones. The flora here is essentially Malayan, with palms, cycads, lianes, and epiphytic ferns and orchids. The open Eucalypt forests with lower rainfall and on poorer soils contain many species, but have not been ecologically separated.

A notable feature of these various forest types is the secondary grassland that results from burning. Even burned subtropical forest gives *Paspalum distichum* in a very short time, hardly a vestige of the original flora remaining. But great areas of natural grassland occur in South Queensland (Darling Downs) and in New South Wales. North-westward there is transition to scrub country such as brigalow-scrub (*Acacia harpophylla*), etc.

Along the south coast (much of Victoria and South Australia) there is open Eucalypt forest of various species but a remarkably uniform facies. Owing to burning, clearing and grazing, grasslands, clearly secondary, are increasing. Further west a sclerophyll type of Eucalypt

[1] Data based on material supplied by Prof. T. G. B. Osborn.

forest occurs, with shrubby hard-leaved undergrowth and no grass. Here the rainfall is from 25 to 40 inches, winter rain, with a hot dry summer. The mallee-scrub which extends over a large area of South Australia includes several species of shrubby Eucalypts and is probably a distinct plant formation, corresponding most closely perhaps with Mediterranean macchia or Californian chaparral. The southern part of Western Australia appears to repeat the features of the south coast, bearing xerophilous forest of various Eucalypts, rather open, with mixed shrubs, sclerophyllous forest with smaller trees, and savanna forest. The northern portion of Western Australia is tropical and is largely arid savanna and steppe.

The arid interior includes, besides scrub areas, great regions of so-called "salt-bush," comprising many communities, in which various species of *Atriplex* and *Kochia* are dominant. Contrary to a general impression, these areas are not halophytic. The rainfall is 10 inches or less, occurring irregularly. Scrub of a very open type, including *Eremophila, Acacia, Casuarina* and *Callitris*, occurs on rocky ridges, while *Eucalyptus* is confined to large watercourses, dry except in flood time. The desert area, with a rainfall of 5 inches or less, has been but little studied. Scattered shrubs of *Acacia, Cassia* and *Eremophila* occur. The "therophyte" (annual) element is very large and contains many valuable grasses such as porcupine grass (*Triodia*), and this partly justifies the Australian objection to calling any of the country "desert." The rainfall is very erratic. Heavy rains in summer or winter produce abundant therophyte vegetation, but there may be intervals of two or three years between effective falls.

Among the earlier pioneers of botanical work in Australia the name of Ferdinand von Müller is the most conspicuous, and the veteran J. H. Maiden's long continued floristic work at the Sydney Botanic Gardens must be mentioned. Maiden also laid the foundation of our knowledge of Australian alpines, while Schombergk in 1876 dealt with the vegetation of South Australia. In his book on the southern part of Western Australia (1906), Diels published the first general vegetation map of the continent, which is excellent and useful and has formed the basis of subsequent maps, *e.g.* that by Griffith Taylor in his useful work on *The Australian Environment* (1918). Cambage has also published valuable work on many aspects of the eastern vegetation, among others the influence of the mountains on climate and forests. In the report of the Ecology Committee of the Australasian Association for the Advancement of Science (1923) there are 212 titles, though the majority, of course, are of floristic works. Most of the specifically ecological work is quite recent. Such work is now being carried on by Miss Collins, by Dr Sutton (secretary of the Committee) and by Prof. T. G. B. Osborn and his pupils at Adelaide.

The following interesting summary of some general conclusions by Prof. Osborn may be added:

1. The secondary nature of the grassland throughout most of extra-tropical Australia is apparent, the natural vegetation being forest or scrub. Settlement requires the clearing of this for agriculture or pasture. The rain forests of the east, whether subtropical or south-temperate, regenerate very slowly. The savanna forests are often prevented from regeneration by fire and grazing. The native flora, which is not naturally close grazing, thus becomes moribund, while sheep and cattle or vermin (rabbits) prevent regeneration by destroying seedlings.

2. Natural regeneration of certain forest types (*e.g.* sclerophylls and red-gum) is comparatively rapid if grazing is prevented. The more complex climax communities of the rain forest probably have a longer series of successional phases, but no data have been published. It is clear that under existing conditions the valuable mixed forests, containing *Araucaria*, *Phyllocladus* and *Dacrydium*, are doomed.

3. The characteristic arid vegetation type known as "salt-bush" is non-halophytic. The plants, however, show marked selective absorption of sodium chloride, and this is possibly connected with water absorption.

4. Unpublished observations show that the vegetation of the arid regions is in delicately balanced equilibrium with its environmental factors. The permanent plant-covering, whether half-shrub (*Atriplex* and *Kochia*) or woody (*Acacia aneura* and *Casuarina*) is not regenerating. Hence grazing represses the succession or produces unstable regressive communities. The secondary prominence of grasses in the salt-bush area as a result of grazing is noteworthy, as is the development of communities entirely dominated by alien plants, *e.g. Nicotiana glauca*.

Work is particularly needed on:

(*a*) Forest successions as a guide to regeneration (if practicable) of the climax rain forests, (*b*) grassland successions with a view to improvement of pasture, and (*c*) succession in arid areas in relation to grazing. The principles of "range management"—resulting from the ecological study of grazing land—as recently worked out in the western United States require investigation on the Australian grazing lands.

New Zealand[1]. The two islands stretch from about 34° (the latitude of

[1] Data taken from the writings of Dr L. Cockayne.

Sydney) to 47° south latitude and are thus temperate in climate. The rainfall nowhere sinks to anything like the low levels reached in large areas of Australia and South Africa, the whole of the North and a large part of the South Island possessing a true forest climate. The forest is indeed mainly temperate rain forest of evergreen character and a considerable number of dominant species. Epiphytes, lianes and the abundance of ferns, recalling tropical rain forest, bear witness to the wet climate. Of the 399 species of the forest communities no less than 162 are woody plants (99 trees and 63 shrubs). Among the most conspicuous dominant trees are the kauri (*Agathis australis*) and species of *Podocarpus* and *Dacrydium*. The *Nothofagus* forests, consisting of various species, are of markedly different character, their dense shade excluding the wealth of epiphytes, climbers and ground species of the mixed rain forests. They are mostly subalpine, though in places they occur at sea-level. Varied shrub communities occur in situations not sufficiently favourable for the development of forest, *e.g.* on windswept coasts, on the mountains and where the soil is very poor. Manuka-thicket (*Leptospermum scoparium*) is one of the most widespread of these, occurring in very varied situations and showing a remarkable plasticity. This species varies from a small tree 30 feet high to a low carpet-forming plant.

On the eastern slopes of the Dividing Range of the South Island, and originally on the lowlands extending to the east coast, where the rainfall is too low to support forest, and on the pumice plateau of the North Island, the very characteristic tussock grassland occupies the ground. This has been grazed for 70 years by millions of sheep, but shows no signs of disappearing. The tall tussock (*Danthonia Raoulii* var. *rubra*) is worthless as pasture: the sheep eat the young leaves of the low tussock communities (*Poa caespitosa* and *Festuca novae-zealandiae*)—the withered foliage being burned off at the end of the winter—as well as other associated grasses and herbs. Many European plants have become established over large areas of these communities, mainly as a result of burning or over-grazing, though its essential nature and physiognomy are unchanged. As a result of cultivation and clearing a large number of "induced" plant communities have been established, some composed of native, others of alien plants from all parts of the world. Thus native rain forest has in part been completely replaced by meadows largely composed of European grasses. The alpine and subalpine vegetation is very well developed, containing a large number of endemics; it shows the usual zonation of plant communities.

Among the early pioneers of New Zealand botany, Banks and Solander are pre-eminent in the eighteenth century; D'Urville and Lesson, Cunningham, Colenso, and above all Sir Joseph Hooker, in the first half of the nineteenth. A number of others followed and the description of the flora culminated in Cheeseman's *Manual* (1906). During the last 30 years it is mainly owing to the work and personality of one man—Dr Leonard Cockayne—and to the colleagues and pupils he has inspired that our knowledge of New Zealand vegetation has become wider and deeper than that of any other Dominion, and this is due very largely to his whole-hearted acceptance of the ecological point of view, though also of course to his unbounded enthusiasm and power of work. Nowhere so markedly as in New Zealand, too, except perhaps in the United States, is the direct importance of ecological work for the proper treatment of forest, pasture-land and sand-dune recognised by the State authorities. Dr Cockayne has often been called in to investigate special economic problems, and ecologists are now recognised officers of the forest and agricultural services. From this point of view New Zealand is undoubtedly a shining example to the rest of the British Empire.

Prof. F. J. Lewis. SCHEME FOR AN OUTLINE BOTANICAL SURVEY OF CANADA

Dr L. Cockayne, F.R.S. NEW ZEALAND ECONOMIC PLANT ECOLOGY

General

It is well known that the New Zealand botanical region offers in its virgin vegetation, and its many endemic species of diverse origin, a peculiarly attractive field for the study of plant ecology. But during the progress of settlement a large part of this vegetation has been either profoundly modified, or entirely swept away and replaced by associations novel to the soil, so that at the present time there is every gradation between the primitive and the artificial. For example, over wide areas the virgin rain-forest of tropical aspect and structure has been replaced by rich pastures, made up of species chiefly European. Other primitive associations of considerable economic value as natural vegetation are being greatly altered, and that value slowly lessened year by year. On the other hand, there are man-made associations, in aspect and members truly primitive, some of high economic importance, which were not present in primeval New Zealand. Evidently then in remembering that agriculture is the mainstay of the Dominion there are economic problems in great abundance for the ecologist to solve.

The papers cited are those in which detailed studies have been made in the field, but not such as record general observations—so frequently inaccurate—the result of superficial study which cannot rank as ecological science. However, the late Mr T. Kirk's suggestive paper (1896) must be mentioned here, since it contains the opinions of that distinguished botanist concerning "natural replacement," and the effect upon certain groups of vegetation, or species, of fire, cattle, rats, insects and particularly introduced plants.

In what follows an attempt is made to trace the development of economic ecology in the Dominion, its commencement being merely a few statements of a practical character in papers purely scientific, and its climax economic researches instigated and paid for by the State. There is also that class of state-remunerated investigations which concern making known the composition of the vegetation of National Parks and Scenic Reserves, and the value of such as pleasure grounds, tourist resorts and natural museums; here also come in the reasons, based on ecological research, for extending such areas, and suggestions concerning their management.

Coming now to the early days of ecology in New Zealand, the pioneer paper was that of Diels in 1896, entitled "Vegetations-Biologie von Neu-Seeland," in which it was sought to correlate the leaf structure of the vascular plants with their environment. Though the author had not visited New Zealand, he was thoroughly versed in the botanical literature, and had received some first-hand information from myself. Certain far-reaching generalisations, which cannot be neglected, were arrived at, but above all the path for future research was opened up by this excellent though daring pioneer work.

Three years later I published an account of the effect of burning a certain kind of subalpine scrub on the Southern Alps (Cockayne, L., 1899), which, though not meant to have an economic bearing, was made use of later by the Department of Agriculture with regard to the question of grassing mountain slopes occupied by that class of vegetation. Indeed, as in so many branches of science, it is not feasible to draw the line between the apparently pure scientific and the applied.

This paper was followed by one more ambitious (Cockayne, L., 1900), which dealt with a wide area extending from the coast to the summit of the Dividing Range of the South Island, and botanical ecology in New Zealand was fairly launched. The paper contained some observations of an economic character, but especially it sought on that head to upset the deep-rooted fallacy that when introduced and indigenous plants came into competition the latter would be wiped out. Since then abundant evidence has clearly shown that the foreign plant invader is unable to gain a footing in almost all primitive plant-associations and that certain artificial associations are rapidly overrun by indigenous plants.

As time went on other workers, especially the younger botanists, came into the ecological fold, and the study of various classes of virgin vegetation prepared the way for work of a purely economic character. In all, up to the present, nearly one hundred ecological publications by New Zealand authors have seen the light.

How quickly these ecological studies have been recognised as of public benefit will be seen in what follows, where it is told how various Government departments, and even one private body, have called in the aid of ecology. It seems, indeed, that economic botanical ecology bids fair in New Zealand to become a profession equally with plant pathology, though necessarily in both branches the demand is limited as yet.

Progress of Economic Ecology in New Zealand

Proceeding in chronological order two papers by myself may be mentioned dealing respectively with the vegetation of Chatham Island (Cockayne, L., 1902) and the New Zealand Subantarctic Islands (Cockayne, L., 1904), which contain sections treating of the effect of introduced animals and fire upon the virgin vegetation. A detailed account is given of how *Phormium tenax* (the Chatham Island variety), a plant originally extremely common in wettish open ground, has come to be now almost restricted to rocks and shallow water. Burning vegetation several times on flat boggy ground led to an enormous increase in the endemic *Poa chathamica*.

From 1906 until 1911, I was employed by the Department of Lands for some time yearly to investigate the vegetation of certain parts of New Zealand which were under its jurisdiction. The result was the series of Reports cited in the List of Literature. Those on Kapiti Island, the Tongariro National Park and Stewart Island come into the category mentioned in the Introduction to this paper as "National Parks and Scenic Reserves." They have certainly aroused considerable public interest and have doubtless played no small part in having the Kapiti Island Sanctuary better cared for, and an Advisory Board appointed, while in the case of the Tongariro National Park, its area has been greatly extended by Act of Parliament and a Board of Control has been set up. All these reports contain information regarding the effect of introduced animals, and the Stewart Island report—easily the most complete of any—deals with the "Agricultural Capabilities," "Sawmilling," "Stewart Island as a Watering-place," the introduced animals, and there is a rather long account of the "bird-life."

The Waipoua forest report, though not intended as a contribution to forestry, is considered such by the State Forest Service and is made use of accordingly. Perhaps its outstanding economic feature is the account of the life-history of a kauri forest, of the relation of the kauri (*Agathis australis*) to light, and of its capacity for regeneration. The relation of various species to aspect and soil moisture is also of fundamental moment.

In 1910 the first purely economic ecological paper appeared from the pen of A. H. Cockayne, then Biologist to the Department of Agriculture, and now also Director of the Fields Division. This paper (Cockayne, A. H., 1910) dealt with the burning of tussock-grassland which is practised from time to time on nearly all the mountain pastures of the South Island in order to cause the tussocks to put forth palatable young leaves and to get rid of dead, useless parts. But in this paper the rôle assigned to burning is quite different. It is to be practised—but not everywhere or at all times—*not* to improve the feeding-value of the original grassland but to alter its composition and to produce "a new type of vegetation better fitted for the purpose required, exactly in the same way as forest is turned into grazing-land by artificial means."

The following year saw another work appear of a purely economic aim, my report on the Dune-Areas of New Zealand (Cockayne, L., 1911). This is of a comprehensive character and it is illustrated with 72 photographs. There are two parts, the first purely scientific telling about the geomorphology of dunes and its relation to the vegetation, the establishment of such and the growth-forms of its members, and the second part, basing its procedure almost entirely on what the first part teaches, deals with dune reclamation. The latter falls under the heads of fixing the unstable dunes by means of marram-grass and *Lupinus arboreus*; preserving them as farmlands; and their afforestation, which it is contended should be the ultimate aim.

The same year Aston's first paper mentioned in the List of Literature appeared (Aston, B. C., 1911). This has been followed by a number of other publications by this author. Though primarily dealing with soil analyses, Aston has never been content to let the matter end there, but he invariably gives reliable details regarding the plant-covering, holding rightly that such should form a part of all soil survey work.

From 1914 to 1919 there appeared a number of important purely economic papers by A. H. Cockayne (see List of Literature), who early on had fully recognised that the improvement of its grasslands was the most important agricultural need for the Dominion and that the only way to approach the question properly was by the application of ecological methods. This becomes of special scientific interest in the light of his remarks (Cockayne, A. H., 1916, p. 421):

Although there are grasslands in the Dominion that have been established over half a century, all the various types of pasture can be traced right from their primeval condition to their present state. All the changes and modifications of vegetation that occur in their development can be studied step by step from the original plant-formation into grassland, the derivation of which it would be impossible to tell from its present condition alone. In all parts of New Zealand, but more extensively in the North Island, the evolution of artificial pastures can be studied in all its varied phases, and the origin of artificial grassland can be investigated with a detail that is impossible in older settled countries, where the origin of many of the types of permanent pasture is wrapped in obscurity.

Perhaps the most important result of A. H. Cockayne's grassland investigations, since the state of knowledge at that date, was summed up in his series of articles, "The Grass-lands of New Zealand," which appeared in 1918. Here he puts forth the dictum that, "Palatability is one of the most important subjects underlying the systematic study of pastures"… "It is also connected with one of the most important practical considerations—namely, the effect of using mixtures consisting of plants of varying palatability."

In 1919 the second edition of my *New Zealand Plants and their Story* was published. It contains a chapter entitled "The evolution of a new flora and vegetation," in which the vegetation is classified into "Primitive," "Modified," "Induced" and "Artificial," and the Induced into "Indigenous induced" and "Adventitious induced," and examples are cited and discussed. This matter is gone into much more thoroughly in *The Vegetation of New Zealand*, pp. 280–292, where also there is a brief account of the agriculture and horticulture.

During the years 1918 to 1922 I was engaged by the Department of Agriculture to make an ecological investigation of the montane tussock-grasslands of New Zealand, a paper published by A. H. Cockayne in 1916 having shown plainly that such a piece of work was urgently needed in order to prepare the way for the improvement of these semi-natural pastures, the carrying-capacity of which was extremely low. Incidentally, too, came the devising and carrying out a series of experiments in order to learn if it was possible to profitably re-grass a certain semi-arid moun-

tainous area of about a million acres, and if so what principles should
guide practice, and what methods should be used. By degrees, the latter
task somewhat overshadowed the former. Some of the results appear
in the series of papers, Cockayne, L., 1919 *b*–1922. Much information
was gained from the re-grassing experiments, and a number of valuable
pasture plants were proved to be capable of cheap and easy establishment. The methods by which this knowledge was acquired were, in the
main, ecological. So, too, with the results of the grassland research.
Important details concerning the relative palatability for sheep of nearly
all the species were secured—a matter of pure guesswork previously,
valuable species being considered worthless and *vice versa*. These results
were secured partly by direct observations of individual sheep feeding
in the open and partly by experiment. Regarding the fundamental
question of introducing better pasture plants into the tussock-grassland
it was found that nature provided excellent seed beds in the numerous
mat-plants and low cushion-plants, themselves valueless for feed. Probably nearly all the members of the tussock-grassland formation first
gained a footing in this manner and not on the bare ground, excepting
certain pioneer mat-plants, species of *Raoulia* mainly, which form a
primary succession on stony river-beds and similar situations.

In addition to this work for the Department of Agriculture, in 1918–19
I carried out a piece of research for the Flax Millers' Association regarding
the yellow-leaf disease of *Phormium tenax* and its relation to the management of the *Phormium* (flax) areas, *i.e.* to variations of its habitat. My
first report strongly supported the view that the disease was largely due
to ecological causes, but in a second report I modified this opinion.
Nevertheless, further investigations by Messrs Waters and Atkinson, of
the Biological branch of the Agricultural Department, and by Mr G.
Smerle, who was continuing the investigation for the millers, seem to
support my first theory.

E. B. Levy, who had for a number of years been attached to the Biological Laboratory of the Department of Agriculture and who, during
that period, had received much instruction from A. H. Cockayne, the
Biologist, in economic ecology, produced a series of articles entitled
"The Grasslands of New Zealand" (Levy, E. B., 1921–23). These give
an excellent and fairly complete general account of the pastures of New
Zealand as a whole. They are the result of wide travel throughout both
islands. The ecological method is used and the ecological principles stated
on which the author bases his practical suggestions. The importance of
succession is emphasised and the plasticity of species recognised in its

economic bearings. Amongst the many subjects dealt with are the "principles of pasture establishment on various habitats," *e.g.* swamp-lands, several types being distinguished and named, peat-swamp lands, steep unploughable forest country and coastal sand-areas. Many more topics are discussed and the whole series forms a notable contribution to economic plant ecology, as well as being eminently practical.

A second series of papers under the same head, but the outcome of more intensive observations over a comparatively small area, are being produced by the same author, the whole of whose time is now devoted to economic ecological pasture research. An attempt is being made to trace the relation between the primitive forest and the pasture by which it is replaced, and each type of forest is considered an indicator of its habitat for grassing purposes. Judging from the first two articles—the one on "Forest Successions" and the other on "Secondary Growth and its Control"—the new series should be welcomed by ecologists far beyond the confines of New Zealand. Most of the numerous illustrations are both admirable and instructive.

The year 1921 was a notable one in the history of New Zealand plant ecology, since there appeared Guthrie Smith's *Tutira*, G. M. Thomson's *The Naturalization of New Zealand Plants* and my long delayed book *The Vegetation of New Zealand*. It is the first-named which chiefly concerns this paper.

Regarding Guthrie Smith's work it is difficult to speak too highly. As an ecological publication it stands alone, nothing like it has been published before. Tutira is a sheep-run situated in hilly country on the east of the North Island, somewhat south of latitude 39°. The topography and climate of the area are described; an account of the original plant-covering and that at the present time is given, together with the changes which have come about through the methods of sheep-farming prac-tised, the arrival and increase of the many foreign plants, not forgetting the paths by which they have journeyed and their methods of travel, and the action of the domesticated and feral animals. Nor is the relation-ship of the ever-changing vegetation to the subtle influence of human beings and even the land laws of the Dominion neglected. It is, in short, an ecological work—the word ecology is never mentioned—in which, as far as may be, by the record of about forty years' trained observations of a born naturalist, most of the factors which have influenced the vege-tation are taken into consideration.

G. M. Thomson's work, though not strictly ecological, must be in the hands of all workers dealing with New Zealand animal and plant ecology.

As a book, it too is the only one of its kind. To botanical ecologists the list of introduced plants is of great value and forms a starting-point for more intensive research.

The Vegetation of New Zealand has been already referred to so far as this paper is concerned. At present I am preparing a second, greatly revised, edition, and propose making the economic trend much more evident than in the original book.

The economic ecological activity of the Dominion is not completely reflected by this account of the published work. The Fields Division of the Agricultural Department is frequently called upon to answer questions which require ecological knowledge. Lectures based on applied ecology are delivered at the "farm schools" carried on from time to time. Mr B. C. Aston is making a soil survey of the region and using the associations and special plants as indicators. The State Forest Service has three ecologists engaged for part of each year studying forest ecology, viz. Mr W. McGregor, the kauri forest; Mr C. E. Foweraker, the rain-forest of Westland; and myself, the whole *Nothofagus* forest area. I am also continuing the re-grassing experiments already referred to. Dr. H. H. Allan is privately carrying out much ecological research both pure and applied. In short, it seems clear that New Zealand is attempting to do its economic ecological duty. But such research is in its infancy, nor are there many as yet with the necessary knowledge.

Ecological Education

The last statement leads me to a brief consideration of the kind of training necessary for a botanical ecologist. It seems almost needless to say that something different from the ordinary botanical curriculum with its "types" and microscopic work in the laboratory is required. It must also be borne in mind—speaking of botanical teaching in general—that as knowledge changes, and new demands make themselves felt, so must the content of any curriculum be altered. But teaching ever lags behind the newer knowledge. I well remember the systematic botanical teaching of long ago, and later the translation of Sachs' textbook which revolutionised the method in vogue. Now that ecology has arrived and is well established is not the time ripe for a branch of botany so full of practical importance to come into its own? Would not a course of quite elementary ecology be an excellent introduction to botanical teaching? But I do not stress that point, the question being considered concerns higher ecological training. Certainly one acquainted only with the ordinary botanical curriculum will be incapable of attacking an ecological problem.

What, then, do I, who am not a teacher but have had some field experience, and have instructed beginners in the field, suggest as a training—in part at any rate—for those intending to pursue economic botanical ecology?

First of all no one, I think, would deny that a thorough grounding in general botanical science is essential, but special attention should be given to physiology and the use of a flora. The latter would need a careful explanation of the conceptions of species therein, for an aggregate species is without meaning in ecological research. Probably "types" as a means of teaching general botany should be abandoned. Anatomy should be taught in conjunction with physiology. A wide knowledge of the local flora is of the utmost moment. Each plant should be so well known that it can be recognised at a glance; such rapid identification is readily acquired and should form part of the field instruction; identifying species from their "characters" can come later. The general principles of ecology should of course be taught, but bookwork alone is useless and misleading. The essential is *practical* teaching in the field by a *competent* man. The material for such instruction is at hand the world over. Where, as in New Zealand, almost virgin vegetation is to be seen within a short distance of all the University colleges the conditions are ideal. But fields, badly-kept gardens, plantations, waste places, anywhere indeed where plants enter into competition, will suffice. All depends on the ability of the teacher and his power to interest his students. In the field the student should be taught to work notebook in hand; there should be no trusting to memory.

A knowledge of practical horticulture is of the greatest value for all ecologists. No other discipline can teach so well the behaviour of living plants as watching them develop to maturity from seed sown by one's own hand. Not in any other way can the relation of plants to ecological conditions be so well studied as by garden practice.

A training in photography is essential both in the use of the stand and hand camera, and the student should do his own developing and printing. Instruction in the use of other instruments seems to me of much less importance, it can be acquired later. The really important training is in observation, the acquiring an eye for perceiving the essential, learning to collect judiciously and to take notes properly. Nor should time be taken up—and here I am on delicate ground—with studying schemes of classification of vegetation, or in learning all kinds of technical terms. When all is said and done, so far as the members of a plant-community are concerned, it is the growing-place of each, their growth-forms and the plasticity of these, and the combinations of the species which count.

List of Literature dealing wholly, or in part, with New Zealand economic plant ecology

ASTON, B. C. (1911). "Some Results of a Flying Soil-Survey." *N. Z. Journ. Ag.* II. 10–17.
—— (1918). "Studies on the Lighter Soils of the North Island. II. Liming of Pumice Soils." *Ibid.* XVII. 259–262.
—— (1920). "Soils of the Manawatu District." *Ibid.* XX. 273–286; XXI. 57–66 and 105–114.
—— (1921). "The Classification of the Virgin Lands of the North Island." *Ibid.* XXIII. 266–270.
—— (1923). "The Ideals of a Soil Survey." *Ibid.* XXVII. 131–137.
COCKAYNE, A. H. (1910). "The Natural Pastures of New Zealand. I. The Effect of Burning on Tussock Country." *Ibid.* I. 7–15.
—— (1911). "The spiked blue-grass of Australia. A valuable plant for denuded lands." *Ibid.* III. 1–8.
—— (1914). "The Surface-sown Grass Lands of New Zealand." *Ibid.* VII. 465–475.
—— (1916). "Conversion of Fern-land into Grass." *Ibid.* XII. 421–439.
—— (1916 b). "Some Economic Considerations concerning Montane Tussock Grass-land." *Trans. N. Z. Inst.* XLVIII. 154–165.
—— (1918). "The Grass-lands of New Zealand." *N. Z. Journ. Agric.* XVI. 125–131, 210–220, 258–266 and XVII. 35–41, 140–142.
—— (1918 b). "Some Grassland Problems." *Ibid.* XVII. 321–328.
—— (1919). "Cocksfoot. Its Establishment and Maintenance in Pasture." *Ibid.* XVIII. 257–271.
COCKAYNE, L. (1899). "On the Burning and Reproduction of Subalpine Scrub and its Associated Plants; with Special Reference to Arthur's Pass District." *Trans. N. Z. Inst.* XXXI. 398–419.
—— (1900). "A Sketch of the Plant Geography of the Waimakariri River Basin, con-sidered chiefly from an Oecological Point of View." *Ibid.* XXXII. 95–136.
—— (1902). "A Short Account of the Plant-covering of Chatham Island." *Ibid.* XXXIV. 243–325.
—— (1904). "A Botanical Excursion during Midwinter to the Southern Islands of New Zealand." *Ibid.* XXXVI. 225–333.
—— (1907). "Report on a Botanical Survey of Kapiti Island." *Parliamentary Paper,* I–C. VIII. 1–23.
—— (1908). "Report on a Botanical Survey of the Waipoua Kauri Forest." *Ibid.* I–C. XIV. 1–44.
—— (1908 b). "Report on a Botanical Survey of the Tongariro National Park." *Ibid.* I–C. XI. 1–42.
—— (1909). "Report on a Botanical Survey of Stewart Island." *Ibid.* I–C. XII. 1–68.
—— (1909 b). "The Necessity for Forest-Conservation." *Forestry in New Zealand,* 85–93. Wellington.
—— (1911). "Report on the Dune-Areas of New Zealand, their Geology, Botany and Reclamation." *Parliamentary Paper,* C. XIII. 1–76.
—— (1919). *New Zealand Plants and their Story,* 2nd ed. Wellington.
—— (1919 b–1922). "An Economic Investigation of the Montane Tussock-Grassland of New Zealand"—a series of papers mostly dealing with distinct topics. *N. Z. Journ. Agric.* XVII–XXV.
—— (1921). "The Vegetation of New Zealand." *Die Vegetation der Erde,* XIV, especially 280–297. Leipzig.
—— (1921 b). "The Southern-beech (Nothofagus) Forests of New Zealand." *N. Z. Journ. Agric.* XXIII. 353–360.
KIRK, T. (1896). "The Displacement of Species in New Zealand." *Trans. N. Z. Inst.* XXVIII. 1–27.
LEVY, E. B. (1921–23). "The Grasslands of New Zealand"—a series of continuous papers. *N. Z. Journ. Agric.* XXIII. XXIV. XXV.

LEVY, E. B. (1923 *b*). "The Grasslands of New Zealand. Series II. The Taranaki Back-Country " *Ibid.* xxvii. 138–156, 281–293.
SMITH, GUTHRIE H. (1921). *Tutira. The Story of a Sheep Station.* Edinburgh.
THOMSON, G. M. (1921). *The Naturalization of Animals and Plants in New Zealand.* Cambridge.

Prof. J. W. Bews. TRAINING FOR FIELD WORK IN THE DOMINIONS

Much of what may be said on this subject must be largely a question of personal opinion and at the outset I may be pardoned if I try to make clear my own standpoint with regard to the elementary course. I have no drastic criticisms to make on present methods. I attach great importance to the maintenance of botany as a single unit science to be taught as a whole. So far as the grave error of teaching pure morphology divorced from the study of function is avoided I regard present methods of University teaching as essentially sound. With regard to the claim that students should be trained as investigators from the outset, this is certainly desirable if it can be satisfactorily managed. But in a new country one meets many men who have been investigators from the outset and nothing else. We honour them for what they have done, but we cannot help being aware of their limitations. There is no doubt that in some respects they would have benefited by a University training whatever faults such training may possess. Possibly it is true they might have suffered in others, and if it is the case that our present methods of teaching tend to stifle to any great extent the spirit of enquiry, then to the same extent the state of affairs is really serious.

It is not so much, however, the teaching of elementary botany that I wish to discuss as the full University course leading to an Honours or M.Sc. degree. In the University of South Africa this is spread over at least four years. Now while there is room even in our newer Dominions for a certain number of specialists, while a few mycologists, a few plant geneticists, a few plant physiologists are required, yet the majority should have as good an all-round training as possible.

Even specialists are all the better for it, and specialisation even to the extent of concentrating on the subject of Botany as a whole should not begin too early. How best to produce the adaptable pioneer type that makes the most successful colonist is a question too wide to be discussed here. It must be left to the psychologists and educational experts. But the successful botanist must include more than Botany in his course. If he is to concentrate on the physiological side, Chemistry must be his chief subsidiary subject. He must know some Chemistry at any rate, but Physics is not so important for the type of work to be done in the Dominions. On the whole, Geography, with a bias towards the physical rather than the ethnological side, is one of the most suitable subjects to be studied with Botany. Geography might with advantage be substituted for the course or courses in Geology now often included in curricula. For the pass degree at the University of South Africa we run two subjects as "major subjects" for three years. One of these "majors" is then carried on for at least one more year, or if a "First Class" is desired students are advised to take two years for the M.Sc. In the last year a fairly thorough training in research methods can be given, but as far as the teaching is concerned the research method can be introduced almost from the beginning. By the time the last year is reached the student should be capable of applying himself to real original investigations.

Out of the whole course the more time spent on experimental plant physiology the better. Unfortunately in the past a good deal of time would appear to have been misspent on this subject. The carrying out of textbook experiments for the purpose of verifying theoretical knowledge does not seem to me particularly valuable except in so far as it gives practice in the handling of apparatus.

The study of experimental plant physiology must be made more investigational. Throughout the course concrete problems should be set, particularly those having a definite bearing on plant behaviour in the field, and the results should not, as a rule, be known beforehand even by the teacher. At the outset the methods and apparatus may be of the simplest. Comparative studies of the physiological behaviour (*e.g.* the water requirements or water loss) of different important ecological types are always useful, and the student has the satisfaction straight away of adding a few relevant facts to the sum total of our knowledge of the subject. Of course at the beginning the student requires very careful guidance. In a new country it is very easy to learn new facts. More mature judgment is necessary to be able to appreciate their relative significance and importance.

The early introduction of what are essentially simple research methods into the teaching of Botany I know from experience makes a strong appeal to all types of student. From the teachers' standpoint I am aware that there are many difficulties, especially where Botany has to be taught in large industrial cities where the chief "limiting factor" is the amount of coal-dust in the atmosphere. The botanist, then, must be as enterprising as his plants. In South Africa we live closely in touch with an extraordinarily varied assemblage of every possible form of plant growing in natural habitats. For those who are not so fortunate the only solution is the greater use of the greenhouse.

I am inclined to agree with the view that the greenhouse is of more importance than the laboratory unless the investigations can be carried on out of doors all the year round as with us. Botany should not concern itself all the time with the study of the "Living Plant" obtained out of a spirit bottle.

When we consider more fully the content of the botanical course as distinct from the methods of teaching the important question to be decided is what emphasis should be laid on different aspects of it, and particularly the place that is to be taken by phyletic morphology. Even in a four-years' course, I am afraid the time that can be devoted to section cutting and drawing and the detailed study of the life-histories of types must be curtailed. However important the results of this study have been during the past fifty years or so, I doubt whether it is or ever has been the best possible training for a botanist who has to become acquainted with the vegetation of a new country. Yet it is surprising how far the system has been carried. When I landed in South Africa and had time to study the syllabus of the Cape of Good Hope University I found that *Pellia* was prescribed as the type of Hepatic to be studied by all first-year students of Botany. On further enquiries I found that *Pellia* was one of the few genera of Bryophyta that occurred nowhere in South Africa. Of course it could be and was imported by Botany teachers from botanical supply agencies in England. In those days before the war the reputation of British products in South Africa was distinctly high, and for that matter, except for motor-cars and hepatics, it still is.

Another case is still more illuminating. There was a rather bright Chinese student whom I assisted in teaching before I went to South Africa. He knew his life-histories remarkably well and he could reproduce almost every word that ever fell from the lips of his teachers. He himself was to become a Botany teacher in China, and when he left here he asked whether it would be possible to obtain seeds of all the British plants he knew. Most of them he realised did not grow in China, but he would like to grow them for his classes. They were the only plants about which he could trust himself to give instruction.

These stories point a moral. I do not think they are exceptional cases. They show the attitude of mind that has prevailed in the past. They show that the intensive study of types has not been effective even for the purpose for which it was intended, the general understanding of the evolutionary story. The various life-histories are remembered as a series of isolated facts so far as details are concerned, and the comparisons only illustrate certain very general theories. What is wanted is less attention to somewhat unessential details and more understanding of the groups as a whole, their phylogenetic classification, and their distribution and ecological behaviour in nature. For the morphologist the spirit bottle has its uses, and one cannot preach absolute prohibition, but if he is to train botanists who will be useful in the Dominions he must combine his detailed studies in the laboratory with more systematic work, which should be carried out as much as possible in the field.

It certainly seems a pity that systematic Botany has to such an extent been divorced from the University study of the subject. If a closer union were brought about the advantages would not be all on one side. If the University teachers are at fault the so-called pure systematists are not blameless. There is far too much species-making in the herbarium and far too little attention to real phylogenetic classification, particularly as regards the Angiosperms. The systematists, like all other kinds of botanist, should also adopt more of the general ecological point of view.

We in South Africa are on the one hand deeply indebted to the patient labours of those botanists who have dealt with the systematic classification of our flora, while on the other hand we are often exasperated at the careless way in which new species are created. I sometimes feel inclined to advocate that when a country has reached a certain stage of development no botanist should be allowed to create new species unless he knows the plant as it grows in the field and can give full information regarding its variations and ecological behaviour. If we are not to allow the physiologist and morphologist to spend all their time in the laboratory still less should the systematist be allowed always to hide in the herbarium.

Above all things I wish to urge that the botanist who intends to proceed overseas should devote more time to the study of the flowering plants. It is appalling how almost utterly ignorant of the dominant plants of the world to-day many botanists are. The young botanist with a good botanical degree, when he lands in a new country should find the rich and varied vegetation a sheer delight. Instead of that he is at first bewildered and then depressed. Then he adopts an attitude of defence and begins to explain that Botany does not concern itself any more with the mere naming of plants. The psychology is interesting but obvious. Such an experience deserves our sympathy, for it is a distinctly painful one.

If, then, we grant that for the sake of training botanists who will know the plants that surround them, it is necessary to devote more time to the Angiosperms, the further question arises of how that time should be utilised. One of the reasons probably why this section of the subject has been so neglected is the real difficulty of teaching it. For those who are not within easy reach of abundant material throughout the teaching year the garden and greenhouse must be drawn upon to an extent far beyond what has been customary at most Universities in the past. I am aware, of course, that certain centres do have most excellent facilities for this work. As a student I had a training in this part of the subject in Edinburgh under Sir Isaac Bayley Balfour that was fuller than most. Yet my experience in South Africa has made it my constant wish that it had been still more complete or that I had had the sense or ability to take fuller advantage of it.

For those who think that teaching is successful only in so far as the so-called great general principles of the subject are grasped there is plenty of scope in the study of Angiosperms. If we re-read such a book as Lindley's *Vegetable Kingdom* we can realise how full of promise it was even in pre-Darwinian days. At the present time the study can be pursued with many advantages particularly in such a country as South Africa. It is an old country geologically, the conditions are very diversified, and the history of its flora has been continuous since Permian times. There has been no interruption by an Ice period since the rise of the Angiosperms. To the North is the great tropical reservoir of plant-life, part of which has migrated southward along the eastern coast-belt. The rest of the subtropical flora of South Africa shows a history of gradual adaptation to drier conditions culminating in the vegetation of the Karroo and semi-desert regions of the Western side. In the South-West (Cape) and along the eastern mountain ranges there is a temperate flora which differs in composition, ecologically and in origin. There is in South Africa a magnificent field for the study of geographical distribution and detailed migrations, and the bearing of these on phylogeny. It is not only floral evolution that can be studied, but also ecological evolution or the evolutionary history of plant forms, a difficult but very alluring subject.

No one surely can be prepared to argue that the study of the Angiosperms is of minor importance and can therefore be neglected in any University botanical course, yet if I judge correctly that is exactly what has happened at a large number of centres.

As to detailed methods of teaching I have not left myself much time for discussion. I should like, however, to give at least an outline of a course which I have found successful. First of all, as part of the elementary course, the student learns how to describe the external morphology of flowering plants in accurate terms. I know that this kind of descriptive work has long been considered old-fashioned and unnecessary. The student is said to find it particularly uninteresting. All I can say is that that has not been my experience. The comparative method is used continually and extensively from the beginning. By the comparison of different types some of the simpler principles of phylogenetic classification are arrived at. It is surprising how soon the student begins to recognise the distinguishing characters of prominent families. After this he is encouraged to commence using a flora. At the same time the intensive study of one or two genera should be undertaken, especially by those intending to go on to the M.Sc. degree. Indeed, any botanist, no matter what his previous training, when he arrives in a new country and is in danger of being overwhelmed by the richness of the flora, will find it an excellent plan forthwith to concentrate on a single genus until he knows as much as possible about it from every different standpoint.

The lectures and more theoretical study of the phylogeny and classification of the Angiosperms are undertaken as near the end of the whole course as possible. The more we study the tropical-subtropical South African flora the more the generally accepted principles of phylogeny support the views I have just outlined, that the tropics to the north contain relatively ancient types, especially if the more essential floral organs (ovary, ovules, stamens, etc.) are considered, while as we follow the steps of migration into South Africa of any of the larger families we follow at the same time the record of their evolutionary history.

The whole study proves rather fascinating, since it is also combined with general ecological studies on the various types of South African plant community. My students always inform me that of the whole course this is the part they find most interesting. Even from the start they find that they can apply the knowledge they have gained to the plants they see around them. I am not sure that I can suggest a better or simpler test than that for any botanist who desires to carry out field-work in the Dominions and wishes to know what Botany he will find most useful for that purpose.

Mr R. S. Hole, C.I.E. The Ecological Aspect of Botanical Training

It is with considerable diffidence that I comply with the request of Mr Tansley and venture to address you to-day. Practical forest officers usually refer to me bluntly as "only a botanist," professional botanists, on the other hand, with the grace and polished manners natural to their profession, generally treat me tolerantly and quite kindly, but, at the same time, they find it difficult to conceal the inner complacency born of the thought that "after all, poor fellow, he is only a forester." On the ground, however, that my position seems to be rather that of the Missing Link, you may perhaps find some of the points derived from my personal experience which I now venture to put before you to be of some scientific interest; if not, you will, I trust, extend your indulgent tolerance to a mere "wild man of the woods."

During 26 years' service in India, first as a Forest Officer and subsequently in the official capacity of a Forest Botanist, I have been the subordinate of hard-headed, practical, forest officers, and in order to obtain the money and facilities necessary for my work I have been continually called on to make it clear that such work is of real importance in practical forestry. I must confess that this has often seemed to me to be very irksome and to be a waste of valuable time which might more profitably have been devoted to botanical research. Looking back now, however, on this period as a whole, I am convinced that close association with practical men is a very desirable thing. On the one hand, it insistently keeps us in close contact with realities and prevents us from straying after what may, in the end, prove to be mere *ignes fatui*, while the solution of problems of practical importance brings with it the satisfaction of feeling that one's work is of real utility. On the other hand, the growing appreciation of practical men secures for us the assistance in the shape of money and other facilities which are essential for the progress of scientific work. In saying this it must not of course be understood that I advocate anything in the nature of a rigid official control of scientific work. The association here contemplated merely implies that, while practical men indicate the problems to be solved and ultimately judge the work done by the actual results obtained, the scientist is left free to tackle his work in his own way.

It may perhaps be thought that this is likely to encourage concentration on problems of minor importance from which definite results may be quickly expected, but my own experience has shown that, judging from the results obtained in minor problems, practical men are quite capable of forming a sound opinion as to the general reliability of scientific work and are by no means averse to assisting work on difficult problems which require prolonged research for their solution.

Again, the opinion is sometimes expressed or implied that there is necessarily something inferior about scientific work, the object of which is to supply the needs of practical men. But is not this due to a confusion of thought? Truth is Truth wherever it is found and our primary object in searching out Truth should surely consist in the desire to apply our discoveries to promoting the progress of ourselves and of humanity generally towards a happier state. To some extent, indeed, this opinion appears to be based on the discredited belief that "every man has his price" and that human nature is too weak to retain its hold on Truth if it leaves the shelter of the cloister and passes out among the temptations, pleasures and material rewards of the busy world of practical affairs.

Surely there is only one science, viz. the search for Truth, and when, on our quest, we push out into the Unknown, we ought not to forget, or lose touch with, the needs and difficulties of those we leave behind us.

Ecology is usually regarded as the study of the relationships existing between plants and the external factors which constitute their environment.

Accepting this as the ecology of plants, we may similarly regard the ecology of the science of botany as including the study of the relationships existing between botany and other branches of knowledge in the household of Science as a whole. Accepting, then, the principle that close association with practical men is desirable, I will now briefly indicate a few of the points of contact between botany and forestry which have a direct bearing on the methods of teaching botany.

The first necessity, in the case of a forest officer, is that he should be able to recognise and identify the species composing his forest. To enable him to do this he must possess a good working knowledge of the morphological terms used in the description of plants, he should be able to use a good flora with facility and should be trained to observe and utilise forest characters which help him to recognise species in the forest at different seasons. In this connection, I agree with what has been said by other speakers at this Conference as to the necessity for greater precision and uniformity in the use of terms than often obtains at present. He ought also to possess a sound idea of what a species and other systematic groups really are. For this purpose his training should, if possible, comprise a comparative study of (1) a good Linnean species as it exists in the field, (2) the same species as it is represented by the material in one or more good herbaria, (3) the same species as grown in an experimental garden with its component minor groups segregated and grown under varying conditions of light, soil, moisture and other factors, and (4) the descriptions of the species contained in standard Floras.

Incidentally it may be noted that, in the training of professional systematists, emphasis should be laid on the fact that sound systematic work largely depends on close co-operation with foresters and other field workers. On the one hand, such co-operation provides the field worker with the foundation he requires of fixed names and a knowledge of herbarium types, whilst the systematist in return obtains valuable herbarium material, a more accurate knowledge of the unit groups on which his schemes of classification must be based, and financial assistance for his work from practical men who realise the value of his assistance.

Having identified his plants, the next object of the forest officer, as a rule, is to obtain from the forest the maximum sustained yield of valuable forest products, and in order to obtain this he must possess a knowledge of the factors of the environment, of the ways in which they may influence the development of his species and of the methods by which they may be regulated and controlled, so that his trees may develop from seedling to maturity as far as possible under the conditions most favourable for the production of those products and so that ample, healthy, seedling growth may be available on the ground to insure the permanence of the forest after the trees have been cut down and removed. In the training of foresters and forest botanists, therefore, it is necessary to pay especial attention to what is called the ecological aspect. Every opportunity should be taken to lay emphasis on the reality and intensity of the struggle for existence, on the fact that no plant growing under natural conditions can be regarded as an isolated individual, and that at no time is such a plant exposed to the action of a single factor injurious or otherwise. In teaching physiology, for instance, instead of studying only in the laboratory or greenhouse the absorption of water and food material under carefully controlled conditions in water-cultures, we ought to extend our work into the habitat of the plant and study also the relations existing between the plant and the seasonal complex of factors existing in the forest soil.

Again, in teaching pathology, instead of limiting our work to the isolation and identification of a particular parasitic fungus, to its culture under the controlled conditions of the laboratory, and to the results of inoculations carried out in the garden and laboratory, we ought to study the disease also in the forest and try to explain why the host plant is susceptible to attack in some localities and relatively immune in others, why the growth of the fungus is favoured in some places and checked in others. Incidentally, this will also indicate to us the necessity of obtaining a clearer idea than we usually possess as to what constitutes a really normal, vigorous, healthy state, and what are the first indications of a departure from that state.

The adoption of this point of view will, it is believed, not only make botanical training more valuable than it often is at present, but will also make it easier and more attractive to the student. It is, I think, a general experience that in so far as facts are treated as isolated entities they tend usually to be uninteresting and difficult to remember; on the other hand, the more they can be linked up with other facts of a different kind, the easier they are to remember and the more interesting they become.

Consider, for instance, the frequent plan of committing uninteresting facts to memory in the form of rhymes, the attractiveness of many vernacular and local names of plants as contrasted with Latin names because the former are said to "mean something," the way in which a long train of recollection may be fired by a sound or a scent, the value of illustrations when associated with verbal definitions and so on. It seems possible that this tendency is due to an innate consciousness that facts which cannot be connected up with other facts are of comparatively little use to us, or is caused by the fact that, our human nature being many-sided, the full force of our personality can only be called forth by appealing to it in as many different ways as possible.

Finally, there is a point of considerable importance which is not, I think, always sufficiently emphasised at present in botanical teaching, viz. the necessity of obedience to lawful authority and what this really implies. I have spoken above of the subordination of botanists to forest officers, but in actual experience this has not worked out as domination on the one hand with slavery on the other, but as an association for mutual benefit and a working together to attain a common object, viz. an increase of knowledge and power to help humanity. Just in proportion as we experience the feeling of working for a common object so do we lose the sense of oppression and of irksome control and understand what is really meant by "service being perfect freedom." Scientists who devote their lives to the search for Truth are especially apt to find control of any kind irksome, but it is surely clear that no battle could be won if isolated detachments insisted on attacking the forces immediately opposed to them regardless of all other considerations and without reference to the strategic scheme of the army as a whole.

Botanists, for instance, who from actual experience realise the great difficulties caused to foresters, wood technologists, forest engineers, paper-makers, timber dealers and a host of practical commercial men by constant changes in the names of plants would, I think, not only readily accept the principle of *nomina conservanda* but would press for an extended application of it. This would be a great help to practical men and at the same time need not, I think, in any way interfere with the search for useful information which has been previously recorded for plants under older names. In our efforts to resuscitate the knowledge of the past we surely must not neglect the insistent realities of the present.

Systematists, also, who are not in close touch with field workers do not, I think, always realise how important it is to avoid changes in comprehensive systems of classification which have been widely adopted and in the limits of the larger systematic groups, so far as this is possible without interfering with the discovery and recording of really new and valuable facts. If practicable, it would probably be an advantage if drastic changes in classification could be carried out periodically by a representative and authoritative group of botanists and if the more important standard floras could be revised in accordance therewith at the same time. This would then provide a secure basis for a long period of years for the preparation of smaller local floras and various technical and popular publications by means of which the results of the labours of systematists are passed into general currency in a form in which they can be easily assimilated by those who are not professional botanists.

In conclusion, the principal points which I wish to emphasise are that, in the teaching of botany:

(1) We should endeavour to bring botany into as close contact as possible with other branches of knowledge, and

(2) We should try to modify current ideas regarding submission to authority and freedom, by promoting an increasing realisation of the conception of working in agreement with a common spirit, the desire for Truth, which, to a greater or less extent, exists in every individual.

Prof. R. S. Troup. VEGETATION STUDY AND PLANT ECOLOGY IN RELATION TO FORESTRY, WITH SPECIAL REFERENCE TO THE TRAINING OF FOREST OFFICERS

1. *Introductory*

Every year we send out to different parts of the Empire numbers of young forest officers whose training forms a suitable groundwork for the prosecution of vegetation studies, and most of whom should have exceptional opportunities of adding to our knowledge of the Earth's vegetation in one way or another. It is not suggested that every prospective forest officer will develop a taste for detailed ecological work, or have special opportunities for carrying it out, but even if only a moderate proportion of trained officers who go out to the various parts of the Empire will take advantage of the opportunities afforded them, they should be in a position at all events to furnish accurate descriptions of forest vegetation in definite regions.

To take one instance; in the preparation of forest working plans in India and Burma during the past half-century, over 60,000 square miles of forest have been described in greater or less detail with the aid of topographical maps on a scale usually of 2 or 4 inches to a mile. The descriptions have been drawn up primarily as a basis for formulating plans of future management, but at the same time they have added very considerably to our knowledge of the distribution and characteristics of the various types of forest in those countries; they have also in certain cases brought to light remarkable instances of woodland succession, since the working plans are revised periodically and fresh descriptions are generally drawn up at intervals of 10 to 30 years, during which time very appreciable changes in the forest vegetation may take place. Working-plans operations must eventually be extended to all those Dominions and Colonies which intend to take up forestry seriously, and much will depend on the lines on which the forest officers of the future are trained to carry out descriptive work.

2. *Objects and Scope of Vegetation Study*

Before considering the lines on which training should be carried out, we must have a clear idea of the objects to be attained by vegetation study. Looked at from the broadest point of view, we may agree that the ultimate aim is to obtain a comprehensive idea of the Earth's vegetation, and of the various factors which operate in producing its many types. Actually the work may be said to comprise anything from a general preliminary description of the vegetation of a large tract or whole country to a highly intensive and prolonged study of environment factors in their relation to the vegetation of a restricted area. It may include an intensive study of the life history of an individual tree, or of some problem in woodland succession. The intensiveness of descriptive vegetation study in any particular region will depend largely on the degree to which that region has been surveyed topographically and geologically, and on the accuracy and completeness of the recorded meteorological observations. Much will also depend on the time and opportunities available. As a rule, every forest officer who is interested in the subject should be able to furnish useful descriptive accounts of forest vegetation, preferably accompanied by well selected photographs, while, on the other hand, detailed and prolonged investigations of an intensive kind must generally be left to those who are given sufficient leisure and opportunity for the prosecution of such work.

But whatever may be the details of the work to be undertaken in any particular case, we should recognise that if vegetation study is to be placed on a scientific basis we must begin with the small units and work up to the comprehensive whole, rather than attempt at the present stage to formulate a comprehensive classification on insufficient data and to fit our units into it. We are not yet in a position to produce a wholly comprehensive classification of the Earth's vegetation, and those classifications which have appeared are open to criticism in that they are based on an insufficiency of characters. Nevertheless, for descriptive purposes, some simple terminology to denote the main ecological formations is necessary, and until something more satisfactory is devised there seems to be no reason at present to depart from the accepted terms employed by Schimper and others.

This question of terminology as applied to descriptive work is one of the first difficulties likely to be encountered, particularly under strange conditions, and at the risk of digressing somewhat from our main subject it will be advisable to consider it a little more fully.

It may be granted first, that although mere descriptive work may be useful as a preliminary, the final aim of vegetation study should be to ascertain the factors operating to produce the various types of vegetation; secondly, that although the association may be regarded as the ultimate unit of descriptive work, nevertheless we must seek for affinities among the various associations with the view of adopting more comprehensive ecological subdivisions. Thus the importance of ecological formations in the wider sense cannot be ignored, and while admitting that our terminology as applied to formations is at present by no means perfect, nevertheless some form of general terminology, even if it is not wholly satisfactory, is better than none at all.

An example may serve to explain my meaning. The terms "savannah" and "savannah forest" are used to describe types of vegetation with certain superficial resemblances, but widely differing as regards origin and environment factors. Modifications of these general forms are loosely designated by a number of different terms, such as thorn steppe, thorn veldt, park steppe, bush veldt, and so on. The prevailing tree-growth may be evergreen, where the rainfall is well distributed, or deciduous, where there is a prolonged dry season: the characteristic grasses, even under the same climatic conditions, may vary greatly with local changes in the soil and topography. Frequently, though by no means universally, such savannahs have been formed by the intervention of man from pre-existing forests of widely differing form, such as thorn forest, evergreen bush forest, monsoon forest, or even rain forest, and they are kept in the savannah form largely if not entirely owing to the occurrence of periodic fires. With the exclusion of fire they may develop in course of time into widely varying types of forest, according to climate and other factors. This would indicate that although the terms "savannah" and "savannah forest" may be convenient general terms to adopt in descriptive work, they are actually representative of widely varying environment factors or stages in succession.

Thus if a particular tract of vegetation of some savannah type is being studied, the important point seems to be not so much whether it should be classified as savannah, or tree steppe, or thorn veldt, as that its physical and botanical characters should be described accurately in relation to all the environment factors, climatic, edaphic and biotic. It will also be of importance to collect evidence as to the stage in succession which the vegetation represents, the factors, if any—fire, grazing, etc.—which operate in maintaining the tract as savannah, and the further stages in succession which may be expected if these factors are eliminated or modified.

Again, the terms "rain forest" and "monsoon forest," as defined by Schimper, are representative of climatic conditions differing from each other in essential particulars. Yet in certain cases these two great formations may be so closely related that the exclusion of fire for a series of years will convert monsoon forest into rain forest, as has actually been the case in some of the moister deciduous forests of Burma and Bengal. There is thus probably a stronger affinity between the rain forests and moister monsoon forests of Burma than there is between the rain forests of Burma and those of tropical Africa.

This question has been touched on at some length in order to emphasise the fact that the detailed study of the various environment factors, not climatic factors alone, is of more importance at present than the formulation of comprehensive schemes of classification or terminology.

3. *Training*

Turning now to the question of training, we may presuppose that the general training of a forester should include, among other things, a sufficient grounding in botany (including systematic botany), chemistry and physics, geology, soil science, surveying and climatology. The more specialised training in ecology might be divided into two parts, (1) the study of works of a general nature such as Schimper's *Plant Geography* and Warming's *Ecology of Plants*, as well as books dealing with the more purely geographical aspect of the subject, and (2) the study of research methods as detailed in such works as Tansley's *Practical Plant Ecology*, Clement's *Research Methods in Ecology*, and, for those who know German, Rübel's *Geobotanische Untersuchungsmethoden*.

As regards the actual lines of work to be followed, it is advisable that each investigator should have full latitude to develop his own methods, but the beginner will be well advised, before undertaking descriptive or investigational work, to read any available papers that have been published dealing with work of the kind he proposes to undertake. Material in great variety will be found in the pages of the *Journal of Ecology*. Among vegetation studies in tropical forest regions pp. 1–48 of Chipp's *Forest Officers' Handbook of the Gold Coast, Ashanti and the Northern Territories* (1922) may be taken as an example of a general account of a wide area on fairly extensive lines, while Shreve's *A Montane Rain Forest* (Jamaica[1]), or W. H. Brown's *Vegetation of Philippine Mountains*[2] furnish examples of somewhat intensive studies of more restricted regions. Hole's work on the Ecology of Sal seedlings (*Indian Forest Records*) may be cited as an example of a detailed and prolonged study of the early stages of development of a single important species of tree—work such as is hardly possible unless carried out by a person specially detailed for research work.

Examples of different lines of work might be quoted to almost any extent, but these will have to suffice for our purpose. It is hardly necessary to emphasise the importance of practical work in the field as an essential part of the training, and in this connection a good grounding in Systematic Botany is a necessary preliminary; for while it may be admitted that every forester cannot be a highly trained systematist, it will hardly be denied that descriptive work carried out by anyone who does not fully appreciate the importance of correct identifications must be of little value.

4. *Some Difficulties and Suggestions*

We may now suppose that a student, say in Great Britain, has made himself conversant with the general principles of ecology and vegetation

[1] F. Shreve, *A Montane Rain Forest* (Jamaica), Carnegie Inst. of Washington (1914).
[2] W. H. Brown, *Vegetation of Philippine Mountains*, Bur. Sci. Manila (1919).

study, and that he has had some opportunity for practical work under conditions prevailing in this country. When he begins work under very different conditions, as most of our forest officers do, he will encounter numerous difficulties, particularly in tropical regions, where the factors are generally more varied and complicated than they are in temperate regions. I propose, therefore, to devote the rest of this paper to the consideration of a few of the more important of these factors, namely rainfall, atmospheric humidity, temperature, light, soil, and the action of man; finally I propose to indicate in a few words the importance of a study of factors bearing on regeneration. Anyone who begins to deal with forest investigations will soon discover that even if climatological and other data can be recorded with accuracy, the interpretation of results may be extremely difficult owing to the size of the trees, which precludes the use of laboratory methods except in the case of young plants. For this reason the elimination and isolation of individual factors may become impossible, and we must often be content to estimate at best the effect of combinations of factors. So far as climatological data are concerned, one of the main difficulties encountered, particularly in more or less undeveloped countries, is their insufficiency. Recording stations are necessarily situated as a rule in convenient centres where observations can be recorded regularly; these centres are generally outside the larger forest tracts, and they are chosen according to the locations of civilised man rather than according to the presence of forests or to topographical conditions.

The difficulty of extending climatological stations into uninhabited forest regions is obvious. It is desirable, however, that at all events rainfall data in forest regions should be amplified, and this should not be impossible in places which can be inspected frequently. If the intervals between readings are too long, errors due to evaporation may be appreciable, particularly in a dry climate or a windy situation. Self-recording raingauges are said to be defective in some respects, but even so they have advantages over gauges of the ordinary type where frequent readings are impossible.

In selecting stations for rainfall and other observations, it is important that the main types of forest, scrub, and grassland should be severally represented as far as possible. For extensive work rainfall measurements made in the open should suffice, but for intensive investigations in the forest it may be necessary to place raingauges not only in the open but also at different points under the forest canopy, in order to ascertain the amount of rain that actually reaches the ground.

Rainfall is rightly regarded as one of the dominating factors in determining the nature of the forest formation, though the type of vegetation appears to be influenced more by the distribution than by the actual amount of rainfall, provided the total amount is not very small. In the tropics it is generally accepted that a monsoon climate, characterised by a long dry season alternating with a decided rainy season, produces deciduous forests, while a climate with a well distributed rainfall produces evergreen forests. The deciduous monsoon forests of India and Burma, for instance, furnish a distinct contrast to the evergreen types prevailing over a considerable part of Africa, even where the rainfall is only moderate in amount.

The question of seasonal rainfall is of importance in connection with the introduction of exotic trees. E. H. Wilson[1] has pointed out that in southern latitudes (South Africa and Australia), other conditions being favourable, certain species of pine, such as *Pinus longifolia* and—so far as results are available—the Mexican pines, thrive only where summer rainfall prevails, while others, such as *P. Pinaster, P. Pinea* and *P. canariensis*, do not thrive in a summer rainfall but flourish where the rainfall is a winter or all the year round one. It is interesting to note that the seasonal rainfall requirements of the two closely allied species *P. canariensis* and *P. longifolia* are diametrically opposed, and these species may be regarded as supplementary to each other in similar latitudes where the seasonal rainfall varies.

The relation of atmospheric humidity to forest growth is a matter regarding which our knowledge is far from perfect. Relative humidity has a direct influence both on evaporation and on soil moisture, and may yet prove to be a more important factor than is generally supposed. In the moister regions of the tropics the copious formation of dew, which pours down like rain from the trees on clear cool nights, must add appreciably to the total precipitation, and to the quantity of water which reaches the soil, often at a season when there is little or no rain.

The effect of increased atmospheric humidity due to the ascent of air in tropical mountainous regions is demonstrated in a marked degree by the richness of the epiphytic vegetation in the mist belt at the higher elevations, where the air may be at saturation or nearly so for months at a time. The branches of the trees are covered with a rich growth of mosses, ferns and other epiphytic plants. The trees themselves are sometimes dwarfed, a condition which cannot under such circumstances be

[1] "Northern Trees in Southern Lands," *Journ. Arnold Arb.* IV. 61 (1923).

attributed to desiccation, though the precise reason for the dwarfing has yet to be determined.

Humidity and temperature observations can best be made simultaneously with the aid of recording thermometers, maximum and minimum thermometers, and recording hygrometers. Assuming that readings can be taken at comparatively short intervals—say weekly—the records can be presented in tabular form for each period: (1) maxima, minima, and means, and (2) averages of daily maxima and minima.

As regards temperature records, it is rightly held that mean maxima and minima for each month, and mean annual extremes, are of importance: absolute maximum and minimum figures for a long series of years are also of interest as showing the extremes likely to be encountered in the habitat of a species. The worthlessness of mean annual temperatures as a criterion of tree-growth can be demonstrated by many examples, but perhaps one of the most striking is the fact that in the cool but remarkably equable climate prevailing near the equator at the higher elevations of the East African plateau no species of *Pinus*, among many introduced, has yet been found to flourish, although the mean annual temperatures at appropriate altitudes are similar to those in the natural habitat of several species. Doubtless the explanation lies in the fact that owing to the equableness of the climate the trees experience no resting period such as they are accustomed to in their natural homes.

There are few factors the effect of which on forest growth it is more difficult to assess than that of light. Even assuming that the relation of light to photo-synthesis can be measured accurately, such measurements, if made in the forest, are complicated by other variable factors which may operate at the same time, such as soil moisture, evaporation, or degree of protection from wind, frost or insolation. Hence it is unsafe to attribute to light conditions alone any difference in growth under the shade of a forest canopy as compared with that observable in the open. Only in the case of young plants in controlled garden or laboratory experiments can the effect of light on tree species be determined with any degree of accuracy. In the case of tree-growth, measurements of light intensity in relation to photo-synthesis should strictly speaking be made in different parts of the crowns of the trees, where the light intensity may vary greatly from point to point. But average figures of any value would be unobtainable under such conditions, and hence for comparative purposes, for instance in comparing light intensities at different elevations and in different situations, we must assume that measurements taken under similar conditions in the open over a fairly long period of time are sufficiently comparable.

The question of light intensity at different elevations is one which requires further study in its relation to tree-growth and regeneration. Brown[1] found that the light intensity at the top of Mount Maquiling in the Philippines was only 44 per cent. of that at the base and suggests that this may be one of the factors accounting for the smaller size of the trees at the higher elevations. It is possible that the occurrence of mist at the higher altitudes may have been partly accountable for the diminution of light intensity in this case.

Soil factors have perhaps hardly had the attention they deserve in descriptive vegetation studies, particularly in the tropics. Caution is certainly necessary in interpreting the effect of geology and soil on natural vegetation, and under uniformly favourable climatic conditions the soil factor may become of minor importance. Some well-known tropical soils, however, such as laterite and black cotton soil, are often characterised by very distinctive types of vegetation. In Burma laterite characteristically supports a form of dry dipterocarp forest with a well-marked floristic composition. Possibly the type of forest in this case is determined by the rapidity with which the rock dries, whether owing to its permeability in the case of disintegrated laterite, or the rapid run-off of water after rain in the case of hard laterite. This marked differentiation of forest type owing to the presence of this rock, however, does not appear to be universal in other countries where true laterite occurs.

The black cotton soils are unfavourable to forest vegetation, and both in India and in Africa support at best an open and often stunted growth consisting largely of thorny acacias and other leguminous species. On the East African plateau small patches of open acacia scrub on black cotton soil may be seen inside tracts of evergreen high forest on red loam, clearly indicating that the soil factor is here the determining one. In this case the precise action of the soil on the vegetation is a question requiring further elucidation.

Nearly 30 years ago the Russian investigator Sibirtzev drew attention to the relationship between soil and climate by showing that the different soils of Russia could be separated into broad zones corresponding approximately to climatic zones, and that over regions where a similar climate prevails the soil, during the process of rock-disintegration, acquires more or less identical characteristics even though the underlying geological formations may vary widely. The same principle has been found to hold

[1] *Loc. cit.* p. 233.

good generally in the case of the soils of North America, while more recently Marbut has traced a similar relationship between soil and climate over broad regions throughout Africa. The work of Shantz and Marbut on *The Vegetation and Soils of Africa*[1] deserves more than passing notice. Adopting Sibirtzev's principles, Marbut has compiled a soil map of Africa based to a considerable extent on climatic data alone, with only a limited number of soil analyses over a portion of the Continent. This procedure may appear daring, and the map is admittedly tentative, but it should form an interesting basis for testing the relationship between soil and climate, and particularly rainfall, when more detailed local examinations of the soils of Africa have been carried out. In the same work the influence of soil on vegetation has in certain cases been clearly demonstrated.

There can be little question as to the need for the more intensive study of soil conditions in their relation to forest vegetation, and soil survey work may be regarded as an essential part of the training of a forester who is to undertake vegetation studies. This work will generally involve co-operation between the local forest investigator and a more centrally situated soil chemist.

Problems of succession are of particular interest to the forester, since they are associated with the important question of natural regeneration. Many of these problems are directly connected with the action of man in clearing and burning forest. Civilised man, generally speaking, exercises a greater influence, destructively and constructively, than primitive man: thus the inroads of civilisation in new countries are accompanied by extensive clearing of the natural forest vegetation, while civilised man may also be responsible for an almost complete alteration of the primeval forest to more or less artificial types, as in Great Britain.

The destructive propensities of primitive man vary with his mode of life. Tribes which live entirely by hunting and fishing, and do not cultivate, produce little or no effect on the forest otherwise than by causing occasional fires during the collection of wild honey or other minor escapades. Pastoral tribes make extensive and constant use of fire in order to convert forests into grasslands and to maintain the vegetation in that condition. In thinly populated countries primitive tribes which cultivate permanent gardens and fields round their villages are as a rule innocent of any extensive destruction of forest. On the other hand, tribes which practise shifting cultivation entirely upset the processes of Nature, and the actual result will depend largely on climatic and other

[1] H. L. Shantz and C. F. Marbut, *The Vegetation and Soils of Africa* (1923).

factors. Shifting cultivation has completely altered the original forest vegetation over extensive tracts in India, Burma, Ceylon, tropical Africa, and other parts of the world where this form of cultivation is practised.

In some regions such temporary clearing of forest is followed by a re-growth of forest, sometimes after a temporary occupation by grass; this appears to be the general rule in forest regions of non-seasonal rainfall or of seasonal but heavy rainfall. Often the species of trees which invade the clearings are different from those of the original forest, and the subsequent progression towards the pre-existing climax type may be rapid or slow according to circumstances. In many tropical regions, however, when forest is cleared grass takes its place and occupies the ground permanently owing to annual fires. This is most noticeable in regions of seasonal rainfall with a prolonged dry season during which the grass becomes highly inflammable, but it may also take place in regions of moderate rainfall where there is a less well-marked dry season but where periods of drought are liable to occur at times. Under such conditions tree-growth establishes itself with difficulty or not at all, whereas the grass escapes permanent injury owing to the immunity of its underground rhizomes. There is abundant evidence to show that many of the savannah tracts in the tropics were once clothed with forest, and that they have been brought to their present condition by the agency of man.

A particular form of retrogression is that in which tree forest, when cleared for shifting cultivation, is replaced by bamboo jungle. In certain tropical countries extensive areas of pure bamboo forest have originated in this way. This condition is probably due more to the aggressive habits of the bamboos than to anything else, for the culms which ascend from the underground rhizomes grow rapidly and quickly suppress any transitory tree species that may appear on a cleared area.

Among the most important of all problems to be solved by the forester are those connected with regeneration. These problems involve the study of factors bearing on seed production, dispersal and germination, and the establishment and development of the seedling. The seedling represents the most critical stage in the life of a tree, and once this stage has been successfully passed the problem of regeneration may be said to have been solved; an accurate knowledge of the conditions under which the various seedlings are capable of establishing themselves, therefore, is of extreme importance. Seedlings can, fortunately, be studied under more exact methods than trees in the later stages, by means of controlled experiments in which the various factors—light, temperature, soil texture and moisture, soil-covering, etc.—in their relation to germination and seedling development can be regulated.

The close study of regeneration factors as they affect different species is a necessary preliminary to the determination of the origin of any particular forest association. In this respect there is a similarity between silviculture and geology. In order to understand the processes by which the rocks as we now find them were built up, the geologist studies similar processes taking place at the present day; he is thus able to interpret the various signs revealed by the rocks and to attribute their origin to definite causes. Similarly, by studying the conditions under which individual tree species are able to regenerate and establish themselves the forester may succeed in explaining the origin of different woodland associations. But whereas the geologist carries his mind back to remote geological ages, the forester looks back through mere centuries, or only decades. Nevertheless, the forester's task is often the more difficult owing to the number of factors to be taken into consideration.

Dr T. F. Chipp. VEGETATION SURVEY OF THE CROWN COLONIES AND THE TRAINING OF PROFESSIONAL MEN AS FIELD OBSERVERS

The need for an inventory or stock-taking of the vegetational units in the various parts of the Empire is apparent, and yet it has not been attempted in the Crown Colonies. The reason appears to be that this duty belongs to no one authority. Forestry and Agricultural Departments each work at and chronicle such areas and vegetation as interest them, but there is no co-ordinating scientific body that is in a position to correlate this work and present a combined result. Our Colonies in this respect may in fact be likened to a large store where the head of each department is interested in the goods under his charge but where a general manager is lacking. Such a state of affairs in the business world cannot of course be contemplated, and yet in the Crown Colonies it exists so far as the study of vegetation is concerned. This cannot be considered as anything but most unsatisfactory.

Not only are such appreciations of a country necessary if a correct idea of the exploitation and development of its economic resources is to be gained, but such records would prove of the greatest value to workers in the future who would be aware of the natural state of the vegetation at a given period. Were we, for instance, to be in possession of data of the period before the present era of agricultural development changed the face of so many countries, it would be of the greatest help in solving many problems with which we are confronted.

It has been argued that our chief object should be to study small units and work up to a comprehensive whole rather than to attempt to formulate a general classification on insufficient data and to fit our units into it.

I venture to submit, firstly, that we have now sufficient data available on which to draw up the broad lines of vegetational zones and chief units of these countries, and, secondly, that students of small units cannot appreciate the relative value of their work without a knowledge of surrounding and ancillary factors and conditions. This knowledge they cannot obtain personally as they are not in a position to travel sufficiently extensively, nor have workers in the field access to libraries where are recorded the results of previous investigations. In fact you cannot build a house with bricks alone, you must have foundations. These must be prepared beforehand and we are, I submit, in a position to effect such preliminary preparations.

Summaries should be produced in booklet form for each Colony, or natural group of Colonies, on the basis of an outline account of the vegetation, and should be uniform in design. They should be presented according to a definite plan so that the completion of each one of the series will mark a definite advancement in our study of each particular group, in a manner rather similar to the method of publication adopted by Engler for *Die Vegetation der Erde.*

Until such outlines are provided the field worker can only record his observations in the limited area or field of work under his consideration, without indicating the relation of his work to the whole surrounding unit.

Efforts so spasmodic and uncorrelated cannot lead to any result, and unless some steps are taken we shall find that our knowledge of the vegetation of the Crown Colonies is as diffuse and intangible in ten years' time as it is at the present day. It behoves us, then, to discuss some definite scheme which we can approve and recommend for consideration to the Conference. The aim of this scheme must be to present the potential worker in the field with a concise account of what is known about his surroundings, and also put him in touch with the institutions at home that can assist him in his studies.

If, at the next Botanical Conference, we can point to the completion of this stage in our studies I am convinced we shall at any rate have reached a common starting-point from which all may join with a knowledge of what has been done.

As to the means by which such compilations may be effected. Many suggestions may be offered, but a central permanent organising Committee is essential if the discussion is not going to end with this Conference. I would put forward for your consideration, therefore, that this Conference approach certain Societies and Institutions with a view to the subject being taken up permanently. In the event of these bodies entertaining such a proposition favourably a directing Committee of their representatives might ensure the inauguration and continuation of the work.

In such a Committee we have our directing force. They would consider the suitable groups into which the Crown Colonies may be gathered, and approve a definite form for this series of Outlines.

The next step is to secure the service of gentlemen with a field knowledge of a Colony or Group and opportunity to co-operate in this scheme, and who would make themselves responsible for compiling these Outline Accounts. This should not be difficult for such a Committee, for only some half-dozen compilers will be required and we know of many gentlemen in retirement who have the ability and opportunity to assist.

We have, however, yet to consider the question of publication, for without publication the work is useless.

The preparation of these Outlines on the lines indicated should not be an expense except for the final typescript and publication. I estimate that such a work as one of these Outlines, suitably illustrated, could be produced at from £200 to £250, with 500 copies as a minimum. Such a work will appeal not only to the ecologist but to all scientific and technical officers in the Colonies concerned, and in addition will be of use to Political and Administrative Officers. It is to be expected, therefore, that Colonies concerned may either advance a sum, receiving a number of copies afterwards for distribution to their officers, or may undertake to buy so many copies. In the case where several Colonies are concerned in only one publication the amounts solicited from each would be small.

In addition, there are various organisations which have funds at their disposal for aiding research work, and I am of the opinion that were the case to be argued before them they would feel disposed to aid in some measure.

With the exception of Ceylon, the Crown Colonies would be considered most conveniently under natural groups such as the following, and this would materially lessen the number of Outlines to be prepared and at the same time arrange for a number of Colonies to contribute to each work.

The *Eastern Group*, comprising, principally, the Malay Peninsula, Straits Settlements and Borneo.

Ceylon.

The *North-East African Group* of the Anglo-Sudan and Somaliland.

The *East African Group* of Uganda, Kenia and Tanganyika.

The *Central African Group* of Nyasaland and Northern Rhodesia.

The *West African Group* of Nigeria, Gold Coast, Sierra Leone and Gambia.

The *West Indian Group* of the West Indies and British Possessions in Central and South America.

The preliminary survey of these countries or groups on which so much work has been done and which I am urging should be consolidated, may be considered as including the linear surveys and area summaries, the latter including the general physical or physiognomic characters, and the floristic analysis, connection and enumeration. This necessary preliminary work can in the first instance be carried out only in the field, but it is only rarely that the field worker has had the opportunity to correlate his work with that of others, and hence we have no one presentation or appreciation of the vegetation of the countries under consideration. This is the work that can be carried out by compilers at home who have had some previous knowledge of the countries.

The second and more thorough survey we have hardly yet begun. Such detailed work has been approached by Colonial teaching staffs, visiting botanists and research officers of scientific and technical departments, who have a suitable botanical education and a knowledge of kindred sciences. Under such a heading I would class detailed studies on the habitat, on biotic, physiologic and physiognomic factors, on crops, floristic surveys and complete ecological studies.

As to the assistance we may obtain from professional men as field observers. Recent history indicates clearly what is required. At the end of the eighteenth and beginning of the nineteenth centuries men, with scientific and technical qualifications, were beginning to travel abroad to the Colonies and lay the foundations of those institutions around which the more modern spheres of scientific and botanical enquiry have centred. These men were keen naturalists both professionally and because of their personal likings and enthusiasm for their studies. At the same time they had a thorough all-round scientific training and were in a position to understand the various problems of life as evidenced in the countries they visited. They were trained essentially on broad lines and were eminently capable of taking a wide outlook in all their bearings on the problems they encountered, but above all they had a sound grasp of systematic botany. They were independent to the extent that they were sent out by Societies or Institutions at home, or they were the head and, at the same time, the whole, of the Government scientific departments in themselves. Their work was not localised neither were their tasks specialised.

As their systematic study of the flora enabled the various essences of the vegetation to be recorded we find that concomitantly the types of the vegetation were also remarked, until eventually we are confronted with the fact that the only useful early descriptions of the vegetation of the tropical colonies are those drawn up by men who are regarded as the great leaders of systematic botany.

Within a few decades, however, a change occurred. Specialists trained in Agriculture and Forestry began to arrive in the Colonies and gradually to take over the particular work of these newly created Departments.

Thus at the present day we have specialised Departments, staffed by specialists who have definite and localised work which demands the whole of their time and energy, in place of the old botanic institution in charge of an all-round naturalist. In none of these Staffs do we find anyone approaching the old naturalist with his broad outlook on life and general training in systematic botany.

It is this personnel, then, to whom we must look for assistance in promoting surveys of the vegetation. Amongst them there are many men, who have had sufficient education and possess a natural interest, who would be prepared to work on this subject if only some instruction could be imparted and their research directed. One has only to think of the botanists of the first rank that have arisen from the Medical Services both at home, in India and the Colonies to realise the assistance that might be looked for in this direction. It is, however, in the Agricultural and Forestry Departments of the Colonies where we ought to expect the bulk of the workers in the field. The ordinary work of the officers in these Departments makes them familiar with definite tracts of country on which they write reports in the routine of their duty. Moreover, at the present day, practically all the senior grade officers of such Departments are men with University science degrees or equivalent training and experience. On examining the staffs of these Departments in the Crown Colonies of the Empire, we find a total of 352 officers who are situated favourably for and quite capable of noting and conveying the information required to complete a Vegetational Survey, at any rate in its broader aspect.

It must be admitted with regret, however, that the majority of young men preferred to these appointments nowadays have turned their thoughts in this direction of earning a livelihood only at the end or towards the end of their University career. The old naturalist was marked out from his boyhood by his passionate interest in all that pertained to Nature. In the present age of examinations boys and young men are rarely directed to first-hand studies in the field after they leave their preparatory school. Moreover, in an industrial age, it is difficult for schools to allow for definite teaching in elementary biology when the examinations which boys must pass at various ages and stages in their careers require knowledge in so many and different subjects.

The situation, then, is that the majority of foresters, agriculturists, and others of kindred professions, know only the bare rudiments of classification and determination of plants, and have very little idea how to set about acquiring a working knowledge of this subject once they are launched into their professional activities abroad. How often do we hear them express their regret at their previous neglect of the study of systematic botany, when they arrive in the field without anyone at hand to help them or any library or herbarium available for consultation. They endeavour to learn the plants from the vernacular names, which is difficult, often misleading, and in the end leads to very little definite knowledge. I have met keen young men in the Colonies who, after several months' residence and work, do not know the botanical names of their most frequently occurring and important trees and plants, and after some years manage to learn only a very few. It may be suggested that I have been discoursing on systematic botany, whereas the subject this section is discussing is Ecology and Vegetation Surveys. But how can anyone describe a unit or piece of vegetation when the essences composing it are quite unknown to him? Such men are not in a position to carry out any investigations in ecology, for they are too scattered for anyone to teach them, and they have not been given sufficient basic knowledge to help themselves. Were they put in possession of a sound systematic education they could then learn to identify their plants and trees and so follow up their distribution and associations.

It is quite evident that the field study of botany and its kindred sciences, such as climatology and geology, is not presented to the student at a sufficiently early age in a tangible and living form so as to turn his thoughts to such study and direct his pursuits into a line that he can follow in after life with a natural inclination, turning what for years should be his hobby into his chief work and pleasure in life. Only when this is done shall we again secure the worker who absolutely loses himself in his daily bread-and-butter work and acquires and retains that spirit to seek into and work out unaided and in isolation the problems of nature.

Important as is the preparation of a summary of the work that has already been done, unless some permanent scheme of registration of current work is organised, we shall in a few years find ourselves in the same difficulty as we are at present.

Publication of current observations takes place in various periodicals wherever publication can be effected, but it is necessary to have them recorded and available for consultation. Bibliographies of some Colonies have already appeared, but the scientific references are very meagre.

What is required for the work of keeping these records up-to-date is an organisation of collaborators at various centres. Centres are already in being at Singapore, Ceylon, Amani and Trinidad, whilst West Africa can be worked from England owing to its proximity. It is necessary to enlist the sympathy and support of the personnel at these institutions to forward these ends. Here again some controlling and directing organisation is necessary to put such work in hand and to maintain it. Publication of such registers as are required could be arranged as appendices to local Government publications and thus be made available for the public generally.

The conclusions to which I would draw attention especially are as follows:—

There is a body of potential workers in the field who need instruction and encouragement by means of Outline Accounts of the Vegetation of their countries if they are to be of real assistance in the prosecution of this study.

We should endeavour to see that future workers are better prepared by fostering the proper spirit and turning their thoughts and energies to the study of Nature at an early age in their school career and maintaining their interest throughout their school and university life.

There is a mass of information that is awaiting examination and publication which should be undertaken at once. Steps should be taken for future work to be noticed publicly under an organised system.

Dr O. Stapf, F.R.S. THE RISE OF PLANT-ECOLOGY IN GERMANY

(*Abstract*)

My paper, which deals with a prominent phase in the development of plant-ecology, is intended to join on to the President's general review of the subject and to form a background to Dr Chipp's contribution on "A Vegetation Survey of the Crown Colonies and the Training of Professional Men as Observers." The time at my disposal obliges me to concentrate on the causes of the great progress plant-ecology has made in the German countries, and to leave the application to our own needs to the audience.

As in other countries, plant-geography in Germany was originally and inevitably directed towards the statistic and topographical branches of the discipline, but almost from the beginning it was permeated by the Humboldtian spirit which had been nurtured in the sight of a vegetation unrivalled in wealth and diversity. This spirit fertilised the thought of German travellers and explorers and it also taught the home student to glean from their accounts and equally from those of foreign travellers whatever had a bearing on the problems that arose from the Humboldtian conception. Grisebach's *Die Vegetation der Erde* (1872) was the first comprehensive digest of the knowledge thus amassed. Parallel with this line another developed in which the geological or historical aspect stood foremost. It had its source in Unger's work, drew its life from the rising Darwinism and had its solid foundation in J. D. Hooker's famous essays. It found its fullest expression in Engler's *Entwicklungsgeschichte der Pflanzenwelt* (1879). Although divergent in their aims, the two branches of plant-geography represented by these lines have a common basis and they are intimately connected. Therefrom arose beneficial reaction on the common base work and interaction between them. Deepened by the application of the knowledge gained through modern botany and by the introduction of more scientific methods, fostered by careful work in the homeland (Kerner) and broadened by familiarity with foreign literature, the earlier of the two branches developed soon into plant-ecology as we know it. Drude's *Handbuch der Pflanzengeographie* (1890) is a markstone in this direction, whilst Schimper's *Pflanzengeographie auf Physiologischer Grundlage* (1898) signifies the completed transformation. When in the 'eighties of the last century Germany became a colonial power, the soil for more intensive and systematic ecological work in the new colonies, all of them in the tropics, was prepared. Guiding literature was, or at any rate soon became, available, whilst the intimate contact which had always existed between the Botanic Gardens and Herbaria on one side and the Universities on the other guaranteed a wide diffusion of the taste and knowledge for and the appreciation of work contributive to plant-geography in the tropics. Collectors, travellers and administrators were equipped for it. Their efforts were sure of support from the home government, whilst the centralisation of the botanical exploration of the colonies under the aegis of the Director of the Botanic Garden and the Botanic Museum at Berlin ensured a useful amount of co-ordination and the publication of the results achieved. Botanic Gardens and Botanic institutes under scientifically trained heads were established in some of the colonies. Although generally a much wider field was assigned to them than ecology and in some cases they were intended in the first place for practical work, yet they became promising centres for intensive field-work and experiment without which progress in ecology is impossible. A development like the one described cannot remain restricted within political boundaries. As Nature is independent of them, so must ultimately be all attempts to interpret her. The recognition of this principle by the German school of plant-geography has influenced the teaching of the discipline throughout the world and gained for it a considerable following and many a collaborator. Nor has such success proved a one-sided advantage. It has given a powerful stimulus to the development of the discipline generally, not least within the boundaries of this empire. Rival schools have sprung up; problems are viewed from a different angle; methods are changed as the national genius leans more this way or that way; but the movement that has been started will go on and we who are in it cannot afford to neglect or disregard either the head-waters or the great tributaries which lie outside our own area.

Mr J. S. Henkel. Proposal for a Uniform Method of recording Ecological Observations with special reference to Africa

It is important that forest officers and other botanical observers when travelling through a new country should have some uniform method of recording observations of the vegetation and the general conditions under which it exists, so that comparable data, available for generalisation, may be obtained. The following points are of importance. They can be recorded on a map from which, later on, "floral regions" can be constructed.

(1) Direction of route.

(2) Geology: samples of rock and soil should be collected.

(3) Altitude, determined by aneroid readings, both in stream-beds and on ridges.

(4) Notes should be made on climate and other meteorological points.

(5) Maximum and minimum daily temperatures observed during the trip.

(6) The flora in as great detail as possible. The occurrence of known plants should be recorded, and as many specimens as possible should be collected, locality and habitat being noted as fully as possible.

(7) *Timber specimens.* The following has been found a practicable and convenient means of securing specimens of timber. Young trees of about 4 inches diameter 4 feet from the ground are chosen and sawed through at this height. A vertical slice $\frac{1}{2}$ inch thick and 4 inches in depth is then cut across the stump, so as to include pith and bark, giving a specimen 4 inches $\times$ 4 inches $\times$ $\frac{1}{2}$ inch, small and convenient for transport. The age of formation of heartwood, if any, should be noted.

It is suggested that the Conference should prepare an outline of the data to be noted by observers. The resulting series of uniform observations could afterwards be co-ordinated by a central body appointed for the purpose.

Dr W. Burns. THE APPLICATION OF ECOLOGICAL TERMS IN THE TROPICS

The recent papers of W. T. Saxton[1] and L. Dudley Stamp and Leslie Lord[2] voice a difficulty in applying certain terms, particularly as regards plant communities, to Indian and Burmese conditions. I imagine that similar difficulties have been encountered by other workers in similar places.

In India we are only at the beginning of ecological investigation. Some excellent work has been done, but we have not yet quite cleared our points of view nor standardised our methods. It seems to me that the Imperial Botanical Conference may help us greatly, particularly in regard to the matter of terminology.

Taking Western India, which I know best, we have an area under the influence of the monsoons, except in the extreme north, where desert

[1] W. T. Saxton. "Mixed Formations in Time, a New Concept in Ecology," *Journal of Indian Botany*, III. 2, 3, 4, pp. 30–33.

[2] L. Dudley Stamp and Leslie Lord. "The Ecology of the Riverine Tract of Burma," *Journal of Ecology*, XI. 2, pp. 129–159.

conditions prevail. In the rest of the area there exists therefore a habitat with a marked wet season and a marked dry season about twice as long as the wet season. The highest temperatures come just before the wet season. The amount of rainfall varies on the same parallel of latitude from (in the Bombay latitude) 80 inches on the coast, through 100 inches on the Western Ghats, tailing off rapidly thereafter through 25 inches down to the famine-stricken eastern area where anything down to 5 inches per year may occur. The monsoon periodicity, the varying amounts of rain received, plus considerable edaphic variations (including mangrove conditions, inland salt areas, sheer trap rock, etc.) produce an environment of considerable complexity. Man and animals have also much modified the vegetation.

Is the terminology of any ecological school suited to the description of plant communities in such conditions? The difficulties of application seem to me to be:

(1) The existence of plant communities growing in highly specialised conditions;

(2) the existence of a considerable number of communities which are neither grassland nor woodland;

(3) the monsoon condition, which brings about several marked changes in the appearance of the same area within one year;

(4) the existence of certain species which live right up to both extremes of a great range of rainfall and soil conditions.

I have a feeling that we should at present concentrate on the accurate description of small units of vegetation, without too much care as to their higher classification. I do not wish to minimise the importance of larger surveys, but we must not generalise too soon. I offer the whole subject of the nomenclature of tropical plant communities for discussion.

RULES OF NOMENCLATURE

(CHAIRMAN: DR A. B. RENDLE, F.R.S.)

The Chairman explained that a discussion on Nomenclature had been included in the programme of the Conference in response to a general desire expressed by overseas botanists and others. The working of some of the International Rules of Nomenclature adopted by the Vienna and Brussels Congresses had not proved so satisfactory as had been hoped. The Executive Committee of the Conference had therefore appointed a small sub-committee to report on the operation of the International Rules. Certain resolutions for the modification of the International Rules were being put forward by the sub-committee for the consideration of the Conference and to serve as a basis for discussion. Any resolutions adopted by the Conference would be forwarded to the approaching International Botanical Congress in America.

The discussion was opened on behalf of the sub-committee on nomenclature by Mr T. A. Sprague, who stated that in their opinion the International Rules had proved satisfactory on the whole, and had brought about greater stability and uniformity in nomenclature. In the course of the 19 years during which the International Rules had been in operation, however, certain defects had become evident, and differences of opinion existed among experts as to the interpretation of some of them. The sub-committee was therefore proposing certain alterations in the Rules in order to remedy these defects and to remove ambiguity (see *Journ. Bot.*, March 1924, pp. 79–81). They considered that the object of taxonomic nomenclature was to provide a means of indicating with certainty the identity and precise circumscription of groups, and that the Rules of Nomenclature should comply with the following conditions:

1. The Rules should be reasonable, otherwise they will not be accepted. They cannot be enforced.

2. The Rules should be simple, so that the average taxonomist may be able to interpret them correctly.

3. They should secure certainty of application of names.

4. They should secure precision of application, otherwise records will be relatively valueless.

5. They should secure stability of nomenclature. So long as the classification is unaltered the name of a group should remain unaltered, unless new facts come to light regarding its history.

6. They should secure uniformity of nomenclature. This can be achieved only by general adherence to one set of Rules. Hence the desirability of agreement between the adherents of the International Rules and those who accept the American Code.

7. The Rules should secure, as far as possible, the conservation of old established nomenclature. This necessitates the recognition of a list of specially conserved names.

Mr Sprague pointed out that Art. 36 of the International Rules (invalidating names of new groups published on and after Jan. 1, 1908, without Latin diagnoses) did not comply with condition 1 (above), and was not in accordance with one of the leading principles on which the Rules were based (Art. 3). It was arbitrary and imposed by authority, and was not founded on considerations forcible enough to secure general acceptance. The ideal that each name of a new group published should be accompanied by a Latin diagnosis was admirable: the attempt to enforce it had failed. From Jan. 1908 to July 1924 at least 10,000 new species of Flowering Plants and 1500 new species of Fungi had been published without Latin diagnoses. These were all invalid according to International Rules, but in actual practice they were treated as valid. Art. 36 had become a dead-letter, and should be revoked.

He also advocated the revocation of Art. 55, 2° (rejecting duplicating binominals, *e.g. Linaria Linaria*). Such names were rejected at Vienna because they were distasteful and appeared ridiculous to many botanists, but it seemed desirable to accept them for the following reasons:

(1) Their rejection prevents the *first specific name* from being retained.

(2) Their rejection often necessitates a long investigation in order to discover the next available name.

(3) Even after another name has been found and adopted for the species it may have to be superseded on some technical ground. Owing to the rejection of duplicating binominals 18 species have borne 43 names during the period 1900–1923 (see *Journ. Bot.*, Feb. 1924, pp. 42–44).

(4) Even when the name is finally fixed it is often unsatisfactory, *e.g. Calamagrostis canescens* is an albino form.

The following considerations were brought forward in support of expunging the "principle of nomina abortiva" from the Rules:

(1) It has introduced unnecessary complications into nomenclature. What we want to know is whether a particular name can be used *now*. Under the theory of nomina abortiva we have to find out whether at the date when a name was first published it contravened the International Rules (ed. 2, 1912). *Inula squarrosa* Bernh. (1800) duplicated

I. squarrosa L. (1763), which is now reduced to *I. spiraeifolia* L. (1759). If Bernhardi regarded *I. squarrosa* L. as a synonym of *I. spiraeifolia*, then he was at liberty to use the combination *I. squarrosa* for *Conyza squarrosa* L. If he did not regard *I. squarrosa* L. as a synonym, then he broke the Rules (of 1905–1912) by using the combination for another species. Thus we have to ascertain Bernhardi's views on *I. spiraeifolia* before we can say whether *I. squarrosa* Bernh. was a valid name or not. This introduces purely subjective questions into nomenclature.

(2) The principle is so complicated that even its authors made mistakes in applying it.

(3) The question of what is and what is not a *nomen abortivum* is not yet settled. Schinz and Thellung hold different views from Briquet and other authorities (see *Journ. Bot.*, Feb. 1924, pp. 43–46).

The instability and uncertainty resulting from the "principle of nomina abortiva" were further illustrated by the case of *Blitum virgatum* L.

In the opinion of the sub-committee three conditions were necessary in order to secure certainty of application of names: 1, each group must bear only one valid name; 2, each name must be used for only one group; 3, provision must be made for the application of a name in the event of the group to which it was originally applied being divided.

1. *One group one name.* When there are two or more competing names for the same group, the one which has *chronological* priority of effective publication is to be accepted. The rule of priority is merely a means to an end, a convenient method of deciding between competing names. Hence—if sufficient cause is shown—it may be set aside in individual cases.

2. *One name one group.* When the same name (or combination of names) has been used for two or more different groups of the same rank, the later uses of the name (or combination) should be treated as invalid. Later uses of a name (or combination) are called *homonyms*. Under International Rules homonyms may be accepted as valid if the previous use or uses of the name have become synonyms. As it is frequently a matter of opinion whether the first use of a name is a synonym or not, the conditional acceptance of homonyms leads to instability and uncertainty (see *Gard. Chron.* 1924, I. p. 92; *Journ. Bot.* 1924, pp. 46, 80).

2 A. *All combinations which are homonyms should be rejected.* The existence of *Chenopodium virgatum* Thunb. (1815) should invalidate the combination *Ch. virgatum* (L.) Ambrosi (1857). *Ch. virgatum* Thunb. is now treated as a synonym of *Ch. album*, and the combination *Ch. vir-*

gatum Ambrosi is used for *Blitum virgatum* L. But if *Ch. virgatum* Thunb. is at any time in the future regarded as specifically distinct from *Ch. album*, then it will be necessary to give up the name *Ch. virgatum* Ambrosi. The *name* of a European and Indian species thus depends on the *taxonomic value* of a Japanese species, which may be a matter of opinion. The result is uncertainty and instability.

2 B. *Generic names which are homonyms should be rejected unless they are specially conserved.* Two genera have been called *Kickxia*, namely *Kickxia* Dum. 1827 (Scrophulariaceae) and *Kickxia* Blume 1828 (Apocynaceae). Under International Rules, Art. 50, the *name* of the apocynaceous genus depends on the *taxonomic treatment* of the scrophulariaceous genus. For many years practically all botanists included the scrophulariaceous *Kickxia* in *Linaria*. Some botanists now regard it as an independent genus, and the result is that the well-known apocynaceous *Kickxia* has now to be called *Kibatalia* G. Don.

3. *Names should be applied according to the so-called "type-method."* "Standard-method" is a better name. A so-called type-specimen of a species may not be at all *typical* of that species, but it does serve as a *standard* with which other specimens may be compared in case of doubt. In seeking to apply a name correctly one naturally turns to the original description, but this may have been insufficient or inaccurate, so that from the description *alone* it may be impossible to apply the name with certainty. Hence it is desirable to have a *standard* to which the name is permanently attached. A standard-specimen is accepted for each specific name, and a standard-species is accepted for each generic name. If a species was described from a single specimen, that is the standard-specimen. If a genus was described from a single species, that is the standard-species. In such simple cases most botanists follow the standard-method as a matter of course. If a species originally included more than one specimen, a standard-specimen is *selected*. Similarly if a genus originally included more than one species, a standard-species is selected. It is desirable that the original author should indicate a standard-specimen or standard-species when he publishes a new species or genus. If he does so, the specific name is permanently attached to the specimen indicated, and the generic name is permanently attached to the species indicated, so that if the group is divided there is no doubt as to which part should retain the name. If he neglects to indicate a standard, difficulties may arise, as different authors may retain the name for different elements of the group. The application of a generic name in such a case may be determined either by the "residue-method"

or the type-method. According to the residue-method the name is applied to what is left in a genus after the removal of one or more species to other genera, old or new. It is unsatisfactory for the following reasons:

1. It does not finally fix the application of the generic name.

2. It requires an investigation not merely into the circumstances attending the publication of the genus, but into its whole subsequent history.

3. It frequently results in the most characteristic and best-known elements being excluded from the genus.

4. It frequently results in the generic name being transferred to a different genus, *i.e.* to one which did not form part of the genus as originally published. Thus the generic name *Gesneria* was applied (in the form *Gesnera*) to the genus *Rechsteinera*, which was not included in *Gesneria* L. (1753); and the name *Banisteria* was applied to the genus *Banisteriopsis*, which did not form part of *Banisteria* L. (1753) (see *Gard. Chron.* 1924, I. p. 104).

The type-method, on the other hand, has the following advantages:

1. It fixes the application of the generic name once and for all by attaching it permanently to a particular species.

2. It usually requires an investigation only into the circumstances attending the publication of the genus.

3. It automatically prevents the transference of the generic name to another genus.

A provisional set of Regulations for fixing generic types was published by the Botanical Society of America in *Science*, April 4, 1919, n.s. XLIX, pp. 333–335; and a type-basis Code of Nomenclature appeared in *Science*, April 1, 1921, n.s. LIII, pp. 312–314. In accordance with these Regulations the type-species of 100 Linnean genera have been ascertained by Hitchcock (*Amer. Journ. Bot.*, Nov. 1923, X. pp. 510–514).

Rigid adherence to the type-method in every case would, however, cause serious disturbance of nomenclature by changing the application of certain well-known generic names. This may be avoided by specially conserving such names, and attaching them to a standard-species which will preserve the generic name in its usual acceptation. The type-species of *Erica* is certainly *E. vulgaris* (*Calluna vulgaris*). The generic name *Erica* may, however, be retained in its present sense by conserving it with *E. Tetralix* as a standard-species (see *Journ. Bot.* 1921, p. 291).

Before taking the Resolutions the Chairman invited those who were not members of the sub-committee to express their opinions. Mr James

Groves gave the desirability of coming to an agreement with the adherents of the American Code as his principal reason for supporting the alteration of the International Rules. He advocated the setting up of machinery to deal with any further suggestions which might be made.

Resolution 1. Certain alterations should be made in the International Rules of Nomenclature. (Carried.)

Res. 2. Art. 36 (invalidating names of new groups published on and after Jan. 1, 1908, without Latin diagnoses) should be replaced by a strong *Recommendation* to supply Latin diagnoses. (Carried.)

Mr Groves suggested that names of new groups published in languages not employing Roman characters should be treated as invalid. Mr Sprague referred to the discussion on this point at the Vienna Congress (see *Act. Congr. Bot. Vienne*, pp. 129–131). Dr Rendle, Mr Wilmott, Miss A. L. Smith and Mr Ramsbottom also spoke, Mr Ramsbottom mentioning the case of descriptions of Fungi originally published in Japanese and ignored until they were subsequently translated into Latin in *Mycologia*. The conclusion reached was that descriptions in out-of-the-way languages could not be treated as invalid, but would in practice be ignored until some means of identification presented itself. Dr Foxworthy advocated the retention of a compulsory Latin diagnosis on the ground that it was a precaution against careless work.

Res. 3. All combinations which are homonyms (*i.e.* later homonyms) should be rejected. (Carried.)

Res. 4. All generic names which are homonyms (*i.e.* later homonyms) should be rejected except such as may be specially conserved. (Carried.)

Res. 5. The principle of the type-method of applying names should be formally accepted. (Carried.)

There was general agreement among the speakers that the type-method should be followed in the case of new groups published *in the future*. Several speakers, including Messrs Britten, Groves and H. N. Dixon urged that the type-method should not be applied retrospectively. An amendment to this effect was lost, as also a further amendment to the effect that the type-method should not be accepted at present. Mr Wilmott illustrated the type-method by the simile of an elastic attached to a fixed pin. Whatever the boundaries traced by the end of the elastic, the pin remains as a fixed point which must be included within them. Similarly, however much the limits of a group may vary, they must be so drawn as to include the type. He argued that the so-called retrospective application of the type-method was really work done for the future.

Res. 6. Art. 55, 2º (rejecting "duplicating binominals," *e.g. Linaria Linaria*) should be revoked. (Carried.)

Several speakers found "duplicating binominals" distasteful, but there was a general feeling that their rejection produced undesirable consequences.

Res. 7. The "principle of *nomina abortiva*" should be expunged from the Rules. (Carried.)

Res. 8. The list of *Nomina generica conservanda* should be revised. (Carried.)

Mr Dixon pointed out that a revision of the list would open the way to an agreement between the adherents of the International Rules and those who followed the American Code. Mr Ramsbottom enquired why the proposal that *nomina conservanda* should be retained as against all competing names (which appeared in the leaflet summarising the suggested changes in nomenclature) had not been embodied in a Resolution. Mr Sprague explained that the wording of the proposal had proved to be open to objection. If it were passed in its original form it would lead, in the event of re-union of *Mahonia* with *Berberis*, to the supersession of *Berberis* L. (1753), which is not a *nomen conservandum*, by *Mahonia* Nutt. (1818) which is on the list. Subsequent discussion led to the proposal of the following Resolution.

Res. 9. It should be made clear how far each of the *Nomina generica conservanda* is conserved. (Carried.)

Two resolutions proposed by Mr Groves were next dealt with.

Res. 10. That for the future the name of a group shall not be regarded as effectively published when the description is issued only with exsiccata. (Carried.)

Res. 11. That for the future the name of a group shall not be regarded as effectively published when the description is in autograph or typewritten documents. (Carried.)

Considerable difference of opinion was shown during the discussion, and with Mr Groves's consent the Resolution was subsequently omitted from those submitted to the plenary meeting of the Conference.

The following new Recommendation by Dr F. W. Foxworthy was next brought forward.

Res. 12. Publication of new genera and species should be only in scientific publications and, if possible, only in such as habitually reach systematic botanists. (Carried.)

EDUCATION AND RESEARCH

(CHAIRMAN: DR H. WAGER, F.R.S.)

(a) THE POSSIBILITY OF PROMOTING AN INTERCHANGE OF STAFF AND POST-GRADUATE STUDENTS BETWEEN THE OVERSEAS AND HOME UNIVERSITIES AND RESEARCH INSTITUTIONS

and

(b) THE DESIRABILITY OF PROVIDING FURTHER FACILITIES FOR BOTANICAL RESEARCH IN THE DOMINIONS, COLONIES, AND PROTECTORATES

Dr H. Wager, F.R.S.

We have met to discuss education and research, but first I regret to say that Prof. Farmer has been unable to come to-day. In his place Prof. Bower has kindly consented to introduce the discussion. Before calling on Prof. Bower I should like to mention that Prof. Bower intends, I believe, to open the discussion on a somewhat wider basis than is implied by the propositions (a) and (b) in the programme which you have before you. It is very important that these two propositions should be discussed as fully as possible. It seems to some of us that it is also important to broaden the discussion so as to include the wider interests of education and research. In fact, Prof. Farmer, had he opened this discussion, would have taken the same lines as Prof. Bower now intends to take.

Prof. F. O. Bower, F.R.S.

I am sorry Prof. Farmer was not able to come to-day. I met him the other day and we discussed the intended material of his speech. I was glad to find that the lines which he proposed to take were the same as my own. The work of this meeting has been to lay stress upon the applied side of botany; anybody not knowing the inside life of botanical Britain may have thought that we were all applied botanists, and that pure botany did not exist in this country. Of course we do not agree with that conclusion—what we shall have realised all through these discussions will have been that the applied side of botany can only be pursued on the basis of a sound knowledge of the science as a whole. That immediately leads our minds to the subject of education because everyone of those who pursue the applied lines of botanical work must proceed upon the basis of the pure foundation. It is only a question of how far that foundation of pure science shall be extended before the application takes place, and that is the question that we have for discussion this morning. We realise, I think, the cleavage which is bound to occur, and I think should occur at the post-graduate period. The pre-graduate period should not be a period of specialised work within the science, but rather of generalised work. This cleavage between the two branches that one sees so plainly increasing as the years go on should be a post-graduate cleavage between the educational and administrative side, or I should say perhaps, between the academic and the non-academic side. That is not using "academic" in any narrow and restricted sense of the word, but I mean merely that it should include the work of those who remain in academies and who carry on the processes of education and research as against those who go out into the field and whose duties are not didactic as a rule, and who work especially on applied lines.

One point which I think we should realise is that the cleavage, when once it has begun, is difficult to recede from. I do not say that this cannot be done, but in general a man should not go back, after years of specialisation in one branch or the other, from applied science to academic work, or *vice versa*.

The subject for this discussion is education and research. I do not recognise that there is any distinction between education and research if education be properly conducted. A few of us have actually seen the new start made by Huxley, Michael Foster, Sir William Thiselton-Dyer, and others. I do not suppose the present generation in the least realises what a tremendous change has come about since we gave up the didactic method which merely "stuffed" into the mind of the student as much as that mind could hold and more. That was the tendency before Prof. Huxley's time. Prof. Huxley was the man who at South Kensington initiated those practical courses which are now so well established. He introduced what is now the modern method, by which each student is set to find out things for himself; he is put down at a table where he can see the things, and if he does not see them, it is his own fault. Of course the demonstrator is responsible for helping him, but no demonstrator can give the "seeing eye"; he can only direct the eye and teach it to see better. It was a most vital and important change which thus made even the elementary student an investigator at once. If he really is so, there is no distinction between education and research, and the most elementary student who is of the proper type will be an investigator from his first entry into the laboratory. The difficulty in pursuing this method is the want of time and the number of students. The larger the number of students the more difficult it is to carry this out; the more limited the time the less opportunity the student has of seeing the things he ought to see. These are the ultimate factors which have to be considered in the practice of a method which is theoretically so excellent. The ideal which we should have is that of leading the student to work on investigating lines by himself. We should not merely inform our students, but teach them to think and to see for themselves. Success really turns on the material—I mean the material of the student. There are students and students. Their ideals are very diverse. There are two classes. One is the student who goes into the science because he cannot help it. He has loved natural history from his boyhood. That is the type we want —the botanist who is a botanist by conviction. Then there is the other student who enters into the profession because he thinks he could make it a paying proposition. He sees before him a career in which he will be able to make some hundreds a year in a short time. That is the student we do not want—at least I do not want. And if you look over the lists of those students who have become distinguished you almost always find that they were students of the subject by conviction, and not by any ulterior motive.

The next point is how to treat this material. Supposing we have our ideal class before us—real enthusiasts—one must provide first of all a general grounding. I do not want to revert to the subject of recent discussions; but we should, in my view, have a clear co-ordination of all the branches of the subject for the elementary student. He should study forms and functions and the method of applying the study of function in the field. He should be introduced to histology and palæontology, and all the other branches. These should be presented to him in a comprehensive course of study so as to get his mind awakened to the clear basis of the branch-subjects from which he may later select. The other way of studying might be through early specialisation, but this is a thing to be avoided: for the man who may be driving a lonely furrow in a distant land will have to depend upon the years he has spent in one of our big institutions for the knowledge necessary for driving that furrow well. Of course the tendency would then be along utilitarian lines. That is the reason why early specialisation is so dangerous; and why in the interests of the individual, and certainly in the interests of science, it should be inhibited as far as possible till later in the course. Otherwise we should have to have one course for those students who are likely to be working in far distant lands, and one for those who are likely to work in home billets for educational purposes.

We ought also to consider very carefully indeed the preparation of the teacher. Those who are going to be teachers should have some special course in didactics offered to them. In my own case I was only taught one thing in didactics, but I was taught it by a great master. I gave my first lecture at South Kensington in 1882. Prof. Huxley sat in the front row. When it was over he came up to me and said "You have told me a lot of things I never knew, but I want to tell you one thing: Address your audience and not the blackboard. Cultivate the attitude of looking over your shoulder." I think we ought to have instruction by an artist on blackboard drawing with chalks: I should say a single lecture would suffice, together with further suggestions how to present what you have got to say to your audience. Method of presentment is still one of the things we have got to acquire for ourselves; nevertheless a kindly Providence has led many of us into safe channels. But think of the audiences we have mangled by our inexperience. I do not wish to be unpleasant but we have seen in the course of this Conference all the laws of didactics violated. A few hints at an early period in the education of the teacher may probably put him on the right track.

Research along the ideal lines I have sketched would be merely carrying further what has been already done by the elementary student in the laboratory. One sees to-day the gradual development of the practice of what is called team work. Team work has its dangers, but there are also splendid results as we have seen during these discussions, but again I repeat that it has its dangers. One is, that there is thereby established an uncertainty as to who has done the work; you may deal with that by a dual publication; otherwise who is to know who has done what stands in the published print. Further there is in joint publication the danger of producing what may be called a blend. Those who are interested in wines will know that a vintage wine is a different thing from a blended wine. A blended wine has its merits but in my view they cannot compare with those of the vintage wine. The personal publication of a thing which you yourself have produced is a vintage wine: but the thing that is produced by team work is very much in the nature of a blend. The second danger which I see is that by team work we should tend to produce an increasing number of persons who, on becoming investigators, would be dependent on the stimulus and suggestions from the head of the team. As the years go on we should find a deficiency of those who have the ability to launch out into lines of their own. That is why I never advise a promising beginner to indulge in team work. I will help a student, but then I leave him to make what he can of his enquiry. That is one way of producing capable investigators. I believe that it is a far better way than by allowing students to become the subordinates of team work.

Then comes the question of facilities abroad. In 1885 I went to Peradeniya and greatly appreciated the advantages of a tropical visit even for a short time. The result was that next year a motion was brought forward by the British Association for giving a grant to facilitate the journey to Peradeniya. That journey had been a great advantage to me, though now of course the thing is becoming a much more ordinary event in the life of a botanist. We may say, I think quite justly, that no teaching botanist should consider himself properly equipped for the carrying out of his duties as a University teacher unless he has had practical experience extending over a reasonable time among a flora quite different from his own. Of course we are aware that this country is a very difficult one for ecological study. There are only certain parts of the country that can really serve for ecological study. To gain experience in ecology on a really natural Flora one must go outside these long established islands; so that there is an additional reason for welcoming and encouraging journeys to other countries than our own.

Now there is one other point I want to bring forward, and that is the presentation of the results of research. It is not everyone who has ability to write clearly. The practice of expressing himself in an intellectual and stimulating way is the thing which makes the writer able to commend what he has before him to the reading public. I think that a great deal of time, trouble and attention should be given to the clear and well-expressed presentment of what we have for publication. What I always do myself in writing is to read my copy over again and again, thinking all the time of the "Advocatus Diaboli" who is looking over my shoulder, and hearing him say, "What a badly expressed sentence that is," or pointing out other blemishes of a faulty composition. If you only take yourself in hand in that matter and think of the other person who is going to read what you write I think you will find your writings become much more interesting, and carry very much more weight than they would do otherwise. One distinguished man who writes in rather an involved style once said to me, "Oh well, I write it down so that I can understand it, and other people must make the best they can of it." I submit that that is essentially wrong. You all ought to make your communications as nearly fool-proof as possible, like motor-cars; and if you will only do that you will secure the attention of your readers.

Dr H. Wager, F.R.S.

I am very glad that Prof. Bower has begun on these lines in the discussion, because I hope that in discussing these two propositions concerning education and research the first will always be kept in mind. In all research a well trained mind is of fundamental importance, and the educational methods by which this can be achieved should receive the most careful consideration. Before we go on I have to bring to your notice a proposal from the Commonwealth Government of Australia, put forward from the Mycological Conference which was held last week, that this Conference should consider the interchange of mycological workers in different parts of the Empire for periods such as six or twelve months.

Prof. F. J. Lewis

When I think of all the things which have been said by Prof. Bower I am bound to say I am in complete agreement with them. Two of his statements struck me very strongly: (1) the classification of the two types of students, and (2) the danger of early specialisation, both of which are fundamental. Further, in regard to the classification of the two types of students we must bear in mind, even with those who take up Botany for the pure love of it, that science should offer them reasonable opportunities of advancement. I think it is an important thing that reasonable opportunities should be offered them when they want to travel without the fear of financial embarrassment. Now I presume that we have a choice in regard to our discussion this morning, and it is my wish to speak on the general report presented by the sub-committee which we have before us this morning. I do not know the names of this sub-committee and so it is quite easy for me to say that I am sorry to totally and thoroughly disagree with the conclusions which they have presented. I feel that the matter which we are discussing this morning is a very important one. It is impossible to exaggerate the importance of it both from the point of view of ourselves as botanists who are discussing the progress of botanical science and from the point of view of the general welfare of the Universities at home and abroad.

In regard to the second paragraph of the report the conclusion is that it is neither desirable nor practicable to encourage the exchange of teachers and research workers who go out from the laboratories and temporarily take positions in places overseas. I have felt during the last two years—it has been impressed upon me strongly—the desirability of arranging these exchanges. Now there are two types of people for whom these exchanges can be arranged. There is what one might call the junior worker who has graduated at home or in some overseas University, perhaps some three or four years ago, who has shown ability to do research work, and who is evidently the person who will make his mark by and by in some field of botany. Then there is the senior worker who has had the full charge of a large department for some years and who has settled down in some particular place with its own field of work. I think the problem of exchange between those two types is somewhat different. In the first case I see no difficulty in the way of exchanging the young worker who has perhaps done good research work and got his M.Sc. and put in three years' teaching work in his own University. I see no difficulty in arranging an exchange between such a person in an overseas University and a similar person in the home University. Sometimes it is apt to be supposed that the advantages are most numerous for anyone overseas of that type to come and work in one of the home Universities. That is quite true. The advantages are numerous. Some of these people—I speak from experience—are extraordinarily keen about their work and they have very wide interests so that they would in that way compare favourably with most of the people trained in this country. Again and again I am told by men who are going on to do scientific work "I want to work on really scientific lines; I want to take up practical problems and find their solution." And one knows—living in distant countries—the moral responsibility that scientific men have in tackling these problems. The point I wish to lay stress upon is the desirability and entire practicability of such exchanges.

Prof. J. W. Bews

I agree with what Prof. Lewis has said and with the introduction by Prof. Bower, which I think forms an excellent general introduction. But when one has come six thousand miles to attend a Conference what one looks for particularly is practical results, and however interesting the general discussion on education may be I think it is secondary to definite proposals. In passing, I might even remark that while one agrees that early specialisation is entirely undesirable yet the real trouble is to decide what is specialisation and what is really the fundamental aspect of the subject. I need not add any more to the discussion which we heard yesterday. Now with regard to the question of interchange. I think that is a thing which should be approached with a proper appreciation of the great differences of outlook between the home worker and the worker overseas. This is a real question, and it is only after a few years' residence in the Dominions that one can see the fundamental differences which exist. The outlook in the Dominions is entirely different, but with the material at one's disposal it is an easy matter to open up new lines, and results can surely be obtained. The answer can be obtained by coming back to the older and more established centres. We should undoubtedly benefit by such an exchange. But on the other hand the other party to the exchange stands to benefit in my opinion even more; the whole usefulness of education is gained by the enormous advantages of a period of travel. Now I should like to suggest that that period should not be a short one but one that can be obtained by such a system of exchange. I am sorry that the sub-committee has confined it to research workers. I do not think that the difficulties would be insurmountable for members of the staff—certainly for the junior members and for some seniors. The actual teaching duties demanded overseas are not so heavy, and where assistance is supplied in departments only a limited number of lectures would be demanded. At a Conference at the South African University where I had the honour of being a representative this very question was discussed at length and our Minister of Education expressed himself heartily in agreement with the scheme proposed. As far as possible we should look forward to a system by which each partner to the exchange would draw the salary from his former institution. Now the second proposal was the desirability of increasing research facilities overseas and in the tropics; in that connection I should like to draw your attention to what is done by Holland in the Dutch East Indies. The use Holland is making of her small colonies is I think an example to this Empire. I think we should impress upon the powers-that-be the fact that the dependencies and Dominions overseas would benefit enormously if greater facilities were given, but I cannot enter into details with regard to this. However, I would point out that it is a very important thing for a careful investigation of the use that is being made of the different Botanical Gardens to be made. We have Botanical Gardens in South Africa but every possible use is not being made of them. If we decided to look into the question we could see what further use could be made of the already established Botanical Gardens in the different parts of the Empire. They could be made centres of research, and it would be a very easy matter in my opinion to prosecute work with more immediate economic results. I do not want to say more because I wish these most important points to be left in your minds as fresh as possible.

Prof. Dame Helen Gwynne-Vaughan

I find myself in very cordial agreement with what Prof. Lewis has said in regard to the report of the Sub-Committee, and I find my views in accordance with those of Prof. Bews respecting the difference of outlook at home and abroad. I want to suggest, however, that there are two points of view from which we ought to regard this question of exchange. There is the point of view of the botanist who travels and comes to a new place and there functions for a certain number of months, and there is also the point of view of the environment in which he finds himself, and it is from this point of view that I hope very much that this Conference will pass some sort of resolution suggesting the possibility that these advantages shall be not merely for research workers but for the actual members of the staff with longer or shorter experience. I say that because I think it would do students a world of good to have some of their lectures from a botanist who has an entirely new outlook. It is desirable that they hear their ordinary routine subjects from someone who has worked on a flora entirely different from their own. I do not think it is at all impossible if we put our minds to it to effect an exchange for a year or so between members of the staff from both sides. This could easily be done with regard to research students and in the same way with many of the junior staff, and I think it could also be done with relatively senior members. We all know perfectly well that junior members of the staff would be pleased to run a large department for a year or so. It will be for them to see that continuity has been maintained, to see that the visiting lecturers and professors do not stray entirely from the curriculum that is essential for examination purposes. At the same time I think it is desirable that some fund should be obtained for the extra remuneration that might be given to the juniors, so that the visiting staff may be free to visit other Universities in the neighbourhood. In this connection we should benefit enormously by having as our colleagues for a time, botanists whose ordinary work is carried on overseas. I wish every botanist here would turn over in his mind the effect upon our colleagues if half a dozen of them went to South Africa, Australia, Canada, or New Zealand to widen and develop their academic knowledge. They would then return with a freshened and new outlook to their original work; while at the same time someone from overseas would be among us going through our ordinary routine, dealing with our different problems, and helping to examine our students. I feel that it would be very desirable that this Conference should pass a resolution that something should be done to produce the funds to bring about this desirable consummation.

Colonel French

I cannot tell you very much in the brief time at my disposal about the Cotton Growing Corporation and its work and history. I will merely say that the Corporation, which is financed partly by the trade in Lancashire and partly by a grant from the Government, is an enthusiastic supporter of research. We are amongst other things trying to produce not only more cotton from the Empire but better cotton, but it is absolutely essential that we should have the assistance of skilled experts and skilled botanists. Amongst other things which we do is to select a dozen or so students every year for one or two years' post-graduate training. They can go to Cambridge or the Imperial College and most of our senior students go to the Imperial College of Tropical Agriculture in Trinidad. On the advice of our experts we do not attempt to create experts but merely try to send out well-educated young men who have got a sound basis of education which will enable them when they have got their local and practical experience in cotton growing to deal with the problems which may arise. Then as regards research, the Corporation has decided upon a definite policy for they are considering the possibility of establishing a central cotton research institution somewhere in the cotton-growing areas of the Empire, and they have also considered the possibility of having a pool of research workers employed under the Corporation that can be sent out to different parts of the Empire as and when their services may be required to deal with local problems. We are still trying to discover how best research can help the cotton grower and how the Corporation can help the research worker. There are one or two points in connection with the administration of research work on which I should like to express a purely personal opinion. One is that I believe research should not be a pensionable civil service. I think that to be in civil service has many merits, and I have very many great friends and people whom I admire enormously in the civil service, but I think you will probably agree that if it were so the research workers' freedom would be bound to be limited by necessary departmental regulations and their intellectual activities would be decreased. The second essential I think for research work of this sort is that there should be absolute co-operation between the research worker and the Agricultural Department. That I consider absolutely essential and I think that there must be an absolutely clear definition of duties between the research workers and the Agriculture Departments. During the war I was associated with a great many departments, and in all cases the inter-departmental warfare in which I was engaged was caused by somebody over-stepping his duties or not minding his own business; therefore I think it is essential that there should be a clear definition of the duties of each.

Dr W. L. Balls, F.R.S.

I have very little to say in addition to what Colonel French has already said about the work of the Cotton Growing Corporation, but I would say that we have come to the conclusion that we are satisfied with the result that outside technical training in the study of botany at any rate is not worth while. The man whose foundations are well and truly laid does not need to have anything but a good general education, and again I repeat that we have come to the conclusion that the applied science of botany is not worth teaching though it is worth learning. Among the many points raised by Prof. Bower I should like to carry one point a step further and that is early specialisation; I visited one of the public schools a short time ago and was impressed by the fact that they were trying to turn out specialists at about the age of fourteen or fifteen, the argument being that owing to the severe competition in the scholarship requirements of Cambridge and Oxford it was possible for a boy to devote himself to only one branch of science. With regard to team work that is a subject on which I am extremely keen myself. We have got to learn how to avoid the difficulties already existing but I think we must recognise at any rate that team work is becoming inevitable and will be completely so in the future. It is up to the scientists to find some cure for the evils of team work and the disadvantages which exist in it. One man alone cannot do effective work for the industrial and agricultural side of botany, so what we must do is to recognise this and to devise some method for dealing with its disadvantages.

Dr E. J. Butler, C.I.E.

In discussing this matter with various members of the Conference I find the idea widely held that while it would be an advantage to promote the interchange of staffs from the Universities and other institutions the obstacles would be almost insuperable. Now I do not think this is the case everywhere. One can easily conceive of a certain number of institutions where it would be practicable to arrange for an interchange of the sort mentioned—between the research worker in the research institutions overseas and the worker in either research or teaching institutions at home. Some years ago this case was put before me. You can easily imagine in systematic mycological work what an advantage it would be to a member of the staff of one of the larger herbaria, interested in a special group that is widely represented in a distant part of the Empire, to spend a year in the country where he could find species of that group all round him. And you might find in that country a man equally interested in systematic mycology to whom it would be a great advantage to be able to study the collections at home by taking over the duties of the other. Also it would be an advantage to us to have a man whose knowledge of a particular group was very extensive working in England. Yesterday we had another case in connection with the applied side of the work; we were discussing an important disease of the rubber tree, investigations of which have been carried on in the Imperial College and are now going on in nearly all the rubber-producing countries. It would seem to me to be an advantage if an exchange of staff was arranged in such cases. Some years ago Mr Brooks spent some time in the Malay States working on rubber tree diseases. The influence of that visit was still very strongly marked when I visited the Malay States several years afterwards. The chief obstacle to such proposals is the provision of the passage money, but I cannot conceive of a more profitable way of spending, say a thousand pounds annually, than in promoting the exchange of men from one part of the world to another.

Prof. R. H. Yapp

I feel some diffidence in saying what I am about to say because I want to move a resolution that has no bearing on what has gone before. I shall not have time to discuss any of the points I should have liked to have discussed. But I would like to add that the matter I have in mind is not really so far removed from the point as might be thought at first sight. If you wish I will first read my resolution, which in my opinion deals directly or perhaps indirectly with research workers (see *Res.* No. 2, p. 383).

Dr A. W. Hill, F.R.S.

I have only time to make a very few remarks. But I feel I must say one thing about research workers going overseas. I would like to remind the members here present that in the Botanical Gardens at Kew we could show them a good many things they would find abroad. There was a case where a botanist went to a foreign country to study a particular subject. When he came back he spoke about his researches to the Director who said "Well, very interesting, but as it happens you could have studied the whole thing in the palm house at Kew."

Prof. D. Thoday

I have no particular points to bring before the Conference this morning, but I should like to say from my experience, even for a few years in South Africa, that time spent among a different flora is invaluable for widening one's outlook and instilling into one's mind a new point of view. Anything we can do to encourage the home botanist to see new vegetation should be done. In visits home one of the most difficult points is the serious personal expense which devolves entirely upon himself. This should be obviated because it is necessary for the overseas botanist to come home in order to meet those who have greater opportunities of going to the root of the matter and developing in a fundamental way the research work which it is not possible for the outlying worker to follow.

Anything that could be done to facilitate the return of the outlying workers to the large centres at home would be of immense advantage to the Empire.

Dr A. S. Horne

Speaking as one who for some years has looked, like Moses, at the land flowing with milk and honey and never been able to enter it, I should like to say that the fact of these visitors from overseas being present amongst us makes one experience that kind of stimulus that I imagine arose when the pioneer botanists—Hooker and others—first brought their work to light. One finds from year to year that visitors from the United States and Japan and from every part of the world come here to get into contact with workers in this country, to find stores of literature and to exchange views. I cannot help feeling that the government which sends over these research workers or senior workers has a most enlightened view. It has been my privilege during the last two years to be associated with the groups of workers who are studying the senescence of apples, and I think that if one had the opportunity of visiting Australia or Canada where apples are produced in great quantities, the experience one would gain would be enormous.

Dr H. Wager, F.R.S.

May I be allowed to say one word. We have had a most interesting discussion and I am glad we have ranged over the wide field of education and research and that it has not been confined to the propositions on the programme. I was glad to hear that Prof. Bower's views concerning specialisation were very like my own, and it was extremely interesting to hear his quite definite statement that applied science is not worth teaching in the earlier stages of botanical education. I think anyone who has been concerned with teaching the elementary stages of any branch of science will agree that it is unwise to specialise too early. We often see how badly prepared some young people are for the work they hope to achieve; badly prepared because they have tried to specialise too early and have never done anything really well. I sometimes feel that the existence of the word "mastery" has been completely forgotten. Many people who have a very interesting and useful amount of knowledge, as Cardinal Newman puts it, may have no real intelligent outlook on the subject they are supposed to have studied and may never master anything. There are two sides to education.

There is the discipline of the mind which is concerned with methods, and there is the knowledge which it is desirable to acquire. The former is sometimes forgotten in the earlier stages of botanical education. The scientific outlook is acquired, not by obtaining as much information as possible, but by the development of a scientific habit of mind. Later on comes the stage when it is important to get as much information as one can, but I think this should be in the post-graduate stage.

LECTURES

(Given in the rooms of the Linnean Society by kind invitation of the President and Council of the Society)

Dr J. M. Dalziel. THE ECONOMIC BOTANY OF WEST AFRICA

(CHAIRMAN: DR A. B. RENDLE, F.R.S.)

Mr F. A. Stockdale. THE PERADENIYA BOTANIC GARDEN

(CHAIRMAN: DR A. W. HILL, F.R.S.)

Professor A. C. Seward, F.R.S. RECORDS OF ANCIENT PLANTS WITHIN THE EMPIRE: WHAT WE KNOW AND WHAT WE NEED

(CHAIRMAN: DR D. H. SCOTT, F.R.S.)

"But I still find it hard to resist the conviction that, from the educational point of view, stimulus is more important than exactness."   A. C. BENSON.

I assume that the intention of the Executive Committee in asking me to give this Lecture is to encourage our colleagues from the larger world beyond the seas to do what they can to promote a wider and more scientific interest in the records of vegetation of the past. Though my title—Ancient Plants within the Empire—is imperial, the department of Botany with which I am concerned embraces the world and belongs to a time when political boundaries were unknown and, strange as it may seem to us, politicians did not exist. Huxley said, "the fact of evolution is to my mind sufficiently evidenced by palaeontology." Here we have the main stimulus: it is, I believe, true that we know less about evolution now than was thought to be known a few decades ago. We have more facts to puzzle us, and wisdom, which is said to increase with years, as the result of longer reflection, tells us that the better way is to avoid dogmatic statement, to remember that in scientific research authority counts as nothing, and to devote ourselves to the discovery of evidence that can be trusted. The present vegetation of the world has been compared by a German writer to a surface—a superficies; and the successive floras of the past to a solid with depth as well as length and breadth in which is contained the material we seek to enable us to follow through the ages the rise and fall of groups, genera, and species. All of us can do something towards the investigation of this material and my object is to suggest, in very general terms, some of the work that needs doing. We may help the science of Palaeobotany both by our own research; or by passing on to others our belief in its importance and its value, both from a scientific point of view and as a recreation.

I will not deal separately with what we know and what we need. It is obvious that what is known of ancient plants throughout the Empire is more than can be presented in an hour's lecture[1]. I can only give a rapidly moving and very indistinct series of pictures of some selected examples of what is known, adding, as opportunity offers, a few suggestions for further research.

The *British Isles* need not detain us: to describe in the most general terms the floras which have left their traces in the home strata, from the uncertain records of Silurian vegetation to the submerged forests of historic times, would occupy a large proportion of the time allotted to the Conference. But though we are fortunate in the abundance of material in the British Isles, we cannot flatter ourselves that we have as yet taken full advantage of the opportunities that are available. The names of the several chapters of geological history with which we are familiar are British or European, and there is a natural tendency to apply these names to series of strata in other parts of the world. One of our aims is to correlate rocks in all parts of the world, to form a picture of each stage in geographical evolution, to determine as nearly as possible the homotaxis or contemporaneity of plant-bearing beds: it is, however, dangerous and misleading to insist on the inclusion of strata in widely separated regions under the same name. A classification suitable for one portion of the earth's surface may be inapplicable to another where conditions were different. The palaeobotanist must not be parochial; he must remember that the plants he studies inhabited a changing world; they are inseparable from the environment. Environment has been defined by an American writer as "the sum of all the contacts which an organism or group of organisms establishes with the forces and matter of the surroundings, either organic or inorganic." "Nature vibrates with Rhythm." The history of the world is intimately connected with the history of the earth's crust: mountain-building and the retreat of the sea, marine transgression and flooding of the land, climatic change, igneous activity are all concerned in supplying the factors which condition the progress or the fate of the organic kingdom.

[1] Lantern slides were exhibited.

One of the essentials of thorough palaeobotanical work which some of us have neglected in this country—I am one of the worst offenders— is the collection and examination of the plant records as they occur in the rocks. Precision in fixing the geological horizon is especially important. By seeing for ourselves the association and method of occurrence of fossils in the freshly exposed rock we can often gain valuable knowledge which may throw light on their interpretation.

Arctic regions. The only part of the British Empire north of the Arctic circle belongs to the Dominion of Canada. From Grinnell Land Tertiary plants were collected by Col. Feilden, the able Naturalist of the Alert Expedition of 1875 and 1876, and described by Oswald Heer, the Pioneer of Arctic Palaeobotany. From Ellesmere Land, to the south of Grinnell Land, a remarkably luxuriant flora was described by Prof. Nathorst of Stockholm: few richer floras of Upper Devonian age are known. From the same Land Nathorst also described some extraordinarily well preserved remains of *Sequoia* and other relics of a Tertiary vegetation. To Heer we owe the description of fragmentary material collected by Sir Leopold MacClintock and other explorers in Banksland, Melville Island, and other portions of the Arctic archipelago. The oldest fossils, from Melville Island, may, as Heer believed, belong to the Carboniferous period; but they are too incomplete to be identified with confidence. If they are of Carboniferous age, with the exception of a few specimens obtained by Nathorst from the north-east corner of Greenland, they represent the most northerly outposts of the vegetation of that period. Several Tertiary plants were described by Heer from a northern locality on the Mackenzie River, but, so far as I know, nothing new has been recorded in recent years. There is no problem in Palaeobotany more interesting than that presented by the Arctic floras of the past; they afford evidence of conditions even more genial than those described by Mr Stefannsen in the "friendly Arctic." One would like to see more money spent in expeditions primarily concerned with geological investigation in relatively inaccessible regions: the discovery of plant-bearing beds is much more to be desired than a successful dash to the North Pole.

Canada. The rocks of New Brunswick bordering the Gulf of Saint Lawrence are classic ground. Here is abundant material; Devonian, Carboniferous, and, as some have maintained, Silurian floras. Sir William Dawson's researches have not in recent years always been estimated at their true value. He was mistaken in some of his conclusions: who has not been? But he set an example of industry and well-directed enthusiasm which is worthy of emulation. Some of Dawson's plants were re-examined by Prof. Penhallow, by Dr White of the United States, and by other workers. The flora of the Fern Ledges of St John, New Brunswick, was fully described by Dr Marie Stopes in 1914, and its Upper Carboniferous age, previously suspected by Kidston, was confirmed. Dawson's *Asteropteris* has been thoroughly investigated by Prof. P. Bertrand of Lille; Mr W. J. Wilson of the United States described a *Lepidodendron*, in 1917, from New Brunswick. But in recent years we have had few contributions to our knowledge of the Palaeozoic floras of this region. The famous Joggins section, where a forest of *Lepidodendra* occurs *in situ* comparable to that at St Etienne in France, previously examined by Logan, Lyell, Dawson and others, has more recently been described by Mr W. A. Bell of the Canadian Survey.

Our great needs are (i) a more precise knowledge of the horizons within the Devonian system of the plant-beds of the Gaspé Peninsula, (ii) more material and its critical investigation, (iii) the discovery of petrifications, especially of *Psilophyton*, *Nematophycus*, and other genera. In the light of the results obtained by Kidston and Lang from their masterly examination of the Middle Devonian plants of Aberdeenshire further knowledge of the Canadian floras is an urgent need. From the interior of Newfoundland Dr Arber, in 1912, recorded a new species of *Psygmophyllum* and the well-known *Sphenophyllum tenerrimum* of Lower Carboniferous age.

Some Permian plants from Prince Edward Island were described in 1913 by Miss Holden of Harvard, whose premature death in Russia when she was working with a British unit during the war was a serious loss to Palaeobotany. She also described Coniferous petrifactions from Triassic rocks in New Brunswick.

Cretaceous floras, some Upper and some Lower Cretaceous, have been described by Sir William Dawson from Vancouver Island, Queen Charlotte Islands, British Columbia, the Rocky Mountains, and elsewhere. Some of these are of special interest because of the apparent identity of certain species with specimens from Greenland and Britain. The political boundaries in North America are responsible for some confusion in nomenclature between Canadian rocks and strata in the United States. It is most desirable that an effort should be made to re-examine the Canadian Cretaceous and Tertiary floras, and to bring them into line with those which are being investigated by American Palaeobotanists. Tertiary floras occur in Alberta, British Columbia, and in other parts of the Dominion, but our knowledge of them is very incomplete. Pleistocene plants are recorded from Ontario, Toronto, Montreal, and elsewhere. Canada is large in area and the available geologists, who have time for work that is not primarily concerned with economic products, is small. I can only urge botanists to do their utmost to see that existing collections are critically revised and to collect additional material. It is only by individual effort and by stimulating amateurs to help us that we can hope to rival the activities of the United States Geological Survey.

Falkland Islands. I will now pass to the far south. In the Falkland Islands we have particularly interesting fragments of disrupted Antarctica which contain many relics of the *Glossopteris* flora and have afforded evidence of an older Devonian vegetation. It is remarkable that many of the fossils, both animal and plant, show a greater resemblance to those of South Africa than to those of South America. Members of Swedish Expeditions (1901-3 and 1907-9) collected many plants, and in 1912 Dr Halle published an excellent account of the Geology and Palaeontology of the Islands. From 1920 to 1922 Dr Baker, of the Imperial College of Science, was occupied in geological observations, and the specimens obtained were described by Mr Walton and myself.

Graham Land, South Georgia, and Antarctica. Farther south on the borders of Antarctica a rich collection of Jurassic plants was obtained by a Swedish Expedition from the rocks of Graham Land. The specimens were briefly described by Nathorst in 1904 and fully described by Halle in 1913. This Jurassic flora is exceptionally interesting; it shows a striking resemblance to the Lower Oolite flora of England and agrees closely with a Jurassic flora of India. Now that a British Expedition is going to the far south I hope that it may be possible to devote some time to a search for more specimens of this remarkable flora.

Reference should also be made to a piece of Gymnosperm stem obtained by members of the last Shackleton Expedition on South Georgia, which Prof. Gordon examined.

Traces of ancient plants on the Antarctic continent were first found by Mr H. T. Ferrar, a member of the Scott Expedition of 1901-4 north of lat. 78° S. The material, too imperfect to be determined, was described by Dr Arber. A well-preserved piece of petrified wood was discovered by Mr Raymond Priestley in 1912 in a boulder on the Priestley Glacier: this was described by me in 1914 as *Antarcticoxylon Priestleyi*, but more recently it has been shown to be an example of the South African genus *Rhexoxylon*. From lat. 85° S., on the Buckley Nunatak on the Beardmore Glacier, Dr Wilson collected fragments of *Glossopteris* fronds and thus demonstrated the spread of this far-flung genus to within

300 miles of the Pole. The age of the rocks is not known with certainty, but they are probably of Triassic or Permian age. Now that the South Pole has been reached one may look forward to the possibility of more favourable conditions for geological investigation in Antarctica. The tragedy of Dr Wilson's death is in some slight degree mitigated by the knowledge that he succeeded in obtaining determinable specimens of a Polar flora, an object which I know he had at heart.

Africa. It is only in South Africa that fossil plants have been collected in any quantity. An outstanding feature of a considerable thickness of the Palaeozoic rocks is their barrenness: from the Table Mountain sandstones, the Bokkeveld series and the still higher Witteberg series only a comparatively small number of imperfectly preserved plants have been obtained. There are traces of Devonian floras, mostly fragmentary pieces of Lepidodendroid stems, which bear a close resemblance to specimens from South America and the Arctic regions; but better material is urgently needed. The Witteberg series from which a few plants have been obtained contains numerous examples of *Spirophyton*, a problematical structure suggesting the action of swirling water in sand or, it may be, as some suppose, the impression of a large Alga. It is interesting to note the occurrence of a precisely similar structure in Carboniferous strata in Scotland. Above the Witteberg series is the Dwyka and Ecca series with widespread, thick masses of glacial deposits similar to those of the same age in other parts of the southern hemisphere and in India. Above the glacial beds *Glossopteris* and other members of the *Glossopteris* flora occur in abundance. The rocks of Devonian and pre-Devonian age have not as yet yielded enough evidence to enable us to chronicle with confidence the events in the plant-world before the advent of the greatest glacial period in the history of the earth, or to correlate the events with contemporary happenings in the northern hemisphere. Many plants have been collected from beds above the glacial series, and I would draw special attention to the discoveries made from time to time by Mr Leslie of Vereeniging. But we need much more information. We want to know whether the glacial period occurred in the Carboniferous, as some of us believe, or in the Permian period as perhaps the majority of geologists believe. We should also like to know much more about the pre-glacial floras and their relation to Lower Carboniferous and Devonian floras of the northern hemisphere. Further search for plants should be made in the immediate neighbourhood of the glacial deposits in order to obtain more exact information about the conditions under which the older members of the *Glossopteris* flora existed. We also require more petrified material which would give precision to comparison between the southern and northern botanical provinces. To mention only one more desideratum: we are still very much in the dark about the real nature of *Glossopteris*, the leaves of which occur in countless numbers in many of the beds.

In 1913 Dr Nellie Bancroft described a remarkable petrified stem, which was sent to Cambridge by Dr Rogers, and made it the type of a new genus *Rhexoxylon*. More recently additional examples—much larger and better preserved specimens—have been contributed by Mr Maufe and other geologists from beds that are probably Triassic in South Rhodesia and Natal. An examination of this splendid material has enabled Mr Walton to add greatly to our knowledge of *Rhexoxylon*, a type of stem, probably a Gymnosperm, which in its unusual method of secondary thickening presents a striking parallel to the anatomical features of modern *Dicotyledonous* lianes. Stems of similar type are known from Antarctica, South America, Russia, and North America. From the same series of strata, the Stormberg series, many impressions have been described which enable us to correlate the South African flora with those of India, Australia, and South America. As data increase it will be possible more accurately to estimate the degree of resemblance between the successive floras of the ancient southern continent with those of corresponding age in the northern hemisphere. Our knowledge of Jurassic plants from the African continent is almost nil. A flora, which is believed to correspond with our Wealden flora, has been obtained from the Uitenhage series of Cape Colony. The extension of the *Glossopteris* flora to Rhodesia and Tropical East Africa has been demonstrated by more than one author: contributions have been made by Dr Arber, Dr Gothan and Dr Brehnen. From the Tanganyika province a few fragmentary plants have been obtained which suggest comparison with Permian or Lower Triassic species of Europe.

In 1914 Dr Marie Stopes described some good specimens of parallel-veined leaves from Nigeria which she named *Typhacites Kitsoni*. Additional specimens of the same type and of other plants were sent to me by Dr Falconer for examination last year, and an account has recently been published. The material, though meagre, is not without interest; it represents the remains of a swamp flora including some large leaves of a *Salvinia* similar to *Salvinia auriculata* of South America, leaves of a *Typhacites* and a fragment of a leaf which may be an *Acrostichum*. The plants indicate a Lower Tertiary rather than a Cretaceous age. The abundance of fossil trees in the Libyan desert is well known: they include

Dicotyledons, Palms, and Conifers and are usually assigned to the Upper Cretaceous period. From a locality 15 miles east of the Gulf of Suez, Dr Ball obtained some casts of Lepidodendroid stems which have not yet been fully described, but an examination of them showed that they are very similar to or identical with Lower Carboniferous European species. In 1907 I described a few incomplete leaves from the Nubian Sandstone. More recently members of the Egyptian Survey discovered some large Dicotyledonous leaves in the Nubian Sandstone of Upper Egypt which are now under examination; they afford a striking illustration of the contrast between the present and past climatic conditions.

India. The peninsula of India represents another part of the ancient continent of Gondwana Land: from a thick series of freshwater deposits that have accumulated in a region little affected by earth-movements a rich harvest of fossil plants has been secured. The oldest species, belonging to *Gangamopteris*, *Glossopteris* and other genera, are usually spoken of as Permo-Carboniferous in age: as in South Africa thick deposits of glacial origin are a striking feature. Few petrifactions have been discovered; the oldest are the *Dadoxylon* stems described by Miss Holden in 1917: some Jurassic Cycadean and Coniferous specimens had previously been described by Dr Bancroft. The plant-beds of India are particularly promising as a source of information on the relationship of the Palaeozoic and Mesozoic floras. For several years Dr Feistmantel laboured with success as Palaeobotanist to the Indian Geological Survey, but in recent years there has been no official representative of this branch of research. There is no lack of material; what is wanted is recognition of the importance of the subject and the appointment of a Palaeobotanist. I have cause to be grateful to more than one Director of the Survey for the loan of specimens, and it is not against them that my criticism is directed; but I feel that to-day, to a greater extent than formerly, the purely economic aspect of geology receives more than its fair share of attention. A policy which neglects pure science is essentially wrong.

Indian Palaeobotany is too large a subject to be treated adequately in the time at my disposal; and the same statement applies with equal force to Australia. I will only add that in 1922 a collection of Jurassic plants, made by Mr Wayland, was described by me from Ceylon: these are the first fossil plants recorded from the Island, and they demonstrate a southern extension of the Jurassic flora of Madras. We know little of the Tertiary floras of India. Petrified woods were described by Miss Holden in 1916 from Burma; in 1912 I described some Dicotyledonous leaves from Assam, and Mr Edwards gave an account this year (1924) of some Tertiary leaves from Burma.

Australia. One of the most difficult questions raised by the older fossil plants of Australia is that of geological age, whether certain species are from Devonian or Lower Carboniferous rocks. The casts known as *Lepidodendron australe* are recorded from many localities, but the age of the beds is often uncertain, and we know nothing of the anatomical structure of the stems. Several years ago a well preserved *Lepidodendron* stem was found in New South Wales, but it has not yet been described. I would suggest to Prof. Lawson that a description of the first petrified *Lepidodendron* from Australia is greatly to be desired. Prof. Sahni's account of a *Clepsydropsis* from New South Wales, following a preliminary note by Mrs Osborne on the same type from another locality, is exceptionally interesting as a contribution towards a correlation of Australian and northern hemisphere floras. During the last few years Mr Walkom has published several important papers on Mesozoic floras, and he is to be congratulated upon his success in a praiseworthy attempt to revise and extend our knowledge of the rich collections from different parts of the country.

I have said nothing of the Palaeobotanical resources of the *West Indies*, of the *Malay Peninsula*, of *Hongkong* and other parts of the Empire; but enough has been said to demonstrate the abundance of the material that is already available. My final word is an expression of hope that younger men may be found who will demand the necessary means for prosecuting palaeobotanical research both from a purely scientific as well as from an economic point of view. The study of fossil plants is not only extraordinarily fascinating, but it offers to those who pursue it opportunities of throwing light upon the vexed question of evolution.

PAPERS

(CHAIRMEN: MISS E. R. SAUNDERS, PROF. H. H. DIXON, F.R.S.,
and DR E. N. M. THOMAS)

Dr E. M. Delf. SPERMATIA OF THE FLORIDEAE

Owing to their minute size and somewhat infrequent occurrence, the spermatia of the Florideae are much less known than either the female organs or the tetraspores, and the details of their development have been described for comparatively few genera. The object of the present communication is to pass briefly in review what is known of the origin of the antheridia and of the course of spermatogenesis in this group and to consider some of the outstanding problems in the light of recent research.

Much of our knowledge of the occurrence of the spermatia in the red algae rests upon the broad basis of Thuret's untiring researches. In 1855, summing up the position at that time, Thuret remarked that there were in all 85 species of European red algae in which the male fronds had been detected, 66 of which he had himself found. These cases were not confined to particular families but were distributed over most of the more important European genera. Subsequent cases reported up to the present day amount to 56 as far as I can discover, making a total of 141 in all. Other cases have been found by Miss V. M. Grubb and myself, and these are now under investigation at Westfield College.

Distribution

The spermatia occur as a rule on special male plants, but most of the Nemalionales seem to be monoecious and amongst the higher Florideae *Rhodophyllis bifida* (Rütz) (= *Rhodymenia bifida* Grev.), and a few species of Callithamnion are said to be monoecious.

The whole surface of the plant may be covered with spermatia as in *Dumontia filiformis* (Grev.); or the spermatia may be localised in patches or sori. In *Rhodymenia palmata* (Ag.) these sori form a pale reticulated pattern on the frond; in *Chondrus crispus* they are mostly limited to the tips, and in *Griffithsia corallina* Ag. and some others the fertile areas form narrow bands around the articulations of the filaments. In many cases, the spermatia are found on short fertile branches as in *Polysi-*

phonia spp. and in *Furcellaria fastigiata* (Lamour.); or they may be sunk in receptacles as in *Laurencia pinnatifida* (Lamour.) and in many species of Corallina. On the whole, however, the distribution of the spermatia in any species corresponds with the position of the female organs on the female plant and with that of the tetraspores on the asexual plant.

The male fronds or their fertile areas are pale in colour, gelatinous in character, especially when ripe, and frequently dwarfed in size. In many cases they are ephemeral and become completely disintegrated when the spermatia are shed. Very little is known as to the seasonal periodicity of the male plants and as to the influence of external conditions on the appearance of the spermatia. Kylin, in Sweden, has collected male plants in the autumn months; Rosenvinge, in Denmark, in the spring and summer. In England, a few examples of spermatia may be found in almost any month of the year, but the recorded cases throw little light on the factors determining periodicity.

Terminology

For the purposes of the present communication, the cell in which the spermatium is developed is called the "antheridium" and is equivalent to the "spermatangium" of Schmitz and others. The structure which has been termed antheridium by Yamanouchi (for example in *Polysiphonia violacea* Grev.) is composed of a central axial filament bearing antheridial mother cells laterally and should be termed an antheridial branch. Any cell which bears one or more antheridia is called the "antheridial mother cell," and is equivalent to the "spermatangial mother cell" of most German writers or to the "cellule anthéridifère" of Guignard. The basal cell (Tragzell), regarded by Svedelius as always present, appears to me to be not always distinguishable and is disregarded in the following account.

Development of the antheridium

According to Schmitz, the antheridia arise as terminal cells on short branch systems which may become more or less compacted into a tissue but always remain superficial. This is the view accepted by Oltmanns in his classic *Morphologie und Biologie der Algen*. In 1917, Svedelius further distinguished many different types of antheridial development. A consideration of those cases in which sufficient evidence is now available, however, impresses one with the essential uniformity rather than with the diversity of the modes of development. Two different types are nevertheless clearly distinguishable.

I. The simpler and presumably more primitive type is that seen in Nemalion, Batrachospermum and possibly also in the other genera of the Nemalionales. In these cases, the antheridial mother cell is scarcely differentiated from the other cells of the filament. It appears to bud off antheridia in an irregular way at any point on its free surface. The mother cell has a well defined chromatophore, and when nuclear division occurs this divides also so that the antheridium itself is at first pigmented. During the development of the spermatium, however, the chromatophore gradually breaks down and the spermatium is finally a colourless cell at the time of its escape.

II. In the commoner and possibly higher types, the antheridial mother cell is often differentiated from the surrounding cells in form and in the chromatophore, which only stains very faintly. The budding is mostly subapical and the chromatophore does not divide or enter into the antheridium. In these types the antheridial mother cells are often compacted together, and the antheridia become covered with a continuous outer layer of the outer cell wall. This is the so-called "cuticle" which is firm at first, but which later either dissolves into a watery gelatinous sheath and is easily ruptured by the escaping spermatia as in *Rhodymenia palmata* (Ag.), or may remain firm and become pierced with little holes through which the spermatia make their way when released. The latter is well seen in the case of *Chondrus crispus* (Lyngb.).

The course of development of the antheridium has been frequently described and may be briefly recapitulated here. The mother cell is always uninucleate and when about to form an antheridium puts up a beak. The nucleus divides, apparently always with an intranuclear spindle, and the upper nucleus thus formed wanders into the beak. A constriction now begins as an ingrowth of the wall at the base of the beak, which gradually cuts off the contents of the beak from that of the mother cell. The cell thus formed is the young antheridium, and even before it is completed a new papilla may be formed near the site of the first one, and in this way two or more antheridia (which may be termed *primary*) may be successively initiated upon the same mother cell. When the first primary antheridium has released its spermatium, the protoplast of the mother cell beneath frequently produces a new papilla which pushes up the floor of the now empty old antheridium and thus a secondary antheridium is formed; occasionally also tertiary antheridia have been observed. The successive production of primary and secondary antheridia seems to be a constant feature of all the higher Florideae in which the spermatia are known.

In most of the higher Florideae, 2–4 primary antheridia seems to be the rule, any or all of these forming secondary ones. In *Rhodymenia palmata* (Ag.) two appears to be the usual number, three being occasionally seen. In *Ceramium rubrum* (Ag.) (as yet undescribed) three papillae are nearly always seen, and in *Laurencia pinnatifida* (Lamour.) recently described by Kylin, four or possibly five is the number observed. The genus Martensia is exceptional according to Svedelius, in that the mother cell always produces a solitary antheridium at its apex, this being followed as usual by a secondary one in the same position. In the Corallinaceae, the antheridia and also the spermatia are considerably modified in form probably in connection with the space relationships of the deep receptacles in which they are contained. The details of development are not clearly understood, but the descriptions given by Guignard are not inconsistent with an interpretation on the lines of the forms already mentioned.

Development of the Spermatium

The young antheridium is colourless and the nucleus at first takes up a central position. Within it, at this time, a faintly staining reticulum can be seen with small granules of chromatin at the angles of the network and a distinct nucleolus. The latter decreases as the granules increase both in size and in staining capacity. The ripe spermatium has a nucleus containing well defined granules of chromatin giving a sharp stain with iron haematoxylin. These are usually about 20 in number, and represent the haploid number of chromosomes in the vegetative cells of the thallus, at least in those cases in which both have been determined (*Furcellaria fastigiata* (Lamour.), *Delesseria sinuosa* (Lamour.), *Laurencia pinnatifida* (Lamour.), and others). These granules are usually regarded as prochromosomes or sometimes as chromosomes, although they remain connected by a delicate linin thread. No spireme stage has ever been seen during their differentiation, and Miss Knight has recently found this to be the case also in the dividing nucleus of vegetative cells of those plants of *Pylaiella littoralis* Kjell. which will form unilocular sporangia. In the majority of the red algae the further behaviour of the nucleus of the spermatium is unknown, but in *Nemalion multifidum* (J. Ag.) it seems clear from the work of Wolfe (1904) and Cleland (1919) that after fusion with the trichogyne, the nucleus completes a division, only one of the resulting daughter nuclei being used in fertilisation. Nuclear division was not actually seen, but was presumed by Kylin to occur in the spermatium of *Batrachospermum moniliforme* (Roth.) after its attachment to the trichogyne, two small bodies resembling nuclei having been found. In all the higher Florideae where fertilisation is reported, no nuclear division has been seen, and the inference is that the so-called chromosomes in the ripe spermatium represent initial stages of an abortive attempt at nuclear division which never reaches more than late prophase. According to Yamanouchi, in *Polysiphonia violacea* (Grev.) the nucleus remains in this state up to the time of fusion with the egg nucleus. Comparison with Nemalion suggests that in all these cases the spermatium in the Florideae is really the equivalent of the contents of an antheridium which formerly divided producing more than one male gamete.

The ripe spermatium is spherical or slightly ovoid in form and encloses a large vacuole, the nucleus taking up an apical position. According to Guignard, the spermatium is surrounded with a delicate cell-wall, distinct from that of the antheridium even before its escape. Kylin and Svedelius, who have more recently denied the presence of such a wall, appear to have relied mainly upon preserved and presumably shrunken material. In my experience and that of Miss V. M. Grubb, a delicate wall can be clearly seen either in fresh or in carefully preserved material. We are unable to say whether this wall is derived from the inner layer of the antheridium wall, but it certainly does not include the whole wall of the antheridium as suggested and figured by Yamanouchi for *Polysiphonia violacea* (Grev.). On the contrary, in our experience, the spermatium always issues from within, leaving behind the empty sheath of the antheridium. In any case, all authors are agreed that the spermatium, when it reaches the trichogyne, is always surrounded with a wall which fuses with that of the trichogyne.

Time does not allow of the consideration of such questions as the homology of the antheridial cell group, the possible relationship between the spermatia of the red seaweeds and those of the Ascomycetes, or the question of alternation of generations. I can, in conclusion, only revert to the question which must ever be in the mind of anyone who seeks to investigate the spermatia of the red algae, namely, the reasons for the comparative rarity of the male fronds. Allowing somewhat for the difficulty of recognising anything so inconspicuous as these often are, and also allowing for their sometimes ephemeral existence, it yet seems strange that in the systematic searches which have been made, male plants are often so rare that only a single one may be seen amongst many female and tetrasporic plants. Whilst it is known that tetrasporic plants are apt to be much more numerous than sexual plants, it is usually assumed that in the haploid generation, both sexes will be produced in equal numbers. One asks whether this is indeed the case, and if so, whether there is a high mortality amongst male plants, either from greater sensitiveness to external conditions or possibly as being an attractive source of food to marine animals. A further possibility is suggested by the work of Sturch on Harveyella (1924), which demonstrates for the first time a diurnal periodicity in the life cycle, the plant being only found in deep water in the summer months. It may perhaps be suggested that the spermatia of the red algae are not often functional, are produced less often than previously, and that the male plants are tending towards extinction. The regular alternation of haploid and diploid generations, in many cases in which fertilisation has not been proved, does not necessarily negative this suggestion. There is, however, no evidence as yet from which any trustworthy conclusions on these points may be drawn.

Miss E. J. Welsford. Diseases of the Clove in Zanzibar

(*Abstract*)

A brief account was given of the clove tree in Zanzibar and of the conditions under which it is grown.

Two diseases were described.

The disease known locally as "sudden death" is the result of the attack of a fungus which causes a root rot. The fungus spreads from root to root chiefly in the older trees. The disease can be controlled by the destruction of the dead trees which otherwise act as centres of infection and by the isolation of groups of infected trees by means of trenches. The fungus has not yet been identified.

Another fungus attacks the leaves and causes a "die back" of the branches. So far only immature pyenidia have been found, the ripe stage developing after the leaf has fallen. Infection was only found on those leaves which had already become attacked by an epiphytic Cephaleuros which affords foothold for the developing fungal spores. At present this disease is of little consequence. Should it become very destructive it can be controlled by spraying with copper acetate or with Bordeaux mixture.

Prof. J. H. Priestley. Vegetative Propagation
(*Abstract*)

If the agricultural produce of the British Dominions is to find a secure place on the world's markets in free competition with the exports of communities whose agriculture is continually more scientifically directed, one great desideratum is that the article exported should maintain a definite standard of quality. Such uniform yield as this demands can never be obtained by agricultural produce if the plants yielding the produce are themselves chance seedlings from parents liable to cross pollination. It is therefore of great importance that the work of selection amongst the numerous seedling strains in plants of agricultural value should be followed where possible by experiments upon the possibility of vegetative propagation of the selected seedling. When the acreage under cultivation is thus filled with plants multiplied by vegetative means, a few selected strains with yield of known quality may be relied upon to rapidly drive out of cultivation the miscellaneous, chance-selected seedlings previously filling the plantation. In Europe, and now in the United States, this process has operated, and in the case of fruit in particular, as also with the potato, the markets are supplied with the produce from comparatively few selected, vegetatively propagated, strains.

In tropical regions vegetative propagation is as yet but little exploited, though recent experiments on the propagation of tea and coffee plants are significant indication of future developments. Whilst it is true that, given sufficient patience and adequate facilities, practically any flowering plant may be propagated vegetatively, it is quite another matter to discover a method, in respect to any particular plant, which is sufficiently reliable and at the same time so simple a procedure that it can readily be carried out by unskilled labour under large-scale conditions. Unfortunately the rules for the guidance of the plant propagator are still largely empirical, and it is the object of this present paper to briefly indicate certain general features of the problem which have developed during its investigation.

The normal healthy plant may be regarded as in a state of equilibrium with shoot and root development commensurate with one another. The first result of an attempt to propagate vegetatively is a separation of one part of this equilibrium system, and then an attempt to get the severed part to restore this former equilibrium. A first essential is a healthy healing of the surface of separation, in itself a subject requiring detailed investigation. The subsequent restoration of the former balance of root and shoot has been studied in the development of roots upon a severed shoot and of shoots upon a severed root system. The result is to show the importance of the existence in the severed part of regions capable at once of further growth, the meristems; anatomical study along these lines explains to a great extent the relative ease or otherwise of propagation. It can be shown too that in the majority of cases the origin of new shoots or new roots is to be found in different tissues so that anatomical study may ultimately throw considerable light upon the principles governing regeneration in different plants.

The results of such anatomical studies were discussed in reference to experimental efforts to propagate plants from cut roots or shoots in a series of cases which supply examples of plants showing a wide range of behaviour as to the facility with which they regenerate the missing portions of the plant system and so restore the normal balanced shoot-root system.

Dr W. Brown. A STUDY OF FORMS OF FUSARIUM OCCURRING ON APPLE FRUIT[1]

A number of isolations of Fusarium from apple fruits gave a variety of forms with different cultural characteristics. Differences between some of these forms were slight, between others much greater, so that according to the usual methods of nomenclature these original isolations would have been divided into at least two and possibly three species, the remainder being grouped as varieties of the latter. With these original cultures an investigation was begun from the following points of view:

(1) To see whether these strains on culture maintained their characteristics or changed in any way.

(2) To study the variability of particular forms under different conditions of culture.

The object of the second investigation, which was carried on simultaneously with the first, was to form a sound basis for the comparison of the different growth forms of the different strains.

The various strains, when grown under identical conditions, differed from each other in respect of such characters as the amount of aerial mycelium, the rate of colony growth, the amount of sporulation, the size and septation and to some extent the shape of the spores, the kind

---

[1] This work was carried out in part in collaboration with Dr A. S. Horne.

of colour produced on certain media etc. All the strains agreed in the absence of a *Cephalosporium* stage, in the absence of typical chlamydospores, at least in the mycelium, and in the colour of the spore masses which was invariably some shade of orange pink.

For purposes of a standard comparison, a synthetic medium, based on the composition of potato extract, was adopted. In its original form this medium contained glucose, starch, asparagin, peptone, neutral potassium phosphate, potassium chloride, magnesium sulphate and ferric chloride, but it was found that this could be considerably simplified without seriously affecting the growth form of the fungus as compared with that on the natural potato decoction. The simplified medium had the following composition:

(a)	Glucose	...	2	grms.	$MgSO_4$	...	·75 grms.
	Asparagin	...	2	,,	Agar	...	15 ,,
	K_3PO_4	...	1·25	,,	Water	...	1 litre

Using this medium as starting point, a detailed study was made of the effect of the quantity of each constituent on the growth characteristics of a number of the strains. By suitably varying the constituents in this synthetic medium, a wide range of growth form can be induced in each particular strain. A number of well defined rules could be made out in this connection, of which the following is a summary:

Effect of concentration of medium. This is a factor of great importance. The effect of this factor will be illustrated in relation to two strains *A* and *D*. The former, when grown on the medium given above, develops very scanty aerial mycelium, shows unstaled type of growth, sporulates freely, the spores having a somewhat high septation mode (4–5); the latter has comparatively persistent aerial mycelium, shows pronounced staling, and while sporulating fairly freely has spores of low mode (practically all 3-septate).

In the case of strain *D* dilution of the standard medium (*a* above) has the effect of diminishing and finally removing altogether the staling type of growth. Simultaneously the aerial mycelium diminishes and finally disappears, while the spore septation rises from three to five. Similarly, on concentrating the medium *a*, strain *A* begins to show more persistent aerial mycelium, while the staling type of growth appears, accompanied by a drop in the septation of the spores to three. Thus in respect of the three characters of amount of aerial mycelium, rate of spread of colony, and septation of spore, each of the two strains may be made to show the characteristics of the other by growing it on the appropriate concentration of the medium α.

The following rule is of general application to all the strains of *Fusarium* studied.

Strains which show the staling type of growth on the standard medium have spores of low septation; those which show unstaled growth have spores of high septation. Strains with staled growth and low spore septation show the converse features when grown on a sufficient dilution of the medium; strains which show unstaled growth and high spore septation on the standard medium give the converse features when grown on a more concentrated medium.

Effect of composition of medium. The important constituents of the nutrient medium are the carbohydrate, the nitrogenous compound, and the phosphate. The magnesium sulphate is of less importance as its concentration may be widely varied without producing any apparent change in the fungal growth.

Carbohydrate. The effects of varying amounts of glucose, saccharose and starch (potato) were determined. For these fungi there appeared to be very little difference as between equivalent concentrations of glucose and saccharose. Glucose and starch however, when added in equivalent amounts, do not produce the same effect, even though the starch in the culture medium disappears with fair rapidity. The cultures to which glucose is added as source of carbon produce much more mycelium than the ones in which the carbon supply is starch. From the point of view of mycelium production, a given amount of starch added is equivalent to a much less amount of glucose. There is also a very pronounced qualitative effect. While increase of carbon compound in the form of glucose increases amount of mycelium formed and simultaneously depresses sporulation, increase of carbon in the form of starch is accompanied by much stronger sporulation with only slight effect on an increase of mycelium.

Nitrogenous compound. The nitrogen constituent in the medium used as standard is asparagin. This also is to some extent a source of carbon. When these fungi are grown with asparagin as source both of carbon and nitrogen, the amount of growth produced is small, and growth is soon checked by the development of alkalinity due to the preponderance of the nitrogen supply.

The effect of reducing the amount of the nitrogenous constituent in the standard medium given above is interesting. From the point of view of surface growth, the effect of reducing the nitrogenous constituent is the same as that of simple dilution of the medium, *i.e.* the removal of the staling effect. A further effect of reducing the nitrogen compound is the appearance of colour in the medium, the colour produced in the different strains being yellow, blue, pink, etc. The yellow colouring matter is contained in granules encrusting the hyphae, while the blue and pink colorations are diffused generally through the medium. The granules containing the yellow colour appear to be of the nature of glycogen.

The factor responsible for colour formation is not simply the amount of the nitrogenous constituent, but is the ratio of the amount of carbon compound to that of nitrogenous compound. When this ratio (C : N) is low no colour is produced and conversely.

For the study of colour formation in these fungi, the following modification of the standard medium was adopted:

(β) Glucose ... 2 grms. $MgSO_4$ ... ·75 grms.
Starch ... 10 ,,   Agar ... 15 ,,
Asparagin ... 0·2 ,,   Water ... 1 litre
K_3PO_4 ... 1·25 ,,

This medium differs from (α) in having increased carbohydrate and reduced asparagin. The additional carbohydrate was added in the form of starch in order to keep up sporulation.

The alteration of the carbon : nitrogen ratio of the medium is also reflected in the microscopic features of the spores. The effects may be summarised as follows:

On the α medium—the septation mode is low (3) or high (5) according to the strain; the spores are hyaline so that the septa are clearly visible in the unstained spores; there is no tendency to constriction at the septa; the spores after a time show a tendency to vacuolation, and later still to atrophy of various segments, and finally to death.

On the β medium—the septation mode is always high (at least 5 and sometimes 6); the spores are granular so that the septa are obscured and in many cases can only be seen properly on staining; there is a tendency to constriction at the septa; the tendency to degenerative changes is much reduced, so that the spores on this medium are much longer lived than on the α-medium.

Phosphate. The effect of increasing the concentration of neutral phosphate is, within limits, to diminish staling, to increase sporulation and to raise the degree of septation of the spores.

In low concentrations, such as in α (ca. ·1 %), there is little difference between the neutral and the acid phosphates. With higher concentrations, however, a difference appears. Whereas increase of neutral phosphate increases sporulation, corresponding increases of acid phosphate lead to increase in mycelial formation and diminution of sporulation.

In addition to the composition of the nutrient medium, certain physical factors influence the growth form of the fungus. The most important of these are temperature and light. Increased temperature enhances the staling effect and lowers the degree of septation of the spores. Light has no appreciable effect on the rate of spread of the colony, but markedly increases sporulation. This stimulative effect on sporulation only requires a short exposure of the colony to light. Colonies which are kept in darkness and exposed at intervals for a few minutes to light subsequently show a strong ring of spores near the place where the growing edge of the colony was at the time of exposure. Colonies grown in alternating day and night show a more or less distinct pinkish colour in the aerial mycelium as compared with the pure white of control colonies grown in the dark.

Zonation in spore formation is shown by certain strains. In some cases these zones are due to and correspond with alternation of day and night. In other cases this is not so; certain strains are characterised on certain media by producing a single spore zone, which is obviously correlated with the general metabolism of the colony and not with the periodicity of light. Even in strains which typically produce day and night rings under certain conditions, fairly small alterations of the composition of the medium will obliterate the zoning effect. The factors underlying zonation in general appear to be complex, and in view of the effect of growth conditions on its appearance, it is a feature of limited systematic value.

The preceding account has dealt with modifications of a particular strain under different environmental conditions. Such variations are of a temporary nature so that when each strain, after passage through a series of media with their resultant growth forms, is returned to the standard medium the original growth form is restored. Variations of a permanent nature, however, have appeared from time to time among these strains and a systematic attempt has been made to separate out the new forms from the originals and to keep them in culture. The result of this has been that the original six strains are now represented by upwards of forty, which, though not all different or at least in some cases only differing very slightly from each other, nevertheless have had a different origin.

The new strains have appeared from time to time in the form of sectors in the parent colony. To guarantee that the parent stocks did not consist of mechanical admixtures of several strains, they were carried on from an early stage from single spores and later from single hyphal tips. During a period of nearly two 'years while these "saltations" were under study, the stock cultures were carried on at each successive sub-culturing from single hyphal tips. This was done irrespective of whether any saltation had happened in the meantime in the parent stock, so that very convincing evidence is forthcoming that these segregations have taken place within single mycelia.

The saltant portion of a culture is in some cases clearly defined from the parent (or original) part, and by taking single hyphal tip cultures it is generally possible to effect a ready separation of parent and saltant in the pure form. In some cases where the saltant was ill-defined, no separation into two distinct forms was effected. Again, on certain media no obvious sectoring was shown, and yet it was possible to pick up spores or mycelium from different parts which gave on further culture a variety of strains.

It happened occasionally that the saltant form tended to crowd out the parent so that a reculture taken at random from the mixed culture would in all probability give the saltant form only, and thus the parent be lost. This would inevitably happen if the saltation was not recognised at once and the parent form recovered, while it was still present. A case of this sort actually happened in the course of this work, and if the particular parent culture had been in process of routine reculture, it would have been lost and its place taken by a saltant from it. This phenomenon is probably responsible for much of the change that fungi are known to undergo in artificial culture.

The strains in themselves appear to be constant. Four of the earlier strains (three of the originals and one early obtained saltant) were grown on various dilutions of the standard medium and under known external conditions, and their growth curves carefully prepared. After an interval of about a year and a half the growth curves of the same four strains were again determined and found to be practically identical with those obtained earlier. In their general growth features also they showed no change. Thus variation within this group appears to take place by sudden jumps and not by gradual change during the process of culture.

The strains which were in cultivation in February, 1924, when the last elimination of duplicates was made, were as follows:

From the original strain A, 7 strains

		B, 10	,,
,,	,,	C, 10	,,
,,	,,	D, 11	,,
,,	,,	E, 4	,,
,,	,,	F, 2	,,

Though all the strains derived from a given parent have up to the present remained distinguishable from each other, some of the strains in *e.g.* the B group are very similar to and practically indistinguishable from members of the C group. Thus the strain B_{111} in the B group is distinctly different from anything else in the B group but is closely similar to C_{11} and C_{211} in the C group; and to certain strains in the D and E groups. In the same way some saltants in the D group are very similar to the original A strain. The same complex of forms is thus being derived out of each original parent. As a provisional measure all the forms have been classified into four groups, based on their appearance on the standard medium:

(1) *Mycelial type.* Characterised by somewhat fluffy aerial mycelium which is more persistent than in the other strains; by sporadic and limited sporulation.

(2) *Sporodochial type.* With very evanescent aerial mycelium and more or less intense general sporulation in the form of sporodochia.

(3) *Pionnotes type.* With faint aerial mycelium only in the early stages; sporulating intensely all along the surface of the culture and also on the hyphae running down into the medium.

(4) *Long-spore type.* With scanty but more or less persistent aerial mycelium; with mode of sporulation similar to that of (3); but differing from all three other types in the abnormal length and high septation of the spores.

These types show further distinguishing features. Thus the mycelial type is the one in which sclerotia are best developed. No sclerotia have hitherto been seen in the pionnotal type. The sporodochial type is associated with strong yellow colours on suitable media; the long-spore type under similar circumstances gives pink or blue shades, and so on.

However, the line between the various types is more or less artificial as some transitional forms have appeared, and there is no doubt but that the remaining gaps in the series would be filled up in course of time. Nevertheless, representative members of each type show very distinct differences between each other and would certainly be accorded specific rank. At the same time it is known that they can all be derived from a single "pure" culture, and that a range of forms intermediate between them exists. It seems necessary, therefore, that they should be reduced to a single species. Furthermore, as the same group of forms is obtainable from each of the original six strains (which came from nature and the antecedents of which are therefore unknown), these original six must on this *à posteriori* evidence be put together as one species.

Similar saltations are known to occur in a number of fungi, *e.g. Helminthosporium, Coniothyrium*, etc., and undoubtedly exist in many more. In consequence fungal genera are probably over-classified; many so-called species are probably saltants. The pronounced effect of nutrient, external conditions, etc. on growth form show how essential it is that descriptions of fungi must have reference to standardised and definitely reproducible conditions. Probably a great deal more use could be made of simple synthetic media than has hitherto been done.

Mr B. Barnes. A New Aspect of the Dung Flora

During the autumn of 1923, a fungus, which from its general characters, as seen by the naked eye and under a lens, seemed to be a species of *Piptocephalis*, appeared on some cultures of *Mucor Mucedo*. It could not, however, be identified with any of the described species of *Piptocephalis*, nor could a parasitic relation with *Mucor* be demonstrated. That in fact the fungus was no parasite was soon shown by growing it in pure cultures on pea and potato agars.

The appearance of the conidiophores was very like what would be presented by a branched *Oedocephalum*, suggesting a *Blakeslea* in which the sporangiola had been replaced by conidia. As it was thought that the conidial stage of a zygomycete was under investigation, attempts were made to induce the formation of sporangia of zygomycetous type by growing the fungus on different media and under various conditions, and to bring about efforts at zygospore formation by growing it against various zygomycetes.

Neither sporangia nor zygospores were obtained, but these cultures led to a further discovery. Among the media used were several dung agars, on which moderately good growth was obtained. None of the common intruders appeared in these plates, but some fourteen days after infection the apothecia of a small discomycete were seen to be present in small numbers. Similar apothecia had already been seen on the cultures of *Mucor Mucedo*, but as these cultures, which were made on dung agar in halves of Petri dishes under a bell jar, were by no means clean, no importance had been attached to the apothecia. In the new cultures, which had been made carefully in whole dishes, and were free from the usual species characteristic of fouled cultures, the regular development of the discomycete led to a suspicion that there was a definite connection between the conidial form from which the agar had been inoculated and the ascus-bearing form which eventually resulted.

Two dozen dung agar plates were made up and inoculated with conidia; five days afterwards all the plates showed a moderate crop of the conidial form, and, after eleven days, small apothecia were observed. All the cultures were free from contamination and remained so until they were rejected some weeks later. Four cultures were then started, each from a single conidium, and these again, whilst showing no intruders, gave, first the conidial, and then the apothecial form. Three cultures, each inoculated with a single ascospore, showed no growth at all, and remained unchanged until the dishes were cleaned up. In spite of this, it was felt that these results justified the conclusion that the conidial form and the apothecial form were stages in the life-history of one fungus, and, incidentally, they also ruled out the possibility that a zygomycete was under examination.

It is not necessary to enter upon a detailed systematic discussion here. It may be said, however, that after a search through the literature the conclusion has been arrived at, that the fungus in its conidial form falls into the genus *Acmosporium* Corda. Of this genus, two species have been described, neither of which agrees exactly with the present one, but the systematics have still to be cleared up. The discomycete is a species of *Lachnea*, but again the specific name is doubtful. It is possible that the characters are modified by culture, so that the reference of the form to a species which has already been described is not an easy matter.

The Fungus

1. *Acmosporium*

Acmosporium is easily obtained by inoculating pea or potato agar with conidia. Growth is visible to the naked eye after twenty-four hours, and proceeds rapidly. In the early stages of growth the mycelium creeps along the surface of the medium, and consists of a comparatively small number of thick ($6-8\,\mu$) and frequently septate hyphae, which dichotomise rather regularly, show little other branching, and cover the surface fairly evenly. A young mycelium, two to three days old, viewed along the surface of the agar, shows a number of short (0·5 mm.) branches rising into the air, and apparent to the naked eye. At the same time, other similar branches develop in an irregular manner from the main hyphae, and lie on the agar; as they are wet, they can only be seen under the microscope. These prostrate branches form H-piece connections with one another, and with the main hyphae, which by this time begin to assume a brownish hue.

By the end of five days, most of the macroscopic growth forms a snow-white cottony ring at the edge of the dish, whilst the central part is occupied by a tangle of creeping hyphae. The dense marginal ring consists mainly of conidiophores, rises to a height of about 5 mm., and has a width of about 1 cm., with the inner edge dropping rather steeply to the surface of the agar on which the creeping mycelium lies. For the most part, conidiophores of this age bear well-formed conidia. It seems likely that the formation of a dense marginal ring is due to the better aeration at the periphery of the dish; when the fungus is grown in half a dish, under a bell jar, there is an even formation of conidiophores over the whole surface of the medium.

A culture seven days old is no longer of a snow-white colour, and after three or four more days a dirty flesh colour is general.

Conidiophores arise as simple branches, with a diameter of 6–8 μ. They contain dense granular plasma. When such a branch attains a length of about 100 μ, it dichotomises, the branches elongate, and dichotomise again. This is repeated four or five times, the branches becoming shorter and shorter. Each pair of branches lies in a plane approximately at right angles to that of the previous dichotomy. The ultimate branches are very short, and consist of a swollen spherical terminal portion, of average diameter 10 μ, united to the general branch system by a short stalk, whose length is usually less than the diameter of the swollen end. This bears on its surface from nine to fifteen narrowly conical sterigmata, and these swell up at their ends to give rise each to one spherical conidium. The conidia have an average diameter of 8 μ. When mature they form a dense mass which completely hides the swollen branch end, and, indeed, the dichotomous arrangement of the whole conidiophore is frequently completely obscured by the mass of ripe conidia, which is of a dirty flesh colour. The conidium wall is smooth; its thickness has still to be determined, as within the wall there is apparently a homogeneous layer of material, the nature of which is obscure.

Single fallen conidia are colourless, and vestiges of the sterigma often remain attached to the wall. The granular contents surround two or three large vacuoles, and in some lights often look greenish.

A few remarks may be made with reference to the germination of the conidia. This occurs readily in potato decoction, pea decoction, dung extract and water. After twelve hours, one or two germ tubes may be sent out—in the latter case, usually from opposite sides—and these may branch almost at once, or elongate for some time before doing so. For about twenty-four hours after germination the young mycelia grow intermingled, but it is not until later that H-pieces are formed. H-pieces have been seen connecting the young mycelia from as many as six conidia. It does not seem that heterothallism can be invoked to explain these unions. Conidia placed in dung extract show a special tendency to form H-pieces after germination. Conidia, which have fallen in a culture over a week old, germinate on the surface of the agar, and become connected with the old mycelium. Usually, when a young mycelium has formed a connection, either with another young mycelium, or with an old one, it gives rise to a specially strong branch containing a noteworthy amount of granular plasma.

Often a young branch will lie parallel with an older one, and form two or three H-pieces with it.

In *Acmosporium*, unions of this sort appear to be quite casual, and at present their significance is wholly obscure.

2. *Lachnea*

Conidia sown on dung agar give rise to a mycelium which in essentials agrees with that developed on potato agar. Growth proceeds at the same rate, and there is the same creeping mycelium, with the peripheral ring of conidiophores. This, however, is much weaker than on potato agar, and is often discontinuous. Sown on sterilised dung, conidia give rise to a good mycelium, but the development of conidiophores is limited and patchy.

The short ascending branches which arise when the mycelium is two to three days old are much more abundant on dung agar than on potato agar, and, when viewed along the surface of the agar, it is seen that they tend to point towards the line of contact of the top and bottom of the dish, so that, in passing from the centre to the periphery, the inclination of these branches approaches more and more nearly to the vertical. This suggests some air relation. In view of the fact that with free access of air conidiophores mature all over the surface of the medium, it is thought that these ascending branches may be abortive conidiophores, arising in an atmosphere where the oxygen supply is not adequate to complete development, and that their inclination may indicate an effort towards an adjustment to the oxygen gradient in the atmosphere of the dish. At the same time, the matter is complex, for it still remains to explain how, under otherwise similar circumstances, conidiophores are developed less abundantly on dung agar than on potato agar, and still less abundantly on dung itself. This matter is under investigation.

Examination of the mycelium some ten days after sowing reveals the presence of spirally twisted and more or less tightly coiled hyphae of the type common in the Ascomycetes in early stages of fructification. Clockwise and anti-clockwise coiling has been seen in material from the same culture; this point is to be worked at. So far, the development of the apothecium has not been traced, but work on these lines, and on the behaviour of the nuclei, which are reasonably large, is in progress.

In from eleven to fourteen days after sowing apothecia become visible to the naked eye. At first they are small and white, sometimes with and sometimes without a fringe of brownish hairs. As they age, the apothecia become pinkish, and finally dirty pink, thus showing the same colour changes as are shown by the conidial stage. Mature apothecia are flattish, or a little convex, and their diameter varies closely around 1 mm. Vigorous "puffing" has been seen from old cultures when the lid of a dish has been lifted; the phenomenon has only been seen to occur once in any given culture.

The hairs arise from a swollen basal cell, either singly or in twos or threes. When more than one arise from the same basal cell, development is not simultaneous. A mature hair is usually less than 100μ long, and contains one or two septa.

The apothecium presents no features of special interest. The asci are eight spored; the spores are ellipsoidal, with a smooth wall, and average $15 \times 8 \mu$. So far, attempts to induce the ascospores to germinate have failed, so that it has not yet been possible to obtain *Acmosporium* from *Lachnea* by sowing ascospores on pea or potato agar.

A definite relation has been made out between the number of apothecia produced and the supply of cellulose in the medium. Dung agars were made up to which scraps of filter paper were added, and the number of apothecia formed in these cultures was compared with that obtained on a plain dung agar. The apothecia developed in positions which had no obvious connection with the position of the pieces of paper. In this respect, *Lachnea* differs from *Sordaria* and *Chaetomium*, the perithecia of which usually show a marked localisation on or near to the paper in the culture.

An even distribution of the paper was obtained in later experiments, by adding to the dung agar a quantity of filter paper which had been thoroughly broken up by maceration in distilled water.

In either case, the addition of cellulose in the form of filter paper led to the production of a better crop of apothecia. In one case, the numbers were:

Dung agar and paper ... 305 apothecia

Dung agar ... ... 78 ,,

This result was one of a set of five, all of which agreed well.

No evidence was obtained that the addition of cellulose had any effect upon conidium formation. Pea or potato agars to which paper had been added seemed to differ in no way from ordinary cultures on these media, and on dung agars the conidia were as abundant in those without as in those with paper.

CONCLUSION

It is no new thing to demonstrate the existence of a conidial and of an ascigerous stage in the life-history of the same fungus. After the present work had been carried almost to the point described here, and the connection of the two stages demonstrated, a paper by Dodge was encountered, giving an account of an almost identical piece of work. His conidial form appears to be an *Acmosporium*, very like the one described here.

The present work shows that the ascospores differ greatly from the conidia in the ease with which they develop, and that, whereas the perfect stage is coprophilous, the imperfect stage, though able to develop on dung, finds its best conditions on a vegetable substratum. In view of the fact that the ascospores of a number of coprophilous fungi have to pass through the body of an animal before they germinate, it would seem that the distribution of such a fungus would be greatly favoured by the possession of a stage able to flourish on vegetable refuse, and to produce conidia requiring no special conditions for germination. It may be that an investigation of *Fungi Imperfecti*, growing on vegetable refuse in meadows, will lead to additions to our knowledge of other common coprophilous ascomycetes.

It is remarkable that so little literature has been found referring to *Acmosporium*. The only original accounts seem to be those of Corda (1839) and Phillips (1884). In view of the fact that to the naked eye, and even under the lens, *Acmosporium*, which has been shown to grow rather scantily on dung, has so deceptive a resemblance to *Piptocephalis*, it is very possible that the form is common but liable to be overlooked.

The details which have been given of the behaviour of this fungus in culture suggest that interesting results may be expected from the more thorough investigation of its growth under controlled conditions.

I wish to express my thanks to Prof. Dame Helen Gwynne-Vaughan for the help and encouragement so freely given, and to Mr J. Ramsbottom, O.B.E., M.A., to whom I owe the suggestion of the filter paper method, and much help and advice.

The work is in progress in the Department of Botany, Birkbeck College, University of London, and the present account is a preliminary one.

SUMMARY

A fungus which appeared on open cultures of *Mucor Mucedo* was at first taken to be a species of *Piptocephalis*, but has since been referred to the genus *Acmosporium* Corda.

Cultures started from single conidia gave on potato agar abundant crops of *Acmosporium*, whilst on dung agar, after a slight crop of *Acmosporium*, a species of *Lachnea* appeared. The addition of cellulose in the form of filter paper caused an increased development of the apothecia of *Lachnea*, but did not seem to affect conidium development.

So far, ascospores have not been induced to germinate.

Acmosporium has a dichotomously branched conidiophore, the branches of which bear rounded conidia. The *Lachnea* stage has small apothecia surrounded by simple hairs, and attaining an average diameter of 1 mm.

It is suggested that the distribution of coprophilous fungi may depend, not so much on the ascospores which require special conditions for germination, as upon conidia produced from a stage able to live on vegetable refuse.

Bibliography

CORDA, A. C. J. (1839). *Icones Fungorum*, III. pp. 11–12 and taf. ii. fig. 32. *Acmosporium Botryoideum.*

PHILLIPS, W. (1884). "*Acmosporium tricephalum.*" *Gard. Chron.* new ser. XXI. p. 317, with 1 text-fig.

DODGE, B. O. (1922). "A *Lachnea* with a botryose conidial stage." *Torrey Bot. Club Bull.* XLIX. pp. 301–305, with 7 text-figs.

Mr G. W. Wickens. EXANTHEMA OF CITRUS TREES

In the following notes on the above disease in Western Australia I cannot claim that I have broken any new ground in the treatment of infected trees, for with one exception—bluestone in solution introduced through the roots—I have followed on the lines of experiments carried out in Florida and published in 1917 in *Bulletin* 140 by B. F. Floyd, Plant Pathologist of that State: but as the cause of exanthema is still not definitely known I am giving particulars of my experiments and observations in connection with it in Western Australia. I cannot definitely state the year when I first saw infected trees, but I believe it was in 1912, and at that time trees showing effects were so limited in numbers that special attention was not drawn to them. This comparative freedom of the orangeries in Western Australia continued until 1920 when suddenly the disease manifested itself in many places, and where previously only an individual tree or two had shown signs of the trouble in particular orchards the numbers in that year greatly increased, as also did the number of affected orchards. In 1921 and 1922 the disease was still prevalent, but in 1923 its decrease was as sudden as its increase in 1920, and this year, 1924, no increase has taken place.

On 1st September 1921 I commenced experiments which comprised:

(*a*) Spraying with Bordeaux mixture using 3 lbs. bluestone, 3 lbs. lime, 40 gallons of water.

(*b*) Spraying with Bordeaux mixture 3-3-40 once only on 1. ix. 21, and on that day also dressing the land around the same trees with 2 lbs. crushed bluestone for a medium-sized tree, and $\frac{1}{2}$ lb. for a small tree.

On 2nd November 1921 I again dressed the land with 2 lbs. crushed bluestone for a medium-sized tree and $\frac{1}{2}$ lb. for a small tree.

The bluestone was spread evenly over the surface of the land from the stem of the trees to the circumference of a circle 2 ft. wider than the spread of the trees' branches.

On the 23rd June 1922 the condition of the trees was as follows:

(*a*) Sprayed trees, some showed a decided improvement in condition, others no improvement whatever.

(*b*) Trees which received ground treatment in addition to spraying showed a very marked improvement.

Control trees in same condition as in 1921.

On 31st August 1922 I continued the treatment on the same lines, excepting that I gave neither spraying nor soil dressing to trees (*b*), so as to ascertain if the improvement in condition would continue without further help.

I examined the trees repeatedly during 1923, but was nonplussed to find nearly all trees showing remarkable vigour, and sound young growth. Some of the badly affected control trees were in just as good condition as the treated trees, while many trees which in 1922 had shown only moderate or slight signs of exanthema were in 1923 apparently completely free and that freedom is continuing to time of writing, March 1924.

The orchard where the above experiments were conducted is privately owned, and is situated in hilly country within 20 miles of Perth, where the rainfall averages from 34 to 40 inches per annum, nearly all of which falls between 1st April and 30th October. The top soil where the affected trees are growing is a gravelly loam inclined to be light and the subsoil is a mixture of gravelly loam and clay. The orchard is well drained with agricultural drain pipes and is irrigated during summer. It has since planting been liberally fertilised with bone-dust, potash and peas as a cover crop for ploughing in, while lime has also been frequently applied.

In 1922 an orchard adjoining the one referred to above came into the hands of the Government through the Agricultural Bank. Many trees in it were badly affected with exanthema. It had received no manure, no irrigation during summer and very little cultivation for several years. The top soil is light, inclined to be "snuffy," and the subsoil is a gravelly clay, rather stiff. In this place I took the opportunity of making experiments which I should have considered too risky to try in a privately owned, well-kept orchard.

These comprised:

(*a*) Spraying once with Bordeaux mixture 4-4-40 on 31. viii. 22.

(*b*) Spraying twice with Bordeaux mixture 4-4-40, once on 31. viii. 22 and again on 7. xi. 22.

(*c*) Spraying once with Bordeaux mixture 6-4-40 on 31. viii. 22.

(*d*) Spraying twice with Bordeaux mixture 6-4-40, once on 31. viii. 22 and again on 7. xi. 22.

(*e*) Applying crushed bluestone to soil around trees, using in one application to medium-sized trees 2, 3, 4, 5, 6 and 7 lbs. per tree, 30. viii. 22.

(*f*) Dissolving 1 oz. of bluestone in one gallon of water and introducing one-sixth of the gallon into the sap of a tree by placing a cut root in a bottle of the solution: 30. viii. 22.

(*g*) Dissolving $\frac{1}{2}$ oz. of bluestone in one gallon of water and introducing one-sixth of the gallon into the sap of a tree by placing a cut root in a bottle of the solution: 30. viii. 22.

(*h*) Dissolving ¼ oz. of bluestone in one gallon of water and intro-ducing one-sixth of the gallon into the sap of a tree by placing a cut root in a bottle of the solution: 30. viii. 22.

(*i*) Dissolving ⅛ oz. of bluestone in one gallon of water and intro-ducing one-sixth of the gallon into the sap of a tree by placing a cut root in a bottle of the solution: 30. viii. 22.

I kept the trees under observation during 1923 and though some re-covery was noticeable both in untreated as well as treated trees, the marked improvement all round which occurred in the well-kept orchard did not take place in this one, and at the time of writing badly affected trees are still in evidence.

In June 1923 the condition of the trees was as follows:

(*a*), (*b*), (*c*), (*d*). All sprayed trees bore fruit, light, medium and good crops being noted on different trees, and practically all of this fruit was free from disease.

There was no difference in the trees sprayed twice compared with those sprayed once, nor in the stronger when compared with the weaker strengths of Bordeaux mixture.

The spraying had no harmful effect on the young growths.

The outstanding features connected with the spraying were these: the trees which I kept for controls did not mature a single sound fruit, but they made better growth than those which were sprayed.

(*e*) On 9. xi. 22 the soil-treated trees showed a marked superiority over the control and sprayed trees in the colour of the foliage, which was a healthy green with hardly any variegated leaves so common on affected trees. This effect was still apparent in June 1923. All the trees made more new growth than they had during the three previous years, but as the control trees also made a corresponding increase, no con-clusions can be drawn favouring the treatment in this particular. Hardly any fruits were borne either on treated or control trees. The growth on the treated trees was practically uniform throughout, and the improved colour was the same in the case of the tree which received 2 lbs. of bluestone as in that which received 7 lbs.

(*f*) A few weeks after this tree was treated I thought nothing could save it from death. The leaves dropped, all the young shoots at the top of the tree died back, the bark split open the full height of the stem and extended 18 inches up the main limbs. The crack in its widest part on the stem measured 2¼ inches. However, in June 1923 this tree showed signs of making a remarkable recovery. The crack had commenced to heal and new growth had taken place all over the tree.

(*g*) A few leaves dropped and some of the young shoots were killed. The bark split open the full height of the stem and extended about 12 inches up one of the small branches. The crack in its widest part measured 1½ inches. In June 1923 the condition of the tree was much better than before treatment: very little fruit was borne but the foliage was of a healthy colour, sound new growth was being made, and the crack had commenced to heal.

(*h*) Very slight injury was caused by treatment to the leaves and young shoots, but the bark on the stem cracked ½ inch in width. In June 1923 the crack had commenced to heal, the tree was making growth and carrying a good crop of perfectly clean fruit.

(*i*) Practically no damage was done to the foliage or young shoots by treatment, and the bark on the stem did not crack. The most noticeable thing about this tree in June 1923 was that on the portion above the root on which the bottle was placed several dozen perfectly sound oranges had matured while on the remaining portion there was not one single specimen. The tree made good growth throughout.

In September 1923 I continued the spraying experiments but did not follow up with bluestone in the soil or direct through the roots, as I wished first to see what those treated in that way would be like after another season.

The spraying results proved negative on this occasion. Treated and control trees alike failed to set fruit and the growth and affection by exanthema is now, March 1924, about the same in both cases. The failure to crop may have been caused by other agencies than disease, for this is a year of only light to medium crops of oranges throughout the State, but at the same time I may mention that other trees in the same orchard are fruiting.

The trees which received soil treatment in 1922 are still making good growth and the foliage is still good, but it does not now stand out ahead of the untreated trees in the same way that it did in 1923. The trees are bearing a few fruits only, but these are perfectly clean. As the control trees have not borne, no comparison can be made in this particular.

The trees treated through the roots direct are interesting:

(*f*) This tree has exceeded all expectations, and is now nearly free from every trace of exanthema. The young growth is strong and healthy, without gum pockets, bark fissures, or stained or abnormally large leaves. The tree is carrying 4½ dozen perfectly sound fruits and the crack is rapidly healing over.

(*g*) This tree has made excellent growth and is carrying a medium crop of perfectly clean fruit. There are no bark fissures in the young wood and only slight leaf staining. The foliage is normal in size. The crack in the stem is healing well.

(*h*) Very light crop of fruit which is quite clean. Foliage not as good in colour as (*f*) and (*g*), and a good many variegated leaves noticeable. The tree has made very little new growth, but there are no excrescences on young wood and no die-back at tips. The crack in the stem is healing well.

(*i*) This tree is only carrying 14 fruits, but again like last year these are on the portion of the tree above the root around which the bottle was placed. About 2 ft. 6 ins. of new growth has been made, but much of it has the typical "S" shape of exanthema-affected trees, and gum pockets are numerous.

With the varying results following on the treatment outlined above it is impossible to say that anything has been definitely proved, but there are grounds for believing that bluestone applied in some form to the tree will serve to control exanthema of orange trees when the best manner of making the application has been solved.

Mr P. S. Jivanna Rao. The Virus Theory in relation to Spike Disease in Sandal

In a communication that appeared over the name of F. T. Brooks on p. 955 of *Nature* for 29th December 1923 a summary is given of the discussion that took place on Virus Diseases of Plants during the meeting of the British Association at Liverpool between the Sections of Botany and Agriculture. Among the contributions referred to in the summary is one by Whitehead in which the impression is sought to be conveyed that spike disease in sandal is definitely established as one of the virus diseases and that in this disease there is neither abnormal accumulation of starch nor phloem necrosis.

As the above statement in respect of a disease which, so far as is known, is prevalent only in India is calculated to influence investigators on virus diseases in general, the present writer, as one who has had some direct acquaintance with this disease, desires to point out that the idea underlying the above statement is contrary to facts that have been recorded on the subject.

In the first place, whatever may be the opinion with regard to phloem necrosis, it is certainly not correct to say that there is no abnormal starch accumulation in the leaves. For it will be clear from the writings

of Barber (1902), Butler (1903), Coleman (1917), the present author
(1921) and others on the subject that leaves of spiked plants are undoubtedly characterised by abundance of starch. Misconception on this
point has probably arisen from the fact that microscopic preparations
of single leaves have been relied upon, whereas the starchy nature will
be more evident if entire twigs are bleached by means of hot alcohol
and treated with iodine. If tested in this manner it will be seen that
the leaves may be starchy in different ways and some of them devoid
of starch either at the top or bottom of the twig. Starch may also be
absent in the final stages of spike and, as the writer has shown in a
previous paper, it may be found in abnormal quantities even in leaves
which morphologically have not developed spike.

Secondly, with regard to the virus theory itself, this rests largely on
the grafting experiments conducted by Coleman, the conclusions drawn
from which are open to objection for the following reasons:

(1) The obligate nature of parasitism of sandal being no longer
doubted, the foremost requirement of the plant is the presence of suitable
hosts, whereas in none of Coleman's experiments is there evidence of the
parasite being fully supplied with hosts.

(2) According to Coleman no sign of the disease appeared in cases
where the scion did not grow on the stock. This really shows that
grafting with spiked scions does not succeed with all stocks but only
with those which are physiologically predisposed to spike. For it is
well known that grafting is itself a phase of parasitism and the success
of such a graft is possible only when the stock and scion are physiologically fitted for a nutritive relationship that is akin to parasitism. One
such condition is that the scion (acting as a parasite) must possess a higher
osmotic pressure than the stock (serving as a host), whereas in spiked
plants the low ash content indicates (if it can indicate) a low osmotic
pressure as compared with the normal plant, and union is, therefore,
possible only between a spiked scion and a stock which is disposed to
spike—not otherwise.

(3) Coleman admits the failure to produce the disease by injecting
extract of spiked plants into healthy ones. It is further known that in
sandal areas plants grow in a perfectly healthy condition side by side
with spiked individuals.

(4) As Coleman says, two of the most striking symptoms are the
accumulation of starch in leaves and branches and the death of the
haustoria and root ends. Regarding starch accumulation there is
experimental evidence that this can be induced by interfering with

the root system of the parasite and its host (Hole, and present author). It is also easy to see that the decadence of the haustoria is bound to arise when the hosts die, being short-lived, or the particular roots attacked are killed by the parasite, which thus destroys the host and suffers in consequence for want of one. (i) Haustoria in a dried-up condition were thus noticed on the roots of *Achyranthes aspera, Cynodon dactylon* and some young roots of *Acacia arabica.* (ii) An Eupatorium hedge was partly destroyed by sandal. (iii) Haustorial scars were found on the roots of *Commiphora Berryi*, whose roots have a tendency to peel. The above represent different types of hosts and conditions which incapacitate the haustoria and bring about their death.

In conclusion the writer wishes to state that an experiment is in progress to test the theory that spike in sandal is due to insufficiency of water to the plant owing to the death or removal of hosts or to the hosts being otherwise unsuitable; the results of this experiment will be communicated in a separate paper.

Papers cited

BARBER, C. A. "Report on Spike Disease in Sandalwood Trees in Coorg." *Ind. For.* vol. XXIX, 1903, pp. 21–31.

BUTLER, E. J. "Report on Spike Disease in Sandalwood Trees." *Ind. For.* vol. XXIX, 1903, App. Ser. pp. 1–11.

COLEMAN, L. C. "Spike Disease of Sandal." *Mysore Agric. Dept. Bull.* 1917.

—— "The Transmission of Sandal Spike." *Ind. For.* vol. XLIX, 1923, pp. 6–9.

HOLE, R. S. "Spike Disease of Sandal." *Ind. For.* vol. XLIV, 1918, p. 327.

JIVANNA RAO, P. S. "The Physiological Anatomy of Spiked Leaf in Sandal." *Ind. For.* vol. XLVII, 1921, pp. 351–360.

—— "The Cause of Spike in Sandal." Indian Science Congress, Bangalore, 1924 (unpublished).

Prof. F. W. Oliver, F.R.S. *Spartina Townsendii* and its Applications

(*Abstract*)

The paper was of the nature of a descriptive report on the position reached by *Spartina Townsendii* as an element in salt marsh vegetation. The gradual, almost total, occupation of the Southampton area (where it was first recorded in 1870) and the remarkable spread and invasion by this plant of other areas on the south coast, and latterly in France, were traced. The peculiarities of its establishment on invaded areas were considered, the part played by seeds and the expansion and massing of the plants established therefrom. Reference was made to experimental cultivations at several spots on the English coast, and the view was expressed that it had yet to be proved that this plant was able to maintain unimpaired its vigour and power of spread under the cooler conditions of the east coast. Alluding to its economic uses, primarily *Spartina Townsendii* should be regarded as an incomparable agent in the reclamation of soft, muddy estuarine ground. Finally, an epitome was given of the current (divergent) views of systematists as to the status of *Spartina Townsendii*, especially those of Stapf, Hitchcock and Chevalier.

Mr J. M. F. Drummond. THE FORMATION OF HERBARIA OF CROP PLANTS

In regard to questions of nomenclature and classification, crop-breeders and other students of cultivated plants are at a great disadvantage in comparison with the regular systematic botanist. Apart from a very few works—such as Percival's *Wheat Plant*—floras and monographs dealing with crop plants are lacking. Herbaria of cultivated "varieties" are non-existent or fragmentary. It is suggested that the time has come when a serious effort should be made to remedy these defects. The problem is large and complex, and its solution calls for a carefully organised and co-ordinated scheme. To be really effective, the organisation should be international; but the setting up of a British Empire Committee for the purpose might prove more readily practicable in the first instance and would provide a sound foundation upon which to erect the more ambitious edifice.

The working out of the details of such a scheme would appear to be a matter mainly for the professional taxonomists, who have the requisite practical experience of the herbarium side of the work. The following tentative suggestions are put forward in order to invite discussion on the broad outlines of policy.

1. A British Empire Committee for Agricultural and Horticultural Taxonomy should be constituted, composed of Taxonomists, Crop-Breeders and other suitable persons engaged in Botanical, Agricultural or Horticultural research.

2. The crops of the Empire should be grouped under certain convenient heads, *e.g.* Cereals, Tuber Crops (Potato, etc.), Root Crops, Herbage Crops, Fibre Crops, etc., with a Sub-Committee to deal with each group.

3. Herbaria should be formed for one or more crops at appropriate centres. Each Herbarium should be supplemented by a Living Collection, maintained preferably at the same or at an adjacent centre.

4. It should be incumbent on all Botanical Gardens or Institutes and Agricultural and Horticultural Research Stations or Departments to supply duly authenticated herbarium type-specimens (or living material) to the proper herbarium centres (or Living Museum). In return, the centres should issue catalogues of specimens in their possession and be prepared to supply small quantities of seed, etc., for research purposes on request, and to undertake identification of doubtful specimens. The data and specimens accumulated in this way should in due course provide the material for monographs dealing with the principal crop plants.

5. A variety put into circulation by a Research Station should not be recognised as new except with the sanction of the Committee, which should draw up a set of rules for the purpose. It would seem to be essential, as one condition of recognition, that specimens of the new variety should be submitted to the proper centre accompanied by a brief but sufficient description clearly stating the differentiae from existing varieties. The official Plant-Registration Stations should also be asked not to accord recognition to varieties submitted as new by private breeders except with the concurrence of the British Empire Committee.

Until it is seen how far the scheme meets with general approval in principle, it would be futile to discuss the details at length. A few remarks may, however, be added in explanation of the five draft suggestions which have just been formulated.

1. The first matters to be adjusted are obviously:

(a) The composition of the Committee.

(b) The definition of its powers and duties.

The most suitable body to deal with these questions would no doubt be the Executive Committee of the Imperial Botanical Conference. It is important that the Taxonomic Committee proposed should include delegates from all the classes of workers directly interested, and, in particular, representatives of the following Departments and Institutions: the Ministry of Agriculture and Fisheries and corresponding Departments in the Dominions and Colonies; Botanic Gardens and Museums; University Departments of Botany; Agricultural Colleges; Plant-Breeding Stations.

2. It is unlikely that it will be feasible to deal with all the crop plants of the Empire at once. A beginning might be made with an intensive study of one or more of the groups named above. All the groups should, however, be defined, and Sub-Committees appointed to deal with them at the outset.

3. Too much stress cannot be laid on the fact that it is useless to attempt the classification of crop plants with the aid of dried specimens alone. A living collection is an indispensable adjunct to a herbarium of crop plants; such a collection should be located at or near the centre selected for each crop or crop-group. For Oats, for example, a suitable arrangement would be the location of the herbarium at the Royal Botanic Garden, Edinburgh, and the maintenance of the associated living collection at the Scottish Plant-Breeding Station, Corstorphine (five miles from the Botanic Garden), which specialises in that crop.

4. Among the ultimate aims of the proposed scheme one of the most important would be the preparation of authoritative monographs dealing with the leading crops of the Empire. There is an urgent and widespread demand for such monographs; but it is difficult to see how that demand is to be met except along the lines of some such co-ordinated scheme as that advocated in the present paper.

5. There can be no hope of finality in nomenclature unless stringent rules governing the issue of new varieties of crop plants are framed and loyally observed. There should be no difficulty in gaining the adherence of scientific plant-breeders to such regulations, which will be favourable to their interests, and in the framing of which they will have a share. Although it is scarcely possible to maintain a strict control over the productions of private breeders, co-operation with the Taxonomic Committee on the part of the official Plant-Registration Stations (National Institute of Agricultural Botany, etc.) would provide a valuable check on careless or wilful multiplication of Synonyms.

That classification and nomenclature of "varieties" of crop plants are at present in a deplorably chaotic condition is very generally recognised; it is, no doubt, the obvious magnitude of the task of reducing this chaos to even a semblance of order which accounts for the fact, that hitherto hardly any serious attempt has been made to deal with the problem. It may be safely assumed that plant-breeders and other workers directly interested in the study of crop plants will be in full sympathy with the aims of the scheme proposed; and it is probable that they will welcome the suggestion of Imperial co-ordination and be prepared to assent to the general principles involved in the policy outlined above, though their opinions may differ in regard to the points of detail which have been indicated rather than discussed in the preceding paragraphs. Criticism of the scheme, in principle and in detail, may be expected on the part of the professional taxonomist, who may quite reasonably object to any proposal which seems to involve an addition to his already onerous duties. In particular he is likely to take exception to a branch of classification which necessarily works in general with units other than Linneons. In this connection it may be pointed out that the bulk of the necessary investigation will have to be carried out on the living collections, which will ordinarily be in the charge of agricultural experts. The smooth working of the scheme, however, and its ultimate success, must largely depend on the willing co-operation of the technical expert in classification and nomenclature, *i.e.* the professional systematic botanist. The taxonomist's faith in the value of the Linneon for the practical classification of plants growing in a state of nature need not be affected in the least by his adhesion to a scheme of the kind proposed. On the other hand, it would seem, at any rate to a geneticist, that participation by the taxonomist in the study of those more elementary units (Jordanons, etc.) which constitute the material of the investigator of cultivated crops should hasten rather than retard the realisation of what is after all the true ideal of taxonomy, namely, the elucidation of the evolutionary scheme of the plant kingdom.

In conclusion it must be frankly admitted that there are grave difficulties of finance and staffing in the way of the scheme. But by means of careful co-ordination and division of labour it should be possible to fit much of the work entailed into the ordinary routine operations of crop-breeding stations, Botanic Gardens and similar Institutions. Until the stage of monograph publication is reached, the cost of the scheme should not be great. Little or no special equipment would be required, provided the "centres" are chosen with due regard to existing facilities; in particular there should be no need for initial capital outlay. As the scheme develops, some expansion of the staff of Institutions selected as "centres" will become necessary. In view of its obvious economic importance, a scheme of co-ordinated taxonomic investigation of the crop plants of the Empire is an object thoroughly deserving of support from public funds.

Dr Arthur S. Horne. FUNGAL DISEASES OF STORED APPLES

The problem of the diseases affecting stored apples had received comparatively little attention although a serious annual loss through disease was experienced in this country when this investigation was started about ten years ago. At the time *Sclerotinia fructigena* (Pers.) Schrot. was recognised as the most commonly occurring parasite of apples both in the orchard and store. Occasionally other fungi were mentioned as

causing, or probably causing, disease. More attention has been given
to this subject in the United States of America where several fungi,
notably, *Glomerella cingulata* (Stoneman) Sp. and von S., *Physalospora
cydoneae* Arnaud, *Penicillium expansum* (L.K.), *Cephalothecium roseum*
(Fris.) Cda., *Volutella fructi* Stevens and Hall, *Sclerotinia cinerea* (Bon)
Schrot., *Sclerotinia fructigena* and *Phytophthora cactorum* (Lib. and Cohn)
Schrot. have been recognised as causing considerable loss to growers.
The cause of fruit spot in apples, a trouble commencing in the orchard
and developing during storage, had been the subject of considerable
investigation. Thus Charles E. Brooks in 1908 found by cultural methods
that *Cylindrosporium pomi* Brooks (at a later date identified as *Phoma
pomi* Passer) was capable of causing spotting in the Baldwin variety.
Besides the fungi found during enquiries more directly concerned with
the spotting of apples, others generally held responsible for rotting or
twig canker, have been recorded as causing spotting, for example *Phoma
mali* Schultz et Saec. (Charles E. Lewis), *Physalospora cydoneae* (Clinton),
Phyllosticta solitaria (John W. Roberts) and *Glomerella cingulata* (Dastur).
Other workers, however, as a result of their experiments inclined to the
view that the spotting was due to physiological causes, thus W. M. Scott,
although obtaining *Cylindrosporium pomi* Brooks and species of *Alternaria* from the Jonathan apple, concludes that the cause of the disease
is unknown.

Various opinions were advanced to explain the occurrence of spotting
in this country. By some observers, it was thought to be due to *Cylindrosporium pomi* Brooks, by others, caused by storage conditions, whilst a
few considered it an effect following upon insect punctures. The study
of the fruit-spot aspect of the apple disease problem was started at the
Royal Horticultural Society's Gardens, Wisley, where an opportunity
presented itself of following the course of spotting and ultimate rot
development in over one hundred varieties of apples. The result of this
work made it perfectly clear that the spots belonged to two categories—
those caused by fungal organisms, and those due to physiological causes,
e.g. bitter pit, etc. Special attention was given to those of the former
category and a general account of the results obtained was published
in 1920(5). A superficial examination of the spotted areas frequently
revealed the presence of black markings of various kinds—fibrous,
dotted, raised pustules, etc.—and a microscopical study of the sub-
cuticular tissues almost invariably revealed the presence of mycelium,
but reproductive bodies were rarely present, the only fungi which were
identified by these methods being species of *Fusarium*, *Valsa* and

Polyopeus purpureus Horne. Accordingly resort was made to cultural methods. Cubical portions of tissue were placed on the surface of apple extract agar in slant tubes and cultures thus made were examined from time to time for the presence of reproductive organs. This medium did not favour spore production. Sporing occurred in only a few cases, notably with species of *Cladosporium, Penicillium* and *Alternaria (A. grossularieae)*; a *Stemphylium* subsequently proved to be the conidial condition of a new species of *Pleospora—P. pomorum* Horne; *Leptosphaeria vagabunda* Sacc., *Coryneum foliicolum* Fuck., and *Myxosporium mali* Bresadola.

In 1918 the unidentified fungi were transferred to potato mush agar. This medium favoured the formation of reproductive bodies. Difficulties of identification now arose especially with fungi belonging to the Sphaeropsidaceae. The diagnostic characters of these fungi have in the past been based on the structures exhibited by specimens developed in or on the tissues of a host plant and not on an artificial medium. In the latter case criteria such as the dimensions and colour of the pycnidia and spores as in the case of *Phoma, Coniothyrium,* etc. are of little value, since the variations in pycnidial and spore characters shown when any specific form is grown in different media would include within their range several of Saccardo's species. In order to attain the object aimed at in the investigation it was necessary to give names to some of the forms obtained. It was thought probable that some of the forms had been described previously as occurring on hosts other than the apple, but it was not possible at the time to identify them with any known organisms. It was found convenient to establish a new genus of Sphaeropsidaceae, *e.g. Polyopeus*—a genus differing from a typical *Phoma*, such as *Phoma hippocastani*, in producing pycnidia with multiple necks when grown in certain media. This genus included four species— *P. purpureus, P. pomi, P. recurvatus* and *P. aureus*. Descriptions of these fungi were published in the *Journal of Botany*(4). Other fungi described at the same time were: *Pleospora pomorum, Alternaria pomicola, Coniothyrium cydoneae* var. *mali, Coniothyrium convolutum, Fuckelia botryoidea* and *Sclerotium stellatum*.

In 1918, Brown, working at the Imperial College of Science for the Food Investigation Board of the Department of Scientific and Industrial Research, isolated several fungi from diseased apples stored under the usual conditions, the most noteworthy being a species of *Botrytis*, *Physalospora cydoneae, Coryneum foliicolum, Alternaria grossularieae, Polyopeus purpureus, Fuckelia botryoidea* and a species of *Fusarium*, since provisionally named *Fusarium Blackmani*, which has been used in investigations into the physiology of fungi carried out by him from time to time.

From 1919 onwards the investigation was carried out for the Food Investigation Board, Department of Scientific and Industrial Research, and special attention was given to incidence of disease in apples stored under low temperature conditions. Several fungi were isolated from cold-stored apples including many of those previously isolated. It was found that many fungal species isolated from apples collected in the orchard and fruit-room, from which the apples destined for cold-storage purposes were obtained, were present also in apples contracting disease under low temperature conditions.

The additional fungi obtained were: *Cytosporina ludibunda*, *Rhizopus nigricans* and *circinans*, *Cephalothecium roseum*, *Fusarium sporotrichioides*, *Ascochyta pirina*, *Asteromella* sp., *Actinomyces* sp., *Oospora* sp., *Podospora* sp. and *Eidamia viridescens* Horne[7].

It is interesting to note that with the exception of *Physalospora cydoneae*, *Cephalothecium roseum*, *Sclerotinia fructigena* and *Penicillium expansum* the fungi obtained are not the same as those described as causing spotting or rotting in America. According to Wormald, *Sclerotinia cinerea*, one of the American rot-producing fungi, is responsible for shoot die-back in certain varieties of apple in England. On the other hand the American fruit-spot fungus, *Phoma pomi* Passer, has not yet been recorded in this country. *Physalospora cydoneae* (*Sphaeropsis malorum*) is not an introduced fungus, having been first recorded for the apple host by Berkeley in 1836. This fungus does not appear to be of common occurrence in this country and causes an inappreciable loss among stored apples. *Leptosphaeria vagabunda* is recorded as prevalent in America on twigs of different trees, and *Coryneum foliicolum* as causing cankers on twigs and branches of American apple varieties.

During the earlier part of the work an effort was made to enumerate the fungi obtained from the numerous varieties of apple under observation, but the effort was abandoned after a time owing to the labour involved in identifying the fungi, and ultimately only named fungi, that is, comparatively few of the total number isolated, were retained in culture. Attention has been concentrated on ascertaining the relative importance of the fungal factor from the cold-storage point of view by determining whether, and to what extent, the fungi are pathogenic. It was also thought that important results might accrue from a study of parasitism in relation to the facts brought to light in connection with the investigations, proceeding concurrently for the Food Investigation Board, at the Imperial College by Dr Haynes and Miss Judd into the chemical changes which take place in the apple tissues during storage at low temperatures.

In the paper on spotting in apples (5) the writer pointed out the almost universal occurrence of spots at the lenticel of the apple. Sometimes over one hundred spots occur in a single apple, each with a centrally situated lenticel. These observations are in agreement with those made by Charles E. Brooks in America, and Barker and others in England. It seemed, therefore, fairly safe to assume that the fungi had obtained an entry *via* the lenticel. It was also clear that some fungi, notably those causing skin troubles, once an entry had been effected, might be unable to make further progress. It was decided, therefore, to test the parasitism of the fungi by introducing the inoculant into the tissues and leave to a later date the problem of the actual penetration and the conditions regulating it.

Parasitism

The first experiments connected with parasitism were made at Wisley in 1917 (5) using *Pleospora pomorum*. With Lane's Prince Albert and Newton Wonder, spotting followed by rotting and the development of sterile perithecia in the diseased tissues was obtained under laboratory conditions. Several apples were inoculated with this fungus while still on the tree. Of 27 varieties of apple inoculated, spots were not formed in four of the firm-fleshed russet type (Court Pendu Plat, Norfolk Beaufin, Ribston Pippin and Cellini). Small spots were formed which remained arrested for a period in seven varieties (Alfriston, Allen's Everlasting, Beauty of Kent, Calville Boisbunel, Cockle's Pippin, King of the Pippins and Lord Derby). The remaining sixteen varieties developed spotting more or less rapidly and rotting ensued (Allington Pippin, Belle de Pontoise, Bismarck, Bramley's Seedling, Cardinal, Crawley Beauty, Duchess Favourite, Early Victoria, Grenadier, Keswick Codling, Lane's Prince Albert, Pott's Seedling, Red Astrachan, Rival, Royal Jubilee and Wealthy). *Pleospora* was isolated from the diseased tissues in the case of Allington Pippin, Rival, Wealthy, Royal Jubilee and Cardinal. The rot which ensued in the case of Bramley's Seedling and Lane's Prince Albert was probably not due to *Pleospora*. A sound Cardinal apple placed in contact with a diseased inoculated specimen rotted within a month. In 1920, apples (variety unknown) affected with rot caused by *Pleospora pomorum* were found on trees in an orchard in

Surrey, and an illustration from a photograph of one of them showing an extensive development of sterile perithecia, was published in the *Gardeners' Chronicle*(6).

In 1920 rots were obtained as a result of placing inoculant on the cut surface of certain apple varieties including Cox's Orange Pippin, using *Botrytis* sp., *Cytosporina ludibunda, Pleospora pomorum* and *Polyopeus purpureus*.

In order to investigate the circumstances influencing the parasitic behaviour of fungi under low temperature conditions it became necessary to carry out inoculations on a large scale and a method was devised with a view to standardisation of operations and rapidity of manipulation. This method was described recently in the *Annals of Botany*(3). The inoculations to the number of some thousands were undertaken by the same person (Mrs Cartwright), and less than 3 % of the inoculated apples were subsequently contaminated with organisms accidentally introduced by the operator. Great care was exercised in the picking and selection of the apples used, they were gently rubbed with cotton wool soaked in absolute alcohol after inoculation, wrapped in sterile grease proof paper, and together with a certain number of controls dispatched to the Low Temperature Station, Cambridge, in boxes specially constructed for the protection of the fruit during transit. The apples were inspected at intervals during storage and returned to London for a final examination, where in all doubtful cases attempts were made to isolate the causal organism.

As a result of this work the following additional fungi have been found capable of causing rotting of apples both under ordinary storage and low temperature conditions (1° C.): *Cytosporina ludibunda, Fusarium Blackmani, mali, sporotrichioides, acuminatum* Ellis and Ev., and certain unnamed *Fusaria; Polyopeus purpureus, pomi* and *aureus; Cephalothecium roseum, Eidamia viridescens, Coniothyrium convolutum* and *Aschochyta pirina*.

All these fungi with the exception of *Fusarium acuminatum* occur in this country. Inoculation of apples (Cox's Orange Pippin) with *Sclerotium stellatum, Physalospora cydoneae* and certain forms of *Cladosporium* failed (1—3° C.). Brown, in 1922(1), showed that the minimum growth temperature for *Physalospora cydoneae* (*Sphaeropsis malorum*) lies between 3° and 5° C. Hence the inability of this fungus to cause rotting under these temperature conditions is in accordance with expectation.

Circumstances affecting the parasitic behaviour of the fungi at low temperatures

It was observed when studying the incidence of disease among apples (Worcester Pearmain) kept under low temperature conditions, in 1920, that the pathogenic fungus flora changed during the period of storage. Thus in November and December, *Sclerotinia fructigena*, *Penicillium expansum*, *P. glaucum* and *Botrytis* sp. were prevalent; in January, various species of *Fusarium*; and later, *Pleospora pomorum* and species of *Polyopeus*. In fact, there appeared to be a succession of parasites; that successional attack occurs has now been demonstrated experimentally. In October, 1921, the variety Cox's Orange Pippin was inoculated with several fungi. The apples were stored at 1° C. With *Botrytis* sp., *Penicillium glaucum* and *Fusarium Blackmani* the maximum attack took place before December 19th, with *Fusarium* sp. (No. 85), *Cytosporina ludibunda* and *Polyopeus aureus* the attack developed during January, 1922; during March it became evident in the case of *Polyopeus purpureus verus* and *Coniothyrium convolutum*: whilst disease due to *Pleospora pomorum* did not develop to any extent until the middle of April. With *Fusarium sporotrichioides* and *Polyopeus pomi torpidus* the attack developed fairly evenly throughout the season.

An effort is being made to ascertain the factors responsible for successional attack. It is thought to be partly due to chemical changes in the cell. Such changes are known to take place during the storage period, notably in relation to hydrogen ion concentration, sugar content, and pectin.

From evidence obtained, parasitic virulence appears to vary in different years. Thus an experiment where *Fusarium Blackmani* was the parasite concerned, was carried out in 1921 and 1922 on Cox's Orange Pippin obtained from the same orchard at Burwell, Cambridgeshire in both seasons. The same strain of *Fusarium Blackmani* was employed and the apples were kept under similar hygrometric and temperature conditions (1° C.). In 1921 40 % of the inoculated apples were severely attacked when examined 39 days after placing in the store and all were rotten after an interval of 90 days had elapsed. In 1922 the disease present was of a comparatively trivial character when examinations were made 165 days (apples gathered early) and 143 days (apples gathered late) after inoculation. Since the experimental conditions were similar in the two seasons and there is no reason to suspect from other experimental results that any marked change in parasitic virulence through culturing had taken place, the difference in parasitism seems to be due to the influence of prevailing seasonal conditions. Indeed 1921 was a hot and dry season whereas 1922 was a wet season and the apples ripened relatively late. In the former year the normal picking took place on September 26th; in the latter, a month later (Oct. 26th). Reference to the analyses made by Miss Judd reveals certain chemical differences between the 1921 and 1922 Cox's, but it is not yet known whether these differences have any direct bearing on the degree of fungal activity at low temperatures.

In 1923 experiments were started with a view to investigating the parasitism of the strains of *Fusarium Blackmani*. As previously stated, this fungus was originally obtained by Dr Brown in 1918, later, five other isolations were made by the writer from different varieties of apple (Cox's Orange Pippin, Allington Pippin, Sweet Alford). From their general growth appearances in potato mush agar the six forms appeared to be identical. When grown in potato extract agar containing 8 % glucose the colour reactions to the medium were found to be inconstant and concurrently Brown found marked differences in the rate of growth when these forms were grown in carefully standardised media (plate cultures). A difference in the septation-mode of the spores was also apparent, with some forms the mode was three, in others, five. By selecting individual spores several strains distinct from the original made their appearance, and later, on repeatedly plating out all the strains obtained in various media, the phenomenon of sectoring appeared, that is to say, sectors of the plate were occupied by a form separable from and differing from the parent in certain details, for example, intensity of sporing, mycelial development, growth-rate, septation, colour, etc. The new races were separated from the parent wherever they occurred and at the present time about 40 new strains or saltants are in culture. A general account relating to these matters has recently appeared in the *Annals of Botany*(2). In 1923, eight saltants were chosen as representative of the different growth types at present recognised, *e.g.* mycelial, sporodochial, pionnotal, etc. Duplicate series of 80 apples (Cox's Orange Pippin) were inoculated and placed at 3° and 12° C. respectively. As a result the parasitism of saltants of the mycelial type was of a more virulent character than that exhibited by the others at both temperatures.

Investigations are in progress into parasitism in relation to the acidity of apple varieties. Series were chosen ranging from apple varieties of the less acid type to those exhibiting a relatively high acidity. Sometimes the series were chosen in terms of total acidity, at other times in terms of pH. In the case of *Botrytis* and *Cytosporina ludibunda* all the varieties were attacked when the inoculations were carried out under laboratory conditions in the autumn, but inoculations of the more acid types, *e.g.* Lane's Prince Albert and Bramley's Seedling with *Pleospora pomorum* and *Polyopeus purpureus* failed. That the difference in behaviour bears some relation to the composition of the apple sap is clear from the cultural experiments carried out at the same time, using extract obtained from the same batches of apples used for inoculation purposes and of which the acidity and sugar content had been determined previously. In the case of *Pleospora pomorum* a striking correlation was found between growth in the extract (0°—5° C.), the parasitisation of the tissues and the acid content of the varieties used. These varieties arranged in order of increased acidity were: Cox's Orange Pippin (5·75 acid in terms of c.c. N/10 sodium thiosulphate), King of the Pippins (8·55), Allington Pippin (11·1), Lane's Prince Albert obtained from Cambridge (14·45) and Lane's Prince Albert from Guildford (15·3). The diameter of the colonies appearing in the extract from Cox's Orange Pippin averaged from 2·5 cm. to less than 3 mm. in the case of Lane's Prince Albert (Cambridge).

In 1921, several apple varieties were inoculated with *Pleospora pomorum* and kept at 20° C. including the following arranged in order of descending hydrogen ion concentration: Bramley's Seedling (pH 3-3·1), Allington Pippin (pH 3·05-3·35), Dymock Red (pH 3·9-4) and Sweet Alford (pH 4·1-4·6). All the apples were attacked with the exception of Bramley's Seedling.

In addition to *Pleospora pomorum*, *Fusarium Blackmani*, which exhibits somewhat similar general reaction towards hydrogen ion concentration, is also incapable of attacking Bramley's Seedling during the autumn under normal conditions. Both fungi are able, however, to invade the tissues late in the season when the hydrogen ion concentration, as evident from the data supplied by Dr Haynes, has fallen. Thus with decreasing acidity the Bramley's Seedling apple appears to lose its relative immunity to fungal disease. It is of course realised that although the altered pH may remove a disability, an acceleration of fungal activity may be due partly to other chemical changes of a more or less complex character, whilst similar or other factors may inhibit or retard fungal activity in other varieties of apple, *e.g.* Cox's Orange Pippin, etc., which should be more or less susceptible from the point of view of pH.

The relation of parasitism to the time of picking was investigated in

1922, using Cox's Orange Pippin obtained from Burwell, Cambridgeshire. The apples were gathered on two occasions—October 2nd and 26th (normal picking). One hundred selected apples from each lot were inoculated with *Pleospora pomorum* and placed at 1° C. After an interval of 168 days, 69 % of the apples picked on October 2nd were diseased whereas after a slightly shorter period (155 days) only 20 % of those picked on the later date manifested fungal attack.

Investigations are in progress with a view to discovering whether and to what extent parasitism is influenced by the conditions under which a particular apple variety is cultivated. Bramley's Seedling apples were obtained from six different localities chosen from the nature of the soil and sub-stratum, *e.g.* from the Old Red Sandstone, silt, fenland, etc. One hundred apples were inoculated in each case, and the experiments have been carried out annually since 1922. The fungus used in the first instance was *Cytosporina ludibunda*, but recently *Pleospora pomorum* and *Fusarium Blackmani* have been chosen in addition.

As a result the fen and silt Bramley's showed a marked difference in the amount of disease caused by *Cytosporina* in two consecutive years. In 1922, the disease in the "silts" was of a comparatively trivial character whereas several fen apples were severely attacked when examined 167 (silt) and 161 (fen) days after inoculation (1° C.). In 1923, nearly all the fen Bramley's were attacked, with 62 % completely rotten, whilst only 3 % of the silts showed progressive fungal invasion when examined at the end of March, 1924 (3° C.). Certain chemical differences between the fen and silt Bramley's have been observed by Miss Judd, but it is not yet known whether they possess any significance when considered in relation to difference in the intensity of fungal attack.

Literature cited

(1) Brown, W. "On the germination and growth of Fungi at various temperatures and in various concentrations of Oxygen and Carbon Dioxide." *Annals of Botany*, xxxvi, p. 276, 1922.

(2) Brown, W. and Horne, A. S. "Studies of the Genus Fusarium. 1. General Account." *Annals of Botany*, xxxviii, p. 379, 1924.

(3) Granger, K. and Horne, A. S. "A Method of Inoculating the Apple." *Annals of Botany*, xxxviii, p. 213, 1924.

(4) Horne, A. S. "Diagnoses of Fungi from 'Spotted' Apples." *Journal of Botany*, lviii, p. 238, Oct. 1920.

(5) Horne, A. S. and E. Violet. "Mycological Studies. 1. On the 'Spotting' of Apples in Great Britain." *Annals of Applied Biology*, vii, p. 183, 1920.

(6) Horne, A. S. and E. V. "On 'Spotting' in Apples." *Gardeners' Chronicle*, p. 216, Oct. 30th, 1920.

(7) Horne, A. S. and Williamson, H. S. "The Morphology and Physiology of the Genus Eidamia." *Annals of Botany*, xxxvii, p. 393, 1923.

Mr E. J. Maskell. THE TECHNIQUE OF PLOT EXPERIMENTS[1]

From any point of view it would be desirable that the data for crop yield, under different experimental conditions, should be of great accuracy: for the purpose of a physiological attack upon the causal analysis of yield extreme accuracy is imperative. For here it is not merely a question of establishing the existence or non-existence of a favourable effect accompanying a specified change in the environment, but of estimating the actual magnitude of the effect. The yield-factor relationship is a case in point. In dealing with the problem of Plot Technique it will, therefore, be most valuable to consider a section of the work in which there seems at present to be the greatest chance of securing an advance in accuracy—that is, the problem of the spatial arrangement of the treatments which are being compared in any experiment. On account of the preponderating influence upon yield of climatic factors, the treatments to be compared must be tested at the same time and must, therefore, occupy different spaces: we have in consequence to meet the problem of differences in yield which arise not from differences in treatment but from differences in the positions of the experimental plots (differences due to chemical and physical soil characteristics, to animal or fungal attack, or other positional factors). The problem is twofold: to devise some control in the arrangement of the plots which shall reduce the magnitude of the error arising from this cause and to secure a valid estimate of the error arising from causes which remain beyond control, *i.e.* of the experimental error.

The nature of the problem is best illustrated by a consideration of the evidence available from "Uniformity Trials." Here we are dealing with land chosen for its apparent uniformity, sown to the same variety and receiving the same manurial and cultural treatment but subdivided at harvest into a large number of sub-plots of equal area, the yields of which have been measured. One of the earliest of such trials is that of Mercer and Hall[2], 1911, at Rothamsted, which gives data for wheat (500 plots) and for mangolds (200 plots). The most important result of this trial concerns the distribution in space of the differences in yield of the sub-plots.

Mercer and Hall found that when adjacent sub-plots were combined to make composite plots the standard deviation of the yields of the composite plots was considerably in excess of the "expected" value, the

---

[1] Paper read in the discussion on "The Physiology of Crop Yield" of which an abstract is given on p. 12.   [2] *Journ. Agr. Sci.* 1911, IV, p. 107.

latter calculated on the assumption that the groups are made up by random choice from among the sub-plots. They rightly ascribed this deviation from expectation to the fact that adjacent plots tended to have similar yields, a fact which is at once evident from any map of the distribution of the yields in space.

The complete formula for the standard deviation of a group of plots (two plots in the simplest case) is

$$\sigma^2 (A + B) = \sigma^2 A + \sigma^2 B + 2r_{AB} \sigma A \sigma B.$$

The third term, which involves the correlation (in yield) between adjacent plots, is a measure of the extent to which the actual distribution in space of the yield variations departs from a chance distribution and the existence of this element of regularity is clearly of great importance.

J. A. Harris[1] has devised a general formula for the estimation of this element in Uniformity Trials. The formula allows of the calculation of a "coefficient of heterogeneity" which expresses as a correlation coefficient the degree of association in yield of sub-plots. Harris has applied the formula to the data from a large number of Uniformity Trials and it is clear that heterogeneity is practically a universal feature of yield variation in the field. Certain physical and chemical soil characteristics also have been found to show the same feature. Table I illustrates a few of the results of Harris's analysis.

Table I

Correlation between adjacent plots

Data from	Crop	No. of plots	Coefficient of heterogeneity	
Mercer and Hall	Mangolds	200	Roots	+·346
			Leaves	+·466
	Wheat	500	Roots	+·336
			Leaves	+·483
Kiesselbach	Oats	200		+·495
Smith	Corn	120	1895	+·830
			1896	+·815
			1897	+·606
Coombs and Grantham	Rice	54		+·344
	Soil feature			
Harris	Moisture content	100	1st foot	+·317
			3rd foot	+·542
Waynick and Sharp	Carbon and nitrogen content	100	Carbon	+·417
			Nitrogen	+·498

[1] *Journ. Agr. Res.* 1920, XIX, p. 279.

It is the general existence of this heterogeneity in the field that makes the problem of plot arrangement so important. If soil variation were perfectly random the securing of any desired degree of accuracy would appear to be merely a question of adequate replication of the treatments which are being compared. A random distribution of soil variation as between the different treatments can of course be secured by a perfectly random arrangement of all the plots, but it is possible to improve upon this method.

The whole experimental area may be subdivided into as many units, of equal area, as there are to be replications (trials) of each treatment, and in each unit each treatment in the experiment made to occur once and once only. For the purpose of estimating the significance of the difference in yield between any pair of treatments (say A and B) we may now, following what is known as "Student's"[1] method, work not with the standard deviations of A and of B, but with the standard deviation of the actual differences $A - B$, between adjacent plots of A and of B. Each trial furnishes one of these differences. An indication of the increased accuracy obtainable in this way is afforded by Table II, which shows the variance of comparisons between adjacent plots and between groups selected at random. The part played in this result by the element of heterogeneity is evident from the complete formula for the variance of a comparison between two groups, which is

$$\sigma^2 (A - B) = \sigma^2 A + \sigma^2 B - 2r_{AB}\sigma A \sigma B.$$

With a random arrangement of plots the last term is of course negligible.

The simplest example of the application of this method is found in Beaven's half-drill strip method of comparing varieties of cereals. Two varieties only are compared at a time and the half-drill strips are arranged as

$$A\ \widetilde{A\ B}\ \widetilde{B\ A}\ \widetilde{A\ B}\ \widetilde{B\ A}\ \widetilde{A\ B}\ B \text{ etc.}$$

The variance of the comparison between A and B is calculated from the series of differences, $A - B$, between adjacent half-drill strips. In this way two varieties have been compared on 1 acre in 27 half-drill strips with a standard deviation of less than 1%[2].

For a comparison involving more than two varieties or treatments, the problem is not quite so simple. In particular it is not altogether satisfactory to base the estimate of the standard deviation of a com-

---

[1] "Student," *Biometrika*, 1908, vi, p. 1; 1917, xi, p. 414.
[2] "Student," *Biometrika*, 1923, xv, p. 271.

parison solely upon the data for the particular pair of treatments involved in any one comparison. A method of analysis devised by Dr Fisher[1] is now, however, fortunately available, which utilises the data from all the plots in the experiment for the calculation of an average variance of comparison. The method depends upon the fact that two of the components of the total variance of the experiment can be estimated directly from the data, namely, the variance due to "treatment" and that due to "trial." These variances are calculated from the series of means of the plots grouped (1) according to treatment, (2) according to position (*i.e.* each unit of the experiment, which involves one plot of each of the treatments, forms one group). The variance still remaining represents that which is beyond control and is assumed to be independent of trial and of treatment. It forms the basis therefore for the estimate of the experimental error.

This method is available for the analysis of "chessboard" arrangements, which can be subdivided into spatial units ("trials") as indicated above. Two samples of such chessboards are shown in Diagrams 1 and 2, and the increased accuracy obtained by the subdivision into trials is indicated by the variance figures for Beaven's chessboard in Table II.

Diagram 1. 10 × 3 *Chessboard*

A	F	C	H	E	K
B	G	D	J	B	G
C	H	E	K	A	F
D	J	B	G	D	J
E	K	A	F	C	H

Diagram 2. *Beaven's* 8 × 20 *Chessboard*

E	B	G	D	A	F	C	H	
D	A	F	C	H	E	B	G	
C	H	E	B	G	D	A	F	etc.
B	G	D	A	F	C	H	E	
A	F	C	H	E	B	G	D	

[1] "Student," *Biometrika*, 1923, xv, p. 283.

Table II

Variance of comparisons between groups of plots

Data	Groups and plots	Variance of Random groups	Adjacent pairs
Mercer and Hall	2 × 250	270	136
	2 × 125	199	94
	4 × 4	252	172
Kiesselbach	2 × 100	72	38·5
Coombs and Grantham	2 × 27	7·6	3·7
	3 × 18	11·4	6·5
Beaven	8 × 20	228	108

There is, however, a further consequence of field heterogeneity which is not met by most chessboard arrangements and which is of increased importance in view of the reduction in the estimate of experimental error introduced by the methods described above. It is best illustrated by the case of a potato experiment by F. C. Stewart of Geneva, N.Y.[1], the results of which have given rise to some discussion by American students of Plot Technique. For the purpose of a study of the effect of missing "hills" upon yield Stewart found it desirable to compare in check rows the yield of two halves of a potato tuber. The halving was carefully done and the half-tubers planted in pairs in nine rows, 50 pairs to a row. For the purpose of the comparison, the first half-tuber of a pair was denoted C, the second C'. The plan of a row is thus

$$C\ C'\ C\ C'\ C\ C'\ C\ C'\ \text{etc.}$$

The brackets indicate pairs of related half-tubers. Of the total number planted 429 pairs of plants were available for yields. The mean yield of the 429 C plants was 21·6 ± ·1856 ozs., of the C' plants 20·5056 ± ·1881 ozs. The mean difference $C - C'$ is thus **1·0944** with a probable error of ± ·2642 as calculated by Stewart by the usual formula, or more accurately one of ± ·1930 calculated from the actual differences $C - C'$ by "Student's" method. There is thus a mean difference between the yield of two halves of a tuber as great as 5·2 % of the mean yield and 5·67 times its probable error. This mean difference is of undoubted significance, and had there been any difference in treatment between the C and the C' plants the experiment would have been considered as establishing a significant difference in yield due to that difference in treatment.

That the results established a significant mean difference between the C and the C' plants cannot be doubted. One looks, therefore, for some consistent difference between them: the only consistent difference dis-

[1] *N.Y. Agr. Exp. Sta. Bull.* 489, 1921.

coverable is one of *position*. The C plant is always the first of the pair in a row, the C' plant always the second—in the plan the C plants are on the average one space to the left of the C' plants. The C plants and C' plants are consequently distributed in space about different centres of gravity. Their yields are likely, therefore, from what we know of soil heterogeneity, to regress towards different mean values. The case would be at its simplest if there were a regular downward slope of soil fertility from left to right of the plan, but the data do not support so simple an assumption. The systematic difference in position between C and C' plants can, however, be overcome by a rearrangement as shown below:

$$C\ C'\ C'\ C\ C\ C'\ C'\ C\ \text{etc.}$$

The C plant is now alternatively the first and the second of the pair, the C' plant alternatively the second and the first. The distribution of C and C' plants is now a balanced distribution, both sets having the same centre of gravity.

Recalculation of the data now gives the mean difference,

$$C - C' = -\cdot197 \pm \cdot1962,$$

i.e. the mean difference is now reduced to about one-fifth of its original value (*i.e.* to $\cdot93\%$ instead of $5\cdot2\%$ of the mean), and is now only of the same order as its probable error. (It will be noted that the rearrangement follows the same lines as Beaven's half-drill strip arrangement, which is similarly balanced, and this case affords a useful test of the value of that arrangement.)

The significant difference between the yields of halves of the same tuber which appeared in the first calculation of the data may fairly be ascribed, therefore, to the unbalanced arrangement of the groups which were compared.

A further example of the effect of an unbalanced arrangement may be taken from the chessboard arrangement shown in Diagram 1. It will be seen that the plots are not balanced laterally, the first five treatments being consistently to the left of the second five treatments. This arrangement was tested by placing it at random on a set of yield data from Mercer and Hall's uniformity trials and calculating the mean differences for all the possible pairs of comparisons. The results, shown in the first item of Table III, point in the same direction as Stewart's result: in unbalanced arrangements statistically significant mean differences appear between groups of plots which differ not at all in treatment but only in position. It will be seen that the average of all the mean differences disregarding sign is still greater than the probable error of a random comparison.

Table III

Mean differences between groups of plots

Data from uniformity trials

Data	Arrangement	Mean differences (as % of the mean yield)	Probable error of a random comparison (as % of mean)
Mercer and Hall	10 × 3 Chessboard (unbalanced)	·03– 9·3 (average 3·5)	2·7
Smith	4 × 4 (balanced)	·06– 1·31	1·76
	4 × 4 ,, 1896	·76– 1·29	1·54
	,, ,, 1895	·90–12·9	13·7
Mercer and Hall	4 × 4 (balanced)	·14– 5·3 (average 2·69)	2·34

The chessboard (Beaven's) shown in Diagram 2 is again unbalanced laterally and "Student" in discussing the data from that experiment suggests the possibility of correcting for position—on the basis of an observed fall in yields of similarly treated plots from left to right. He remarks, however, that such correction is unsatisfactory and that it is better to arrange that no such correction is necessary. Mitscherlich in his field experiments also used an unbalanced arrangement and, moreover, an unsatisfactory method of correction for position.

Balanced Chessboards. The extension of the balanced arrangement to a 4 × 4 chessboard (4 treatments, 4 replications of each) is fairly simple, and is shown in Diagram 3.

Diagram 3

$$\underline{|4} \qquad \begin{matrix} A & B & C & D \\ C & D & A & B \\ B & A & D & C \\ D & C & B & A \end{matrix}$$

The results of testing such an arrangement on uniformity data are shown in Table III and represent a great advance on the unbalanced arrangement. This arrangement is, however, a very rigid one: once the first row is filled up, which may be done in $\underline{|4}$ ways, the position of the treatments among the remaining plots is fixed. It is desirable that within the conditions necessary for securing balance, the element of chance in the arrangement of the plots shall be as great as possible: only so will it be possible to obtain a valid estimate of the experimental error.

The arrangement shown above, as Dr Fisher has pointed out to me, is only a special case of a general scheme of arrangement which fulfils the desired conditions.

The number of replications is made equal to the number of treatments —the plots form a perfect square in arrangement—and each treatment occurs once in each row and once in each column. Within these limits the arrangement of the treatments is determined by chance. There are now in the case of a 4×4 chessboard $\lfloor 4 \times \lfloor 3 \times \lfloor 2$ possibilities of arrangement and the only limitations imposed are those which are to be used in the analysis of variance for the estimation of the "experimental error": thus the estimate of the variance due to "trial" is based upon the fact that both the rows and the columns form "trials." An example of such an arrangement is given in Diagram 4.

Diagram 4

A	B	C	D	
C	A	D	B	$\lfloor 4 \times \lfloor 3 \times \lfloor 2$ possibilities of arrangement
D	C	B	A	
B	D	A	C	

Where it is not possible to make the number of replications equal to the number of treatments, balance may be secured by making the centres of gravity of all the treatments coincide.

Diagram 5 shows such an arrangement of a 3×7 chessboard, an arrangement which is now being tested at Rothamsted. With four replications, however, it is much more easy to secure a balanced arrangement and that should be the minimum number for field experiments.

Diagram 5

	A	D	G
	B	E	C
3 × 7 chessboard	C	F	F
	D	G	B
2 × ⌊7 possibilities of arrangement	E	A	E
	F	B	A
	G	C	D

The foregoing considerations, while developed principally in reference to field experiments in which the effects of various experimental treatments upon yield are being estimated, have an important bearing upon a problem more strictly plant physiological, that of sampling a crop at short intervals for studies of growth. In view of what is known of soil heterogeneity it is clearly unsatisfactory to take plants from one row at a time as has sometimes been done. The standard deviations of the means of such harvests are of no value for the estimation of the accuracy with which the increment in growth between one harvest date and the next has been measured and may be quite misleading.

If the approximate number of harvests to be made is known at the outset the sampling can be planned like a field experiment as just described. Wherever possible the plan of the perfect square with equalisation of rows and columns should be adopted. A simpler plan for barley plants which is now being tested at Rothamsted is shown in Diagram 6.

Diagram 6

1 2 3 4 5 6 7 etc.→	Sampling for weekly harvests
←7 6 5 4 3 2 1	
1 2 3 4 5 6 7 etc.→	1 = 1st harvest
←7 6 5 4 3 2 1	2 = 2nd ,, etc.

This plan had to be subject to the limitations that to avoid damage by trampling the sampling must start from the edges of the plot and subsequent harvests must be adjacent to previously sampled areas. It will be seen that plants are taken from four areas each week, that harvests in two successive weeks are from adjacent areas and that the direction of comparison between successive harvests is alternately left to right and right to left.

There is another method of sampling which may be mentioned here both because of its great possible value and because it links on to another aspect of plot technique which is of great importance. That is, sampling on the basis of measurements previously taken on the growing crop. On the basis of measurements involving about one hundred plants it is possible for instance to arrive at the average plant on a barley plot in terms of such characters as total height, height of shoot (combined leaf-sheaths), number of green leaves, width of highest fully expanded leaf, number of tillers, number of leaves on first and second tillers, and so forth. It is then possible to select from the plot a small number (of the order of ten) of plants corresponding with these mean dimensions for the determination of dry weight, leaf area and related characteristics.

In this way field experiments can be made to furnish data upon growth as well as upon yield: and we are at least on the way to securing greatly increased accuracy in the measurement of both sets of data.

CLOSE OF CONFERENCE
AND CONFERENCE RESOLUTIONS
(SATURDAY, *July 12th*, 1924)

THE PRESIDENT IN THE CHAIR

The following resolution was passed unanimously:

"That the present Executive Committee remain in being, with power to co-opt, to consider matters arising out of the Conference and to take action with regard to resolutions."

Proposed by the PRESIDENT, seconded by Dr J. BURTT-DAVY.

At the suggestion of the President the following resolutions were unanimously referred to the Executive Committee for consideration:

(1) "That it is highly important that botanists for posts of whatever kind, whether of teaching or research, should have a sound scientific training in Botany and cognate science, by means of which alone a real scientific outlook can be obtained."

Proposed by Dr H. WAGER, F.R.S., seconded by Prof. F. J. LEWIS.

(2) "That steps should be taken to prepare as complete a list as possible of names and addresses of Botanists resident in the British Empire, and that the question of preparing an extended list to include the Botanists of all countries be considered."

Proposed by Prof. R. H. YAPP, seconded by Dr A. W. HILL, F.R.S.

(3) "That the Conference desires to call the attention of Overseas Governments to the great lack of scientific literature in some of their scientific institutions, and to urge them to remedy this defect where it exists."

Proposed by Mr F. T. BROOKS, seconded by Mr W. NOWELL.

(4) "That this meeting, fully aware of the necessity for co-ordination of botanical work in all parts of the Empire and recognising that both investigators and undergraduates will benefit by contact with fresh points of view, urges that every effort be made to encourage the exchange both of members of the staff and of research students between the Universities and Research Institutions of the Empire, and recommends the establishment of a permanent committee for this purpose."

Proposed by Prof. Dame HELEN GWYNNE-VAUGHAN, seconded by Prof. F. J. LEWIS.

(5) "That Mycology would greatly benefit by the publication of a list of new genera of fungi such as is already in manuscript at the British Museum."

Proposed by Prof. Dame HELEN GWYNNE-VAUGHAN, seconded by Dr E. J. BUTLER.

(6) "That this Conference strongly urges the need for further facilities for research in Forest Pathology in all its aspects, and for closer co-operation between forest pathologists, executive forest officers, and other silviculturists."

Proposed by Dr MALCOLM WILSON, seconded by Dr J. MUNRO.

(7) "That a systematic effort be made to render the existing knowledge of the natural vegetational resources of the Empire generally available as well as to encourage and promote further research on the vegetation of the Empire."

Proposed by Mr A. G. TANSLEY, F.R.S., seconded by Dr T. F. CHIPP.

(8) "That a sufficient body of knowledge has been accumulated to justify the preparation of a series of outline monographs of the vegetation of the Dominions, Colonies, and Protectorates."

Proposed by Mr A. G. TANSLEY, F.R.S., seconded by Dr T. F. CHIPP.

(9) "That all future work published on the vegetation of the Dominions and Colonies should be registered and abstracted, the abstracts being made generally available by periodical publications uniform with the monographs."

Proposed by Mr A. G. TANSLEY, F.R.S., seconded by Dr T. F. CHIPP.

(10) "That a handbook dealing with the aims to be kept in view in general work on vegetation, and with the best methods for use in the field, should be prepared for circulation to present and future workers in the British Empire."

Proposed by Mr A. G. TANSLEY, F.R.S., seconded by Dr T. F. CHIPP.

(11) "That a central permanent committee be formed to carry out the objects specified in Resolutions 8, 9 and 10, and generally to promote the aims referred to in Resolution 7."

Proposed by Mr A. G. TANSLEY, F.R.S., seconded by Dr T. F. CHIPP.

(12) "That a temporary committee of the following four members be appointed forthwith to arrange for the formation as soon as practicable of the permanent committee referred to in Resolution 11: Prof. F. W. Oliver, F.R.S., Mr A. G. Tansley, F.R.S., Dr T. F. Chipp, and Mr J. Ramsbottom."

Proposed by Mr A. G. TANSLEY, F.R.S., seconded by Dr T. F. CHIPP.

(13) "That an adequate training for work on vegetation should involve a broad botanical course of University standard in which a prominent position is occupied by practical instruction in Systematic Botany and Ecology in the field."

Proposed by Mr A. G. TANSLEY, F.R.S., seconded by Dr T. F. CHIPP.

(14) "That a committee be appointed to consider the various proposals made in the discussion on taxonomic work, and to take such steps as they may think fit."

Proposed by Dr A. B. RENDLE, F.R.S., seconded by Dr A. W. HILL, F.R.S.

The following resolutions on Nomenclature were passed unanimously:

(15) "That the sub-committee on Nomenclature remain in being (with power to co-opt) to receive and collate additional proposals for changes in the International Rules."

Proposed by Dr A. B. RENDLE, F.R.S., seconded by Dr A. W. HILL, F.R.S.

(16) "That the following recommendations concerning the Rules of Nomenclature be brought before the next International Botanical Congress through the Executive Committee:

1. Art. 36 (invalidating names of new groups published on and after Jan. 1, 1908, without Latin diagnoses) should be replaced by a strong *Recommendation* to supply Latin diagnoses.
2. All combinations which are homonyms (*i.e.* later homonyms) should be rejected.
3. All generic names which are homonyms (*i.e.* later homonyms) should be rejected except such as may be specially conserved.
4. The principle of the Type-method of applying names should be formally accepted.
5. Art. 55, 2 (rejecting 'duplicating binomials,' *e.g. Linaria Linaria*) should be revoked.
6. The 'principle of *nomina abortiva*' should be expunged from the Rules.
7. The List of *Nomina generica conservanda* should be revised.
8. It should be made clear how far each of the *Nomina generica conservanda* is conserved.
9. That for the future the name of a group shall not be regarded as effectively published when the description is issued only with exsiccata.

10. Publication of new genera and species should be only in scientific publications and, if possible, only in such as habitually reach systematic botanists."

Proposed by Mr T. A. SPRAGUE, seconded by Miss E. M. WAKEFIELD.

The following resolution was unanimously referred to the Executive Committee for consideration:

(17) "That this Conference is of the opinion that the formation of an Imperial Botanical Association is desirable for the furtherance of botanical work throughout the Empire, that a special committee be appointed to consider and to take the necessary steps for its formation, and that all members of this Conference be considered Foundation Members of such Association."

Proposed by Prof. S. SCHONLAND, seconded by Mr F. A. STOCKDALE.

The President then read a telegram from the Acting Agent-General for Victoria regretting his inability to be present at the closing session of the Conference.

At the invitation of the President, Dr Schramm announced that American botanists were arranging to hold an International Botanical Conference in the United States in June, 1926, and cordially extended an invitation to members of the Imperial Botanical Conference to attend it.

The President announced that a meeting of the Botanical Society of Edinburgh would be held on July 16, 17, to which all members of the Imperial Botanical Conference were invited.

The President further announced that a Report of the Proceedings of the Conference would be published in due course.

The President then proposed that hearty votes of thanks be conveyed to the Excursions sub-committee, the Hospitality sub-committee, the Governing Body of the Imperial College of Science and Technology, the lanternists and the attendants for their help in connection with the Conference.

These were carried unanimously.

The Conference closed with a cordial vote of thanks to the President for his great services, proposed by Prof. F. O. Bower, F.R.S. and seconded by Prof. H. H. Dixon, F.R.S., which was carried with acclamation. The President in acknowledging the vote of thanks accepted it as including all members of the Executive Committee as well as himself.

EXCURSIONS *AND* RECEPTIONS

Monday evening, July 7. RECEPTION BY THE LINNEAN SOCIETY, BUR-
LINGTON HOUSE, W. (President, Dr A. B. Rendle, F.R.S.) During
the evening Dr B. Daydon Jackson gave an account of the Linnean
collections.

Wednesday afternoon, July 9. EXCURSION TO THE ROYAL HORTICULTURAL
SOCIETY'S GARDEN, WISLEY.

Friday afternoon, July 11. VISIT TO THE NATURAL HISTORY MUSEUM.
Reception by the Director.

Saturday afternoon, July 12. EXCURSION TO THE JOHN INNES HORTI-
CULTURAL INSTITUTION, MERTON.

Saturday, July 12—*Monday, July* 14. EXCURSION TO BLAKENEY POINT,
NORFOLK. (Conducted by Prof. F. W. Oliver, F.R.S.)

Monday, July 14. EXCURSION TO THE ROTHAMSTED EXPERIMENTAL
STATION.

Tuesday, July 15. EXCURSION TO THE ROYAL BOTANIC GARDENS, KEW.

Wednesday, July 16. EXCURSION TO CAMBRIDGE.

MEMBERS OF IMPERIAL BOTANICAL CONFERENCE
(1924)

OVERSEAS BOTANISTS AND DELEGATES

CANADA
Dr G. H. Duff.
Prof. F. J. Lewis.
Mr C. W. Lowe.

AUSTRALIA
The Agent-General for New South Wales.
The Agent-General for South Australia.
The Agent-General for Western Australia.
The Acting Agent-General for Victoria.
The Acting Agent-General for Queensland.

NEW ZEALAND
Mrs M. W. Aitken.
Dr L. Cockayne, F.R.S.
Mr G. H. Cunningham.

SOUTH AFRICA
Prof. Bews.
Dr Ethel Doidge.
Mr V. A. Putterill.
Prof. Schonland.

INDIA
Mr Badami.
Prof. S. R. Bose.
Mr R. H. Dastur.
Mr C. E. C. Fischer.
Lt.-Col. A. T. Gage, I.M.S.
Mr R. S. Hole, C.I.E.
Mr A. Howard.
Dr H. M. Leake.
Mr F. R. Parnell.
Mr A. C. Tunstall.

CEYLON
Mr C. H. Gadd.
Mr C. P. Jagawardana.
Mr F. A. Stockdale.

FEDERATED MALAY STATES
Mr B. Bunting.
Dr F. W. Foxworthy.
Mr W. N. Sands.

PALESTINE
Mr F. J. Tear.

EGYPT
Mr M. A. Bailey.

SUDAN
Mr R. E. Massey

UGANDA
Mr L. Chalk.
Mr T. D. Maitland.
Mr S. Simpson.

KENYA COLONY
Mr H. M. Gardner.

ZANZIBAR
Miss E. J. Welsford.

SEYCHELLES
Mr P. R. Dupont.

NYASALAND
Capt. C. Smee.

NIGERIA
Mr O. T. Faulkner.
Mr T. Laycock.

SIERRA LEONE
Mr M. T. Dawe.

GOLD COAST
Mr R. H. Bunting.
Dr J. M. Dalziel.
Mr T. Hunter.
Mr L. H. King Church.
Mr W. H. Patterson.

TRINIDAD
Prof. S. F. Ashby.
Mr W. Nowell.

JAMAICA
Mr C. G. Hansford.

S. VINCENT
Mr L. H. Burd.

BERMUDA
Mr E. A. McCallan.

BRITISH GUIANA
Mr F. A. Stockdale.

CYPRUS
Mr W. Bevan.

IRISH FREE STATE
Mr N. G. Ball.
Prof. C. Boyle.
Prof. Cummins.
Prof. H. H. Dixon, F.R.S.
Mr M. J. Gorman.
Prof. A. Henry.
Dr P. Murphy.
Dr Lloyd Praeger.

HOME BOTANISTS

Miss F. Adams.
Mrs Alcock.
Mr W. W. Allen.
Miss V. Anderson.
Miss Archbold.
Dr W. R. G. Atkins.

Miss A. S. Bacon.
Mr Bagchee.
Mr N. G. Ball.
Dr W. L. Balls, F.R.S.
Dr H. Bancroft.
Dr C. A. Barber.
Mr W. Barnes.
Dr K. Barratt.
Mr W. Barratt.
Mr A. W. Bartlett.
Mr W. Bateson, F.R.S.
Dr L. Batten.
Mr A. Beaumont.
Mr T. A. Bennet-Clark.
Dr E. M. Berridge.
Dr W. F. Bewley.
Dr F. F. Blackman, F.R.S.
Prof. V. H. Blackman, F.R.S.
Miss E. M. Blackwell.
Mr B. D. Bolas.
Mr L. A. Boodle.
Dr A. W. Borthwick.
Prof. F. O. Bower, F.R.S.
Mr W. Braid.
Mr J. M. Branfoot.
Dr W. B. Brierley.
Mrs Brierley.
Mr G. E. Briggs.
Miss D. Bright.
Mr J. Britten.
Mr F. T. Brooks.
Dr W. Brown.
Sir Archibald Buchan-Hepburn.
Dr E. J. Butler, C.I.E.

Miss Carré.
Mr Cartwright.
Mrs Cartwright.
Miss D. Cayley.
Dr S. E. Chandler.
Mr Chapman.
Dr T. F. Chipp.
Mr F. J. Chittenden.
Mr R. N. Chrystal.
Mr L. W. Cole.
Mr W. R. Cook.
Mr A. D. Cotton.
Prof. W. Craib.

Mr J. M. Cranfoot.
Mr E. M. Cutting.

Prof. O. V. Darbishire.
Miss A. J. Davey.
Mr A. N. David.
Dr J. Burtt Davy.
Mr W. R. Day.
Dr E. M. Delf.
Miss O. Dickinson.
Mr S. Dickinson.
Prof. H. H. Dixon, F.R.S.
Mr H. N. Dixon.
Mr W. J. Dowson.
Dr G. C. Druce.
Mr J. M. F. Drummond.
Mr S. T. Dunn.

Mr C. Eley.
Dr Bayliss Elliott.
Mr F. L. Engledow.
Miss Erith.

Prof. J. B. Farmer, F.R.S.
Mr W. Fawcett.
Mrs Ferguson.
Mr C. Ford.
Miss A. M. Frampton.
Prof. F. E. Fritch.

Prof. R. R. Gates.
Mr K. J. George.
Mr A. S. Gepp.
Miss L. S. Gibbs.
Mrs A. B. Gillett.
Mr R. D. O'Good.
Mr N. K. Gould.
Miss M. L. Green.
Prof. P. Groom, F.R.S.
Mr J. Groves.
Miss Grubb.
Prof. Dame Helen Gwynne-Vaughan, D.B.E.

Dr P. Haas.
Mr W. Hales.
Miss A. C. Halket.
Miss M. P. Hall.
Mr R. M. Harrison.
Mr F. Haworth.
Dr D. Haynes.
Miss Hayward.
Mr O. V. S. Heath.
Mr Henderson.
Mr W. E. Hiley.

Dr A. W. Hill, F.R.S.
Dr T. G. Hill.
Mr A. N. Hoare.
Miss Hodgkinson.
Miss I. Hoggan.
Mr R. S. Hole, C.I.E.
Dr A. S. Horne.
Major C. C. Hurst.
Mrs Hurst.
Mr J. Hutchinson.
Mr H. A. Hyde.

Dr B. Daydon Jackson.
Miss D. E. Jaggard.
Sir H. H. Johnston.
Mr T. J. Jenkin.
Prof. W. Neilson Jones.

Prof. Sir Frederick Keeble, C.B.E., F.R.S.
Dr F. Kidd.
Mrs Kidd.
Dr R. Kidston, F.R.S.
Dr R. C. Knight.

Miss M. S. Lacey.
Miss V. Lasbrey.
Miss J. Latter.
Miss Lee.
Miss E. M. Lee.
Mr J. Line.
Miss G. Lister.
Miss F. M. Loader.
Mr G. W. Loder.
Lord Lovat, K.G.
Mr A. G. Lowndes.

Prof. S. Mangham.
Mr R. W. Marsh.
Mr J. F. Martley.
Mr E. W. Mason.
Mrs Mason.
Mr W. T. Mathias.
Mr J. K. Mayo.
Miss R. McIlroy.
Miss Mehta.
Dr J. Cosmo Melvill.
Mr W. A. Millard.
Mr W. C. Moore.
Sir Daniel Morris, K.C.M.G.
Dr J. W. Munro.
Mr A. E. Muskett.

Miss E. F. Noel.

Prof. F. W. Oliver, F.R.S.

Dr S. G. Paine.
Mr J. Parkin.

Mr R. Paulson.
Miss C. Pellew.
Prof. J. Percival.
Miss D. M. Pertz.
Dr G. H. Pethybridge.
Dr R. Lloyd Praeger.
Sir David Prain, C.M.G., C.I.E., F.R.S.
Dr C. A. Pratt.
Mr W. R. Price.
Prof. J. H. Priestley.

Mr J. Ramsbottom, O.B.E.
Miss Rankin.
Miss M. Rathbone.
Mr C. Rea.
Dr A. B. Rendle, F.R.S.
Mrs M. Bristol Roach.
Mr W. A. Roach.
Miss I. Roper.
Sir John Russell, F.R.S.

Mr J. L. Sager.
Dr R. N. Salaman.
Dr E. J. Salisbury.
Mr C. E. Salmon.
Miss K. Sampson.
Mr P. C. Sarbadhikari.
Miss E. R. Saunders.
Mr W. T. Saxton.
Dr D. H. Scott, F.R.S.
Mr G. O. Searle.
Prof. A. C. Seward, F.R.S.
Mr K. Shaw.
Mr A. Simmonds.
Mr B. I. Slaughter.
Miss A. L. Smith.
Mr A. M. Smith.
Mr N. J. G. Smith.
Dr W. G. Smith.
Prof. W. W. Smith.
Prof. W. Somerville.
Mr T. A. Sprague.
Prof. R. G. Stapledon.
Major F. C. Stern.
Dr H. M. Steven.
Miss E. H. Stevenson.
Prof. W. Stiles.
Dr V. S. Summerhayes.
Mr F. Summers.
Mr L. G. Sutton.
Mr M. H. F. Sutton.
Mr E. W. Swanton.

Mr R. G. Tabor.
Mr A. G. Tansley, F.R.S.
Prof. D. Thoday.
Dr E. N. M. Thomas.

Mr H. S. Thompson.
Mr J. Thomson.
Mr M. A. H. Tincker.
Prof. R. S. Troup.

Mr G. Varngis.
Mr M. C. Vyvyan.

Mr S. M. Wadham.
Dr H. Wager, F.R.S.
Miss E. M. Wakefield.
Mr J. Waller.
Mr A. S. Watt.
Dr C. West.

Miss A. Westbrook.
Mr W. A. R. Dillon Weston.
Mr A. J. Wilmott.
Dr Malcolm Wilson.
Mr S. P. Wiltshire.
Mr N. J. Wood.
Dr T. W. Woodhead.
Mr R. C. Woodward.
Dr H. Wormald.
Mr C. Wright.

Prof. R. H. Yapp.
Mrs Yapp.

FOREIGN VISITORS

Professor W. H. Dore (United States).
Dr East (United States).
Dr H. A. Gleason (United States).
Mrs H. A. Gleason (United States).
Professor J. M. Macfarlane (United States).

Professor Peklo (Czechoslovakia).
Dr C. V. Piper (United States).
Professor A. Povah (United States).
Professor W. E. Praeger (United States).
Professor J. R. Schramm (United States).

Printed in the United States
By Bookmasters